Internet of Things

Technology, Communications and Computing

Series Editors

The series Internet of Things - Technologies, Communications and Computing publishes new developments and advances in the various areas of the different facets of the Internet of Things. The intent is to cover technology (smart devices, wireless sensors, systems), communications (networks and protocols) and computing (theory, middleware and applications) of the Internet of Things, as embedded in the fields of engineering, computer science, life sciences, as well as the methodologies behind them. The series contains monographs, lecture notes and edited volumes in the Internet of Things research and development area, spanning the areas of wireless sensor networks, autonomic networking, network protocol, agent-based computing, artificial intelligence, self organizing systems, multi-sensor data fusion, smart objects, and hybrid intelligent systems.

Indexing: *Internet of Things* is covered by Scopus and Ei-Compendex **

Tien Anh Tran · Debashis De · Diganta Sengupta ·
Spyros Hirdaris
Editors

Big Data Analytics for Internet of Marine Things

Springer

Editors
Tien Anh Tran
Department of Marine Engineering
Vietnam Maritime University
Hai Phong, Vietnam

Diganta Sengupta
Department of Computer Science
and Engineering
Heritage Institute of Technology
Kolkata, West Bengal, India

Debashis De
Department of Computer Science
and Engineering
Maulana Abul Kalam Azad University
of Technology
Nadia, West Bengal, India

Spyros Hirdaris
Global Ship Systems Centre
American Bureau of Shipping
Athens, Greece

ISSN 2199-1073 ISSN 2199-1081 (electronic)
Internet of Things
ISBN 978-3-032-28170-8 ISBN 978-3-032-28171-5 (eBook)
https://doi.org/10.1007/978-3-032-28171-5

This Springer imprint is published by the registered company Springer Nature Switzerland AG
The registered company address is: Gewerbestrasse 11, 6330 Cham, Switzerland

Preface

The creation of an Internet of Marine Things platform, including the Underwater Internet of Things, is needed to ensure the interaction and digital navigation of heterogeneous marine robotic agents. It is necessary to combine the following robotic agents: autonomous underwater vehicles, remotely operated vehicles, active and passive marine sensors, buoys, underwater sonar stations, coastal communication posts, and other elements of the platform. To ensure the interaction of all these elements, it is necessary to use a common communication system within the platform, as well as a common navigation and control system to solve complex problems of the navigation and control of the movement of robotic agents in order to implement a joint mission to collect and transmit data, including video information in real time. The architecture of the Internet of Marine Things platform must first be defined in order to use a unified approach to data exchange. This book provides an overview of approaches to determining the architectures of network underwater and marine communication systems based on the concept of the Internet of Things. Additionally, this book provides a comprehensive study of Internet of Marine Things applications, challenges, and architectures. The main contributions of this book are summarized as follows: Introducing the potential Internet of Marine Things applications; Pointing out the challenges of Internet of Marine Things (i.e., the differences between Internet of Marine Things and Internet of Things); and Analyzing the Internet of Marine Things system architecture. In this regard, this book on "***Big Data Analytics for Internet of Marine Things***" serves as a focal point in the Save Oceans movement, along with the Sustainable Maritime transportation systems.

The chapters in this book cover a wide range of topics exhibiting the multi-faceted applications of Internet of Marine Things (IoMTs) and Data Analysis in the maritime transportation and ocean engineering. The overview of Internet of Marine Things and big data technique has been summarized technically to understand its benefits into maritime science and technology. The utilization of Internet of Marine Things and machine learning has been studied and revealed the feasibility into solving the existing issues for ships and underwater vehicles nowadays. Additionally, the Edge-Artificial Intelligence (AI) has been applied along with the Internet of Marine Things to create the greener ocean environment. A hybrid deep learning model has been established to support the Decision Making System (DMS) into improving the Internet of Marine Things. The privacy and maritime security have also been investigated to ensure the Internet of Marine Things system effectively from the external factors. Moreover, the post-quantum cryptography and swarm intelligence have been enclosed to ensure effectively into maritime security from data technique. The processing and data acquisition have been provided to increase the reliability of database from the Internet of Marine Things. Finally, the application of Internet of Marine Things has provided its benefits to create the sustainable maritime transportation by 2050 from these studies of this book.

In a nutshell, this book covers almost all the areas of use of Internet of Marine Things (IoMTs), AI, Edge computing, Machine learning, Data analysis, and underwater vehicles for updating the modern technologies of ocean engineering to create the sustainable maritime transportation.

Enjoy reading!

Hai Phong, Vietnam — Tien Anh Tran, Ph.D.
Nadia, India — Debashis De, Ph.D.
Kolkata, India — Diganta Sengupta, Ph.D.
Athens, Greece — Spyros Hirdaris, Ph.D.

Contents

Contributors

P. Ashok Symbiosis Institute of Digital and Telecom Management (SIDTM), Symbiosis International (Deemed University) (SIU), Lavale, Pune, Maharashtra, India

Amreen Ayesha Manipal Institute of Technology Bengaluru, Manipal Academy of Higher Education, Manipal, Bengaluru, India

V. Vignesh Babu SRM Institute of Science and Technology, Kattankulathur, Tamilnadu, India

Kamalika Bhattacharjya Centre of Mobile Cloud Computing, Department of Computer Science and Engineering, Maulana Abul Kalam Azad University of Technology, West Bengal, Nadia, West Bengal, India

Neepa Biswas Narula Institute of Technology, Kolkata, India

Sitanath Biswas JIS College of Engineering, Kalyani, Nadia, India

Tatayya Bommali Department of Management Studies, Sanketika Vidya Parishad Engineering College, Visakhaptnam, AP, India

Sanjukta Chakraborty Department of CSE, Seacom Skills University, Bolpur, WB, India

Rupsa Chatterjee Department of Information Technology, Maulana Abul Kalam Azad University of Technology, West Bengal, Nadia, West Bengal, India

Kudzaishe S. Chitsenga SRM Institute of Science and Technology, Kattankulathur, Tamilnadu, India

Deepanwita Das National Institute of Technology, Durgapur, India

Debashis De Centre of Mobile Cloud Computing, Department of Computer Science and Engineering, Maulana Abul Kalam Azad University of Technology, West Bengal, Nadia, West Bengal, India

D. Devaraj Kalasalingam Academy of Research and Education, Krishnankoil, Tamilnadu, India

Tiasha Dutta Department of Information Technology, Maulana Abul Kalam Azad University of Technology, West Bengal, Nadia, West Bengal, India

T. Ganesh Department of Ocean Studies and Marine Biology, Pondicherry University-Andaman Campus, Sri Vijaya Puram, India

N. Geetha Department of Computer Applications, Alagappa University, Karaikudi, India

Jincheng Gong School of Computer Science and Artificial Intelligence, Wuhan University of Technology, Wuhan, China

Harsh Kumar Gupta JIS College of Engineering, Kalyani, Nadia, India

Abhisek Hazra JIS College of Engineering, Kalyani, Nadia, India

Spyros Hirdaris American Bureau of Shipping, Global Ship Systems Centre, Athens, Greece

Bala Naga Jyothi Division of Deep Sea Technologies, National Institute of Ocean Technology, Chennai, India

Mustafa Kayyali Maaref University of Applied Sciences, Sarmada, Syria

Milind Kundu JIS College of Engineering, Kalyani, Nadia, India

S. Lakshmi Sridevi Department of Computer Science and Engineering, Chennai Institute of Technology, Chennai, Tamilnadu, India

Huanhuan Li School of Engineering, University of Southampton, Southampton, UK

Suchismita Maiti Narula Institute of Technology, Kolkata, India

Banashree Mandal National Institute of Technology, Durgapur, India

Zhang Mingyang School of Ocean and Civil Engineering, Shanghai Jiao Tong University, Shanghai, P.R. China;
Key Laboratory of Marine Intelligent Equipment and System, Shanghai Jiao Tong University, Shanghai, P.R. China

Ritwika Mukherjee Narula Institute of Technology, Kolkata, India

K. Murali Krishna Department of Electronics and Communication Engineering, Rajalakshmi Engineering College, Chennai, Tamilnadu, India

S. Muthakshi Department of Computer Science and Information Technology, M.O.P. Vaishnav College for Women (Autonomous), Chennai, India

N. Nasurudeen Ahamed College of Information Technology, United Arab Emirates University, Al Ain, United Arab Emirates

N. A. Natraj Symbiosis Institute of Digital and Telecom Management (SIDTM), Symbiosis International (Deemed University) (SIU), Lavale, Pune, Maharashtra, India

P. K. Paul Department of CIS, and Information Scientist (Offg.), Raiganj University, Raiganj, West Bengal, India

Nithyashree Pillai SRM Institute of Science and Technology, Kattankulathur, Tamilnadu, India

Ujan Pradhan Department of Computing Technologies, SRM Institute of Science and Technology, Chennai, TN, India

K. Praghash Department of Electronics and Communication Engineering, Christ University, Bangalore, India

V. Jothi Prakash Department of Information Technology, Karpagam College of Engineering, Coimbatore, India

S. B. Pranesh Division of Deep Sea Technologies, National Institute of Ocean Technology, Chennai, India

J. Preetha Roselyn SRM Institute of Science and Technology, Kattankulathur, Tamilnadu, India

K. Rajalakshmi Department of MCA, C.M.R. Institute of Technology, Bangalore, India

R. Ramesh Division of Deep Sea Technologies, National Institute of Ocean Technology, Chennai, India

Shinde Sopan Raosaheb Department of Computer Engineering, Dr. D.Y. Patil College of Engineering and Innovation, Pune, Maharashtra, India

Ansh Raj Rath School of Artificial Intelligence, Amrita Vishwa Vidyapeetham, Coimbatore, India

J. Preetha Roselyn SRM Institute of Science and Technology, Kattankulathur, Tamilnadu, India

Samarjit Roy Computing and Information Technology, EIU, Becamex IDC Corp, Binh Duong, Ho Chi Minh City, Vietnam

Sapta Rathi Roy Department of Information Technology, Maulana Abul Kalam Azad University of Technology, West Bengal, Nadia, West Bengal, India

R. Saavedra International Programs, Azteca University, Chalco de Díaz Covarrubias, Mexico

Subhashree Sahoo JIS College of Engineering, Kalyani, Nadia, India

Minakshi Sarkar Department of Information Technology, Maulana Abul Kalam Azad University of Technology, West Bengal, Nadia, West Bengal, India

Niharika Sharma School of Artificial Intelligence, Amrita Vishwa Vidyapeetham, Coimbatore, India

S. K. Sharma Texas A&M University, Texarkana, USA

B. P. Sreeja School of Artificial Intelligence, Amrita Vishwa Vidyapeetham, Coimbatore, India

Mahamuda Sultana Guru Nanak Institute of Technology, Kolkata, India

Prabha Sundaravadivel University of Texas, Tyler, USA

Tien Anh Tran Department of Marine Engineering, Vietnam Maritime University, Haiphong, Vietnam;
Department of Electrical Engineering, University of Malta, Msida, Malta

V. Varun SRM Institute of Science and Technology, Kattankulathur, Tamilnadu, India

C. H. Vasanthkumar SRM Institute of Science and Technology, Kattankulathur, Tamilnadu, India

N. Vedachalam Division of Deep Sea Technologies, National Institute of Ocean Technology, Chennai, India

D. Vignesh SRM Institute of Science and Technology, Kattankulathur, Tamilnadu, India

Zaili Yang Liverpool Logistics, Offshore and Marine Research Institute, Liverpool John Moores University, Liverpool, UK

Chapter 1
Data Clustering and Machine Learning in the Internet of Marine Things

Huanhuan Li, Jincheng Gong, and Zaili Yang

1.1 Introduction

With the rapid advancement of digitalisation technologies, the Internet of Marine Things (IoMT) is undergoing a profound digital and intelligent transformation [60]. By integrating maritime sensors, Automatic Identification System (AIS) [11], satellite communications, unmanned surface vessels, and shore-based systems, IoMT enables the collection of multi-source, real-time data, such as vessel positions, navigation routes, and environmental parameters. This data facilitates the improvements of maritime traffic safety, operational efficiency, and environmental sustainability.

IoMT applies the principles of the Internet of Things (IoT) to marine and maritime contexts. Through the integration of sensors, communication networks, and data analytics, IoMT interconnects distributed maritime assets, including ships, buoys, unmanned platforms, and coastal facilities, facilitating efficient information exchange and intelligent system management. These capabilities support real-time monitoring, decision-making, and automation in diverse maritime applications, including navigation, logistics, environmental monitoring [68], and safety management [25].

As IoMT networks expand in scale and complexity, the volume of trajectory and environmental data generated has grown exponentially. Deriving actionable insights from such heterogeneous, high-dimensional datasets requires advanced analytical

H. Li (✉)
School of Engineering, University of Southampton, Southampton, UK
e-mail: Huanhuan.Li@soton.ac.uk

J. Gong
School of Computer Science and Artificial Intelligence, Wuhan University of Technology, Wuhan, China

Z. Yang
Liverpool Logistics, Offshore and Marine Research Institute, Liverpool John Moores University, Liverpool, UK

T. A. Tran et al. (eds.), *Big Data Analytics for Internet of Marine Things*,
Internet of Things, https://doi.org/10.1007/978-3-032-28171-5_1

approaches capable of identifying hidden structures and patterns. In this context, data clustering and machine learning (ML) techniques, core techniques in unsupervised learning and pattern recognition, are increasingly applied to analyse maritime trajectory data, detect typical route patterns, and predict anomalous behaviours [28], thereby enhancing risk management and operational optimisation. Integrating clustering and ML into IoMT enables the discovery of latent trends in massive trajectory datasets, offering more intelligent, precise, and timely decision support for the shipping industry.

Existing studies have explored a wide range of clustering and ML approaches for maritime trajectory analysis. For example, Zhang et al. [82] measured trajectory similarity based on direction and density, applying a trajectory segment clustering algorithm to compressed vessel trajectory data. Gao and Shi [20] developed a multi-step sub-trajectory clustering method to identify distinctive longitudinal navigation patterns, enhancing ML applications in collision avoidance, route planning, and anomaly detection.

Despite these advances, most research on the application of clustering and ML in IoMT has largely remained confined to specific case studies or individual algorithms. This has led to a lack of systematic literature reviews and performance comparisons, limiting the potential for cross-method optimisation and broader applicability. To address these gaps, this paper makes the following contributions:

(1) Systematic literature analysis: Conducts a structured review of IoMT-related studies applying data clustering and ML in IoMT, using CiteSpace to map research hotspots and emerging trends in maritime traffic analysis. This identifies the primary areas of academia and industry interest, as well as notable research gaps.
(2) Method and technology evaluation: Compares the strengths and weaknesses of various clustering and ML algorithms, assesses their adaptability to real-world maritime data, and examines their performance in IoMT application scenarios.
(3) Limitations and future directions: Discusses key challenges such as data quality, real-time deployment, and cybersecurity, and proposes future directions that integrate maritime Artificial Intelligence and digital twin technologies to drive sustained innovation and development.

The remainder of this paper is organised as follows: Sect. 1.2 introduces the proposed framework and provides its analysis, establishing the foundation for subsequent discussions. Section 1.3 reviews methods applied in IoMT, followed by Sect. 1.4, which examines real-world applications. Section 1.5 then addresses key challenges and outlines future research directions. Finally, Sect. 1.6 summarises the main findings and concludes the paper.

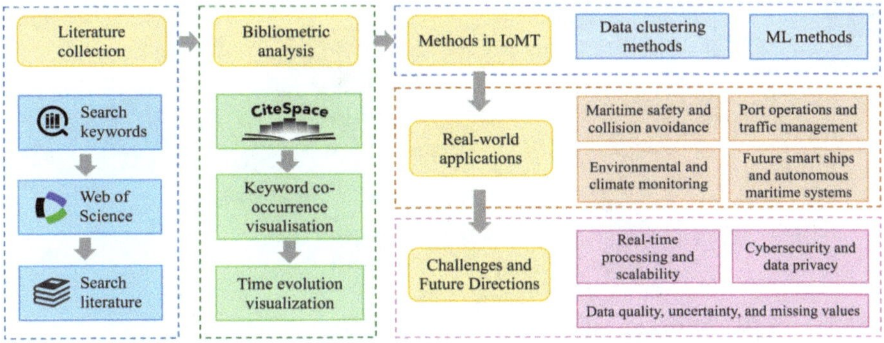

Fig. 1.1 Overall framework of this paper

1.2 The Proposed Framework and Analysis

1.2.1 The Proposed Framework

The overall structure of this chapter is illustrated in Fig. 1.1. The process begins with literature collection from the Web of Science (WoS) database, followed by CiteSpace-based keyword co-occurrence and temporal evolution analysis. The subsequent sections examine IoMT methodologies from clustering and ML perspectives, explore real-world applications, address critical challenges, and propose recommendations for future research.

1.2.2 Keyword Co-occurrence Analysis

A systematic search was conducted on 5 August 2025 in the WoS database [58] using the following search strategy:

TS = (((“Data clustering“ OR “Clustering analysis” OR “Cluster analysis”) OR “Machine learning” OR “Deep learning”) AND (“IoT” OR “IoUT” OR “IoMT”) AND (“Marine” OR “Ocean” OR “Maritime” OR “Offshore” OR “Sea-based” OR “Coastal” OR “Vessel” OR “Ship” OR “Port” OR “Aquaculture” OR “Fisheries” OR “Oceanography” OR “Marine environment monitoring” OR “Underwater sensors” OR “Maritime logistics” OR “Marine equipment”)).

A total of 972 publications were initially identified. After refinement based on article type (e.g. article, review, meeting, and early access) and research topic, 964 relevant papers were retained. CiteSpace software [13] was subsequently employed to construct keyword co-occurrence networks and timeline visualisations.

As shown in Fig. 1.2, the keywords extracted from the retrieved literature can be clustered into nine groups. Based on their technological focus and application

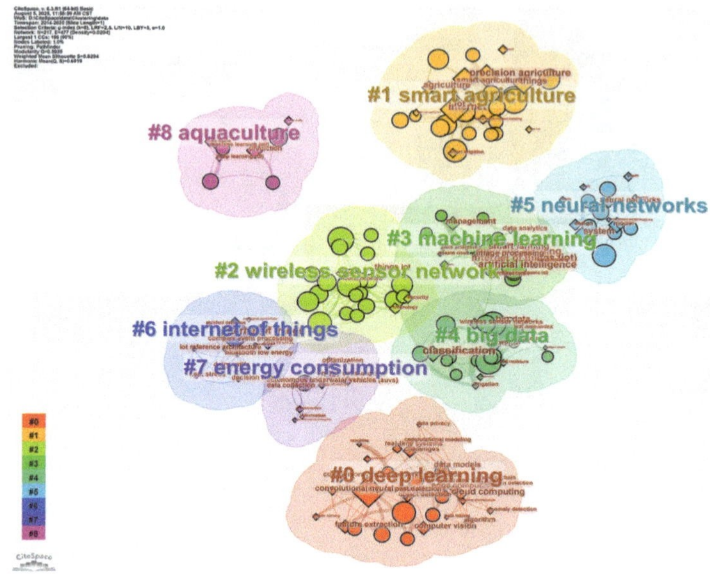

Fig. 1.2 Keyword co-occurrence diagram

scenarios, these groups can be further categorised into three overarching domains: data acquisition technologies, data analysis technologies, and practical applications.

(1) Data collection: In this stage, #2 wireless sensor networks and #6 IoT are the key enabling technologies. Wireless sensor networks (#2) facilitate real-time sensing and transmission of marine environmental data through distributed nodes. For example, Al-Zaidi et al. [5] developed a novel IoMT-based data collection and mapping system that leverages Very High Frequency (VHF) wireless networks. Complementarily, the IoT (#6) provides the foundational architecture for device interconnection and data exchange. For instance, Chi et al. [14] integrated sensor technology, 4G communication, and Zigbee protocols to design an IoT-based marine environmental monitoring system.

(2) Data analysis: Given the large-scale marine data collected, most studies employ Big Data technologies (#4) for storage and management, while leveraging neural networks (#5), ML (#3), and Deep Learning (DL) (#0) to uncover latent patterns. ML is extensively applied in trajectory clustering and pattern recognition. For example, Li et al. [37] proposed a multi-step trajectory clustering approach that applies Dynamic Time Warping (DTW) to calculate trajectory similarity, followed by Principal Component Analysis (PCA) for dimensionality reduction. The reduced distance matrix was subsequently clustered using an improved K-Means algorithm to analyse AIS trajectories. Similarly, Wang et al. [69] performed vessel trajectory clustering based on the Hausdorff distance and the Hierarchical Density-Based Spatial Clustering of Applications with Noise (HDBSCAN) algorithm. DL techniques are particularly effective in capturing

complex, nonlinear data patterns. For example, Li et al. [40] employed offset reconstruction-based representation learning and deep Gaussian Mixture Model (GMM) estimation networks to reveal intricate clustering structures of vessel trajectories.

(3) Practical applications: Clustering outcomes are primarily applied in three domains: aquaculture (#8), smart agriculture (#1), and energy consumption (#7). Marine aquaculture (#8) represents a key IoMT application area. Adiono et al [2] developed an IoT system for seawater aquaculture that supports remote water quality monitoring. Coutinho and Boukerche [16] proposed an innovative four-layer architecture for smart marine aquaculture (#1) that integrates IoT and Internet of Underwater Things (IoUT) technologies with edge and cloud computing, as well as ML, to enable real-time data processing and intelligent farm monitoring. Beyond aquaculture, IoMT applications extend to marine environmental monitoring for reducing vessel energy emissions. For example, Xu et al. [75] reviewed IoT-based applications in marine environmental monitoring, emphasising the potential of IoT and big data to mitigate energy emissions and enhance marine environmental protection.

1.2.3 Analysis of Keyword Evolution Direction

From a temporal perspective illustrated in Fig. 1.3, the evolution of this research field can be divided into three primary phases: the technology accumulation phase (before 2015), the technology development phase (2015–2019), and the application expansion phase (from 2020 onward).

The technology accumulation phase concentrated on constructing fundamental theoretical frameworks and conducting early exploratory applications. During this period, big data technologies (#4) began to emerge, enabling researchers to address

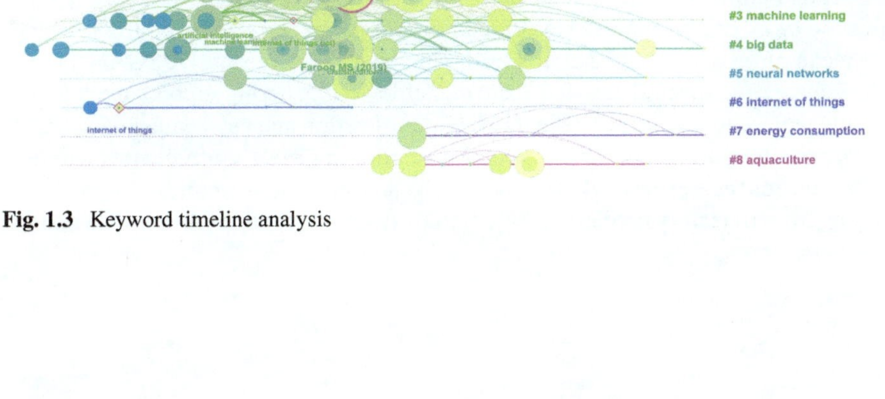

Fig. 1.3 Keyword timeline analysis

the complexity of maritime datasets. For example, Kim et al. [33] proposed a big data–driven approach to evaluate external forces acting on vessels in maritime traffic.

The technology development phase was characterised by rapid advances in DL (#0), ML (#3), and IoT (#6). Research emphasis shifted towards algorithm optimisation and deeper integration with shipping-related scenarios. For instance, Dobrkovic et al. [18] applied the Density-Based Spatial Clustering of Applications with Noise (DBSCAN) clustering algorithm to analyse AIS positional data, extracting key waypoints from navigation trajectories to enhance route analysis accuracy. Similarly, Orgaz Expósito [55] modelled European ports as a directed graph, grouped them using an improved spectral clustering algorithm, and incorporated DL techniques to predict vessels' next destinations, thereby advancing the dynamic analysis of maritime networks.

The application expansion phase reflected a transition from technology-centred exploration to practical implementation, with aquaculture (#8) and energy consumption (#7) emerging as prominent application domains. This stage highlighted the integration of intelligent shipping with sustainable development objectives. For example, Flores-Iwasaki et al. [19] investigated the application of sensors and IoT in monitoring water quality for aquaculture, identifying key knowledge gaps. Similarly, Han et al. [24] compared 18 DL and ML models for predicting marine fuel oil consumption, providing crucial decision support for shipping companies and ports in operational optimisation and green shipping initiatives.

Overall, the progression of ML and data clustering in the IoMT domain exhibits a clear trajectory from foundational technological exploration, through methodological breakthroughs, to widespread practical deployment. As these technologies mature further, their contributions to intelligent shipping and sustainable maritime development are expected to expand significantly.

1.3 Methods in IoMT

Methodological advances in IoMT have evolved along two principal axes: data clustering and ML. Clustering methods provide the foundation for understanding heterogeneous and large-scale maritime data, while ML approaches build upon these representations to enable prediction, detection and optimisation. IoMT data are inherently heterogeneous, noisy and spatially distributed, often generated by AIS, underwater sensors and unmanned aerial vehicles (UAVs). Before high-level prediction or optimisation tasks can be applied, it is crucial to group, structure and simplify raw data into meaningful patterns. Clustering techniques such as K-means, DBSCAN, hierarchical clustering, spectral clustering and model-based clustering support applications in route pattern recognition, traffic hotspot identification and environmental sensing. ML approaches, including support vector machines (SVMs), reinforcement learning (RL), recurrent neural networks (RNNs) and Transformer architectures, support tasks ranging from trajectory prediction and anomaly detection to multi-source data fusion.

As illustrated in Fig. 1.4, the methodological trajectory of IoMT research from 2015 to 2025 shows a clear progression. Early studies relied heavily on relatively simple approaches such as SVM and DBSCAN, suitable for small- to medium-scale datasets. From the mid-phase onward, more sophisticated clustering techniques such as hierarchical clustering, spectral clustering, and model-based approaches were introduced to capture complex spatiotemporal structures. In parallel, advances in ML, particularly DL (RNNs, Transformers) and RL, have been increasingly adopted to address prediction, optimisation, and decision-making in dynamic maritime environments.

This dual evolution, from clustering-based unsupervised structuring to ML-based intelligent decision-making, reflects the field's growing need to both organise complex data and exploit it for higher-level autonomy.

To provide practical guidance, Table 1.1 compares the main data clustering and ML methods in terms of descriptions, strengths, disadvantages and suitability for

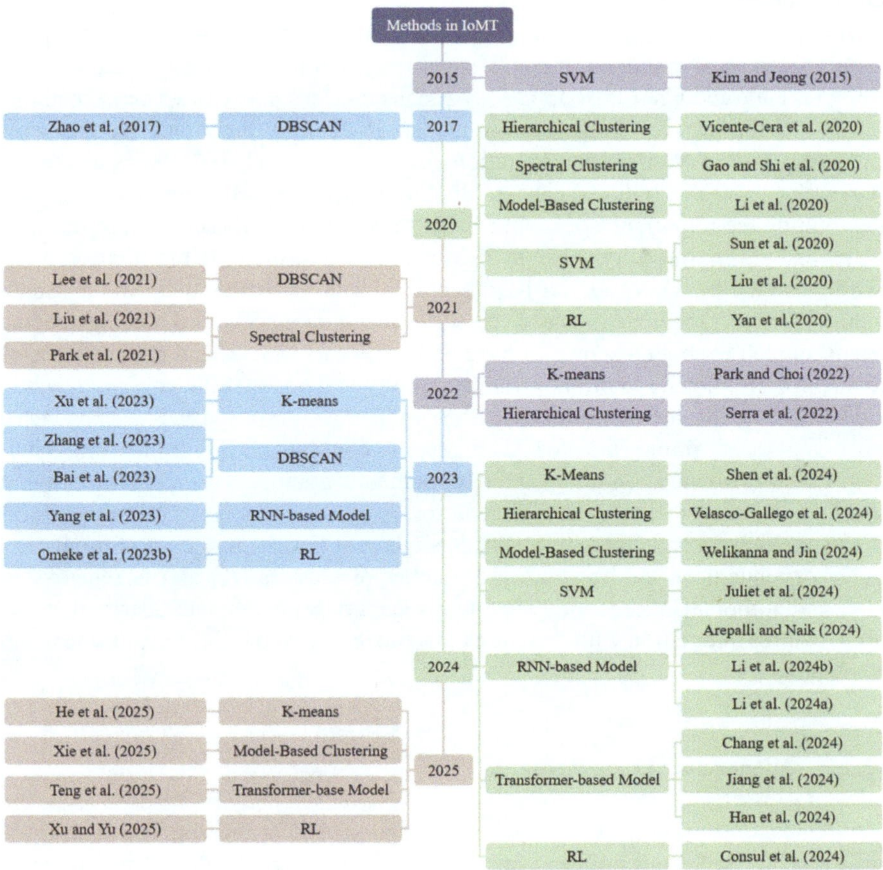

Fig. 1.4 Progression of IoMT methods

IoMT scenarios. This comparison provides researchers and engineers with select algorithms tailored to specific tasks, while also clarifying how clustering and ML complement each other in constructing robust IoMT systems.

1.3.1 Data Clustering in IoMT

1.3.1.1 K-means Clustering

The K-Means algorithm [3] is a widely adopted and computationally efficient clustering technique that partitions data points into K groups by assigning each point to its nearest centroid. In the IoMT, K-Means plays a significant role in optimising vessel scheduling and route identification while enhancing the efficiency and stability of communication networks. Its advantages are especially evident in UAV swarms and underwater sensor networks.

When combined with AIS data, K-Means enables the identification of typical vessel operational patterns, thereby supporting maritime decision-making. For example, Park and Choi [56] applied K-Means clustering to extract representative operational profiles, assisting decision-makers in defining functional requirements for vessels tasked with similar operations. Xu et al. [74] used K-Means clustering to examine how the initial database influences navigational prediction accuracy, highlighting the importance of input feature selection in database construction.

Beyond vessel trajectory analysis, K-Means also contributes to maritime communication networks. He et al. [26] applied K-Means to cluster underwater sensor nodes (USNs) to facilitate cluster-head selection, using a weighted scoring scheme that accounted for both distance to surface sink sensors (SSNs) and residual energy. This approach improved communication stability and energy efficiency in dynamic underwater environments. Similarly, Shen et al. [62] proposed a K-Means-based self-organisation framework for UAV swarm communication, addressing challenges such as signal attenuation and environmental complexity, thereby improving communication efficiency in IoMT scenarios.

Despite these advantages, K-Means has inherent limitations. Its performance is highly dependent on the pre-specified number of clusters (K) and is sensitive to noise and outliers. Consequently, its effectiveness depends on data distribution and often requires integration with complementary techniques to improve robustness and adaptability.

1.3.1.2 Hierarchical Clustering

Hierarchical clustering [51] is a classical unsupervised learning method that organises data into nested groupings, typically visualised using dendrograms. Its interpretability makes it attractive for maritime applications, although its computational

Table 1.1 Method summary

Method	Description	Advantages	Limitations	IoMT application example
K-means clustering	Cluster data by assigning points to the nearest K centroids	High computational efficiency, suitable for large-scale datasets	Sensitive to initial centroid selection, it requires a predefined *K* value	Ship trajectory analysis, common route pattern recognition
Hierarchical clustering	Cluster data based on hierarchical relationships	Able to discover multi-level structures in data	High computational cost for large datasets, challenging for dynamic data	Hierarchical analysis of ship movement patterns
DBSCAN	Density-based clustering that identifies clusters based on point density and can identify noise points	It can automatically detect the number of clusters, robust to noise	Performs poorly with clusters of varying density, sensitive to parameter selection	Abnormal trajectory detection, ship behaviour anomaly identification
Spectral clustering	Utilises eigenvalues and eigenvectors of data similarity matrices to map data to a lower-dimensional space	Excels at handling high-dimensional and nonlinear data	High computational cost, particularly with large datasets	High-dimensional ship trajectory data analysis, ship behaviour pattern recognition
Model-based clustering	Cluster data based on statistical models	Provides probabilities for data points belonging to each cluster	It can be computationally expensive	Complex data structure analysis, such as recognising variable sailing patterns
SVM	Classifies and regresses by finding the optimal hyperplane that separates data	Handles high-dimensional data well, strong ability to classify nonlinear data	Sensitive to noise and imbalanced data, computationally expensive	Ship behaviour classification, such as distinguishing normal and abnormal behaviours
RNN-based model	RNN processes sequential data, ideal for capturing temporal dependencies	Handles long-term dependencies in sequential data	Susceptible to the vanishing gradient problem, computationally inefficient	Ship trajectory prediction, forecasting future route changes

(continued)

Table 1.1 (continued)

Method	Description	Advantages	Limitations	IoMT application example
Transformer-based model	A deep learning model based on a self-attention mechanism	Powerful feature extraction ability excels in large-scale datasets	High computational cost, long training time	Multi-ship coordination analysis, ship behaviour pattern recognition
RL	Learns optimal strategies through interactions with the environment	Optimises decision-making processes, such as route planning, scheduling, etc.	Requires large amounts of interaction data for training, slow convergence, and high computational resources	Ship route optimisation, scheduling, decision-making, and risk forecasting

demands limit its practicality for very large-scale datasets, restricting its use to small- and medium-sized cases.

In the maritime domain, hierarchical clustering has been widely used to address diverse practical problems. For instance, Vicente-Cera et al. [67] combined vessel diversity metrics with agglomerative hierarchical clustering to identify five homogeneous cruise regions. These regions are closely aligned with the boundaries typically defined by cruise associations, providing a detailed segmentation and a deeper understanding of cruise traffic behaviour across European waters.

The method is also effective in maritime safety and failure analysis, where interpretability is critical. Velasco-Gallego et al. [66] applied hierarchical clustering to system failure datasets, enabling the identification and classification of distinct anomalies and malfunctions, thereby generating labelled data for supervised learning. Similarly, Serra et al. [61] analysed 1079 maritime accidents from 2009 to 2019 using hierarchical clustering, uncovering latent causal relationships among different accident types. Their findings highlighted the differing roles of technical and human factors, with human error being particularly prevalent in occupational incidents.

1.3.1.3 DBSCAN

DBSCAN [17] is a density-based clustering algorithm that groups data points by evaluating their spatial density. A data point is assigned to a cluster if it has a sufficient number of neighbour points within a specified radius, otherwise, it is treated as noise. Unlike partitioning methods, DBSCAN does not require the number of clusters to be predefined, making it particularly suitable for irregularly shaped clusters and datasets with significant noise.

In the IoMT, DBSCAN has been widely applied to enhance the security and accuracy of maritime data analysis. For example, Zhang et al. [83] employed DBSCAN to detect routing attacks in Underwater Wireless Sensor Networks (UWSNs) by analysing features such as node energy consumption, forwarding behaviour and link quality. This approach effectively clustered malicious nodes, thereby improving the accuracy of intrusion detection schemes (IDSs).

DBSCAN has also been extensively used in trajectory clustering, where maritime data often exhibit irregular structures. Zhao et al. [84] developed an adaptive hierarchical clustering method based on DBSCAN, which clustered vessel trajectories according to similarity and improved maritime traffic awareness. Lee et al. [35] used DBSCAN to separately cluster vessel arrival and departure trajectories into distinct navigation phases, such as leg and turning segments, offering key data support for autonomous navigation and intelligent port management.

To overcome the limitations of traditional DBSCAN, particularly parameter sensitivity and computational overhead, enhanced variants have been proposed. For instance, Bai et al. [7] introduced an adaptive threshold selection for trajectory similarity and incorporated an improved K-Adaptive Nearest Neighbours (KANNs) method to adjust DBSCAN parameters. These modifications improved both clustering performance and computational efficiency for maritime navigation data.

1.3.1.4 Spectral Clustering

Spectral clustering [52] is a graph theory-based method that represents data as a graph and partitions it using eigenvalue decomposition of the Laplacian matrix. The resulting eigenvectors are projected into a lower-dimensional space, where a traditional clustering method (e.g. K-Means) is applied to generate clusters. This approach is particularly effective for detecting complex, nonlinear and irregularly shaped clusters, while also demonstrating robustness to noise.

In maritime applications, spectral clustering offers notable advantages for trajectory analysis, where vessel movements often display nonlinear and heterogeneous patterns. For example, Liu et al. [45] enhanced the spectral clustering algorithm by optimising the affinity distance function and the k-nearest neighbour method, yielding better accuracy and efficiency when clustering vessel trajectories in the busy ports of Dover and Calais. Park et al. [57] measured trajectory similarity using the longest common subsequence distance, thereby increasing the accuracy of vessel trajectory prediction models and providing robust data support for intelligent maritime collision avoidance systems. Gao and Shi [20] employed spectral clustering on trajectory segments to identify vessel behaviour patterns, offering precise technical support for route planning and collision-prevention strategies.

1.3.1.5 Model-Based Clustering

Model-based clustering approaches assume that data are generated from a mixture of underlying probability distributions, enabling flexible probabilistic modelling of complex datasets. Among these, GMMs [21] are particularly popular due to their ability to represent heterogeneous distributions.

In maritime applications, GMMs have been effectively applied to remote sensing and environmental monitoring. Welikanna and Jin [70] introduced a GMM–Expectation Maximisation (EM) framework enhanced with a stochastic ordering optimisation strategy, successfully segmenting Sentinel-1 SAR imagery into oil spills, open sea areas and vessels. This automated method eliminated the need for manual mask selection and improved both accuracy and efficiency, supporting disaster response and marine environment protection.

Another branch of model-based clustering leverages neural network approaches, such as Self-organising Maps (SOMs). Li et al. [42] developed a SOM-based clustering framework for maritime big data, with results closely aligned with the original datasets, confirming its stability and interpretability. Xie et al. [71] integrated SOM with PCA to perform functional clustering and validation of major ports from the South China Sea to ASEAN regions. The results revealed the hub roles of ports such as Singapore, Hong Kong, Shenzhen and Guangzhou in the regional shipping network, offering data-driven insights for port positioning, planning and sustainable maritime development.

Overall, model-based clustering provides the flexibility to capture the probabilistic nature of maritime data while accommodating both structured statistical models (e.g. GMM) and neural network–driven approaches (e.g. SOM). Its adaptability makes it an indispensable tool in IoMT for applications ranging from environmental monitoring to port function analysis.

1.3.2 ML in IoMT

1.3.2.1 SVM

SVM [78] is an ML method with strong generalisation capabilities. It is well-suited for handling nonlinear problems in high-dimensional spaces and is effective under small-sample conditions. Its core concept is the construction of an optimal hyperplane to separate classes or, in regression tasks, to fit a predictive curve. In the maritime domain, SVM has been applied to integrate sensor and navigational data for accurate modelling of marine environments, vessel operations, and port activities, thereby supporting maritime safety management and environmental protection.

In IoMT, SVM has shown particular advantages. Juliet et al. [31] integrated IoT with SVM to predict pollutant distribution in aquatic habitats, providing valuable insights for long-term pollution control. Sun et al. [64] developed an adaptive kernel selection SVM that automatically matched optimal kernels to environmental factors, improving prediction efficiency and robustness. For trajectory analysis, Liu et al. [44] proposed an online multi-output least squares support vector regression model for real-time vessel trajectory prediction, while Kim and Jeong [32] optimised SVM models for port traffic data to detect abnormal navigational behaviours and support risk management.

1.3.2.2 RNN-Based Model

RNNs and their variants, such as Long Short-Term Memory (LSTM), Gated Recurrent Unit (GRU), Bidirectional LSTM (BiLSTM), and Bidirectional GRU (BiGRU), are effective in modelling sequential data with temporal dependencies by retaining both short- and long-term information. These models are widely applied in IoMT and aquaculture monitoring. For example, Arepalli and Naik [6] employed a lightweight Spatial Shared Attention LSTM (SSA-LSTM) to accurately capture the spatiotemporal variations in dissolved oxygen, achieving efficient early warnings of hypoxia risks. Yang et al. [80] used RNNs with IoT sensors to achieve high-precision predictions, improving aquaculture efficiency.

RNN-based models also hold substantial potential for vessel trajectory prediction, with RNN-based models receiving particular research attention. Li et al. [39] combined BiLSTM and BiGRU in a specific sequence to construct a three-layer information-enhanced architecture, leveraging their complementary strengths in time-series modelling and feature extraction. This bidirectional information fusion cascade network achieved higher trajectory prediction accuracy, enhancing collision avoidance and navigational safety. Furthermore, Li et al. [36] optimised the weights of trajectory features extracted by BiLSTM and BiGRU using an attention mechanism, with adaptive adjustments to produce the final predictions. This approach demonstrated exceptional performance in mixed navigation environments involving both manned vessels and Maritime Autonomous Surface Ships (MASS).

1.3.2.3 Transformer-Based Model

A Transformer architecture has emerged as a key tool in IoMT and IoUT due to its strengths in sequence modelling, multimodal fusion and parallel computation.

In marine life monitoring, Transformers have been primarily used to integrate temporal and visual information for accurate prediction of dynamic behaviours. For example, temporal and visual data are fused for early detection of behavioural changes [10]. This approach enabled early detection of low-activity states, facilitating real-time aquaculture warnings and reducing mortality rates. Jiang et al. [30] introduced an efficient Vision Transformer (ViT) with token-selective and merging strategies (TSMVTs), which can be flexibly integrated into various IoUT systems, substantially enhancing fine-grained underwater classification performance. Similarly, Teng et al. [65] proposed MarineEdgeformer, a lightweight Transformer-based time-series prediction model employing a linear attention mechanism and an optimised decoder architecture. In maritime communication optimisation, Han et al. [23] combined Transformer and Turbo architectures to achieve semantic-level joint source, channel coding, improving reliability and adaptability under dynamic maritime channel conditions. Overall, Transformers extend IoMT research from single-source modelling to multimodal perception, communication optimisation, and edge deployment, supporting the development of smart oceans.

1.3.2.4 RL

RL offers online learning and adaptive decision-making, making it highly suitable for the dynamic environments of IoMT. Omeke et al. [53] reviewed the applications and recent progress of RL in addressing critical IoUT challenges, showing its advantages in managing variability in underwater channels. In network optimization and design, Xu and Yu [76] addressed the complexity and limited resources of maritime and IoMT architectures by proposing a cooperative computation offloading and resource allocation algorithm based on a Double Duelling Deep Q-Network (D3QN). This approach simultaneously optimised task offloading decisions and communication–computation resource allocation, reducing latency and energy consumption while significantly improving resource utilisation and task matching efficiency. Similarly, Consul et al. [15] focused on enhancing communication and energy efficiency in maritime IoT by employing deep reinforcement learning (DRL) to accurately identify faulty or abnormal sensors, strengthen the stability of underwater monitoring networks, and optimise routing strategies from seabed sensors to terrestrial networks, thereby reducing hop counts and transmission delays. In Autonomous Underwater Vehicle (AUV) research, Omeke et al. [54] proposed a simultaneous wireless information and power transfer (SWIPT) scheme for AUV-enabled underwater networks. This method achieved high throughput and wireless charging, extending node lifetimes while reducing environmental impacts. The joint maximisation of throughput and harvested power was formulated as a Markov Decision Process (MDP) and solved using model-free RL, with a tailored reward function guiding AUV trajectory planning to improve communication efficiency and reduce energy consumption. Similarly, Yan et al. [79] developed an RL-based localisation algorithm that applied online value iteration to optimise the positions of AUVs and sensor nodes, achieving higher localisation accuracy and energy efficiency.

1.4 Real-World Applications

With the continuous advancement of the IoMT, an expanding range of practical applications has emerged where ML and data clustering play pivotal roles in enhancing safety, optimising operations and protecting the environment (Fig. 1.5). By enabling real-time monitoring and analysis of marine environments, vessel behaviours and port activities, these techniques allow for more accurate risk prediction, operational streamlining and overall efficiency improvement.

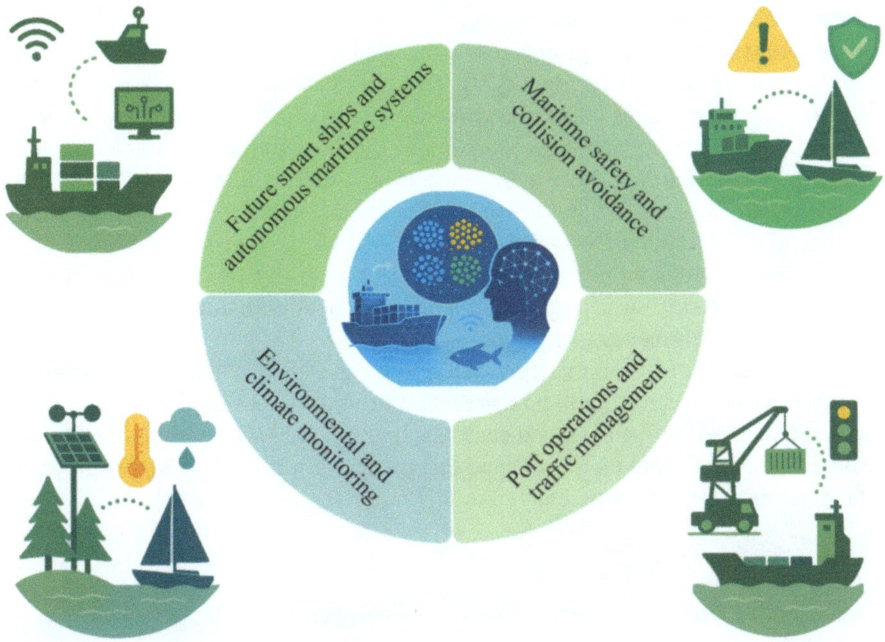

Fig. 1.5 Real-world applications of data clustering, ML and IoMT

1.4.1 Maritime Safety and Collision Avoidance

Traditional navigation methods, although effective, are constrained by manual operation and the inherent complexity of marine environments. The integration of ML addresses these challenges by enabling vessels to analyse their surroundings in real time and predict potential collision risks. Through data clustering and DL algorithms, vessels can identify and track nearby ships, obstacles and weather changes, allowing for early warnings.

A bibliometric analysis by Xue et al. [77] indicates that ML methods in maritime safety are evolving towards more systematic and integrated approaches, with future research expected to converge proactive safety management, advanced technologies and intelligent algorithms, providing the theoretical foundation for autonomous navigation.

Data clustering techniques have also demonstrated significant potential in maritime safety. For instance, Lee et al. [34] analysed AIS data to quantitatively derive vessel traffic routes. Their method employed the Hausdorff distance to segment vessel trajectories, simplified them using the Douglas–Peucker algorithm, and applied DBSCAN clustering to identify waypoints, ultimately producing safe route maps. The resulting safe route maps, complemented with kernel density estimation to analyse route widths, provide a robust basis for optimising major shipping lanes.

In vessel collision avoidance, ML applications are critical. Xin et al. [72] developed a method for extracting, characterising, and sampling maritime traffic navigation scenarios, accurately identifying real-world multi-vessel encounter situations and precisely representing navigational contexts. This method supports the encoding, interpretation, and understanding of dynamic traffic behaviours, systematically selecting representative maritime traffic scenarios through multiple selection principles to provide reliable data for collision prediction. In another study, Zhang et al. [81] applied Dempster–Shafer theory to infer give-way vessel intentions and integrated nonlinear velocity obstacles with the NSGA-II algorithm to optimise manoeuvring strategies. This method balances collision avoidance timing with navigational efficiency.

The application of ML in maritime safety is progressively enhancing capabilities in collision avoidance, navigational risk prediction and proactive safety management. With continued technological advancements, future research is expected to focus on the development of more intelligent and integrated maritime safety systems, further advancing the safety and efficiency of autonomous shipping.

1.4.2 Port Operations and Traffic Management

Clustering and ML technologies become increasingly important in port operations, enabling improved efficiency, optimised resource allocation and enhanced safety. By grouping data on vessels, cargo or port facilities, clustering algorithms can identify patterns and characteristics of different groups, providing scientific support for operational decision-making. For example, Hou et al. [27] improved a port state control inspection decision model using a K-Means clustering algorithm. Their model classified ships into risk categories, improving inspection efficiency and safety outcomes. ML techniques are also widely applied in port management. Mekkaoui et al. [47] used ML and historical vessel trajectory data to predict arrival times at destination ports and provided a structured review on how ML contributes to terminal operations. Addressing the trade-off between operational efficiency and energy consumption in port operations, Cai et al. [9] proposed a three-objective mixed-integer programming model to optimise port scheduling, balancing time energy, and robustness using a dynamic multi-objective algorithm.

ML has also demonstrated significant value in traffic management, particularly vessel traffic flow prediction. Xing et al. [73] employed a Bayesian Gaussian tensor decomposition model with empirical mode decomposition and DTW, improving navigation safety and efficiency. Similarly, Li et al. [43] combined periodic, temporal, and semantic perspectives, applying a Convolutional Neural Network (CNN)-BiLSTM model to capture periodic and temporal features, while using a Graph Convolutional Network (GCN) to extract inter-regional relationships. This method achieves high prediction accuracy and computational efficiency, supporting port planning.

Overall, data clustering and ML have not only advanced the intelligence of port operations but also delivered unprecedented optimisation in traffic management. Their applications in port traffic management enhance safety and economic benefits while offering valuable directions and methodologies for future research and practice.

1.4.3 Environmental and Climate Monitoring

The integration of ML and IoT has significantly strengthened environmental monitoring by improving predictive modelling, real-time analytics and automated decision-making. Miller et al. [48] noted that such integration enhances data accuracy, reduces costs and improves scalability, offering powerful tools for addressing complex marine and climate dynamics. ML also supports environmental performance analysis. Skarlatos et al. [63] applied clustering and PCA to classify vessels into four groups based on 14 features, providing insights into emission-reduction policies and optimised operational strategies.

IoT delivers irreplaceable real-time sensing and data transmission capabilities in environmental monitoring. Addressing limitations in data utilisation and decision support within IoUT, Razzaq et al. [59] proposed an architecture and algorithm integrating predictive analytics, significantly improving the efficiency of converting sensor data into actionable insights. Bates et al. [8] deployed a Long Range Wide Area Network (LoRaWAN)-based marine sensor network in Australia's Clyde River to monitor water and air parameters, reducing oyster mortality from heat exposure and safeguarding aquaculture production.

The combination of ML and IoT in environmental monitoring not only optimises data acquisition, processing, and analysis workflows but also drives the transition from experience-driven to data-driven environmental management. ML is effective in uncovering latent patterns and identifying key features in environmental datasets, while IoT provides high-density, low-latency real-time data streams. Their integration shows broad application potential in water resource management, ship environmental performance evaluation and aquaculture risk prevention.

1.4.4 Future Smart Ships and Autonomous Maritime Systems

ML and clustering analysis play a pivotal role in developing intelligent vessels and smart maritime systems, particularly in route planning, vessel behaviour modelling, and trajectory prediction.

Clustering methods are especially effective for extracting navigational patterns from the vast and complex AIS data, thereby enhancing the decision-making accuracy and adaptability of autonomous systems. For instance, Li and Yang [41] developed a new framework based on AIS data that integrates novel feature measurements, pattern extraction, and route planning algorithms, providing a systematic approach for unsupervised route planning in MASS. Similarly, Murray and Perera [50] examined historical vessel behaviours in specific geographic regions by clustering trajectory segments, compressing them using the Karhunen–Loéve transform, and applying a GMM. Unlike traditional full-area clustering approaches, their method emphasises similarity among relevant trajectory segments, resulting in a more accurate representation of navigational behaviour. Addressing the limitations of redundant AIS trajectory data, Li et al. [38] proposed a multi-stage approach that combines trajectory mapping with an improved DBSCAN algorithm. This method incorporates merged distance metrics and multidimensional scaling to improve trajectory similarity estimation, thereby enhancing the accuracy and robustness of trajectory pattern recognition.

In the ML domain, the introduction of DL has substantially advanced the trajectory prediction capabilities of intelligent maritime systems. Jiao et al. (2023) explored six DL models for vessel trajectory prediction, incorporating factors such as latitude, longitude, course over ground and speed over ground, and provided comparative insights into their suitability across MASS contexts. Gong et al. [22] further extended this field by integrating spatiotemporal features with probabilistic trajectory modelling, using Gaussian heatmaps and ViTs to quantify prediction uncertainty. Their approach addressed long-standing challenges in uncertainty modelling and improved both the safety and environmental adaptability of autonomous navigation in dynamic maritime environments.

Overall, clustering methods in intelligent vessel systems are most effective for pattern recognition and behavioural grouping, allowing unsupervised discovery of navigational strategies and environmental adaptation. ML, on the other hand, serves as the core driver of predictive modelling, delivering accurate, robust decision support for autonomous navigation. The integration of clustering with ML therefore not only enhances route planning and risk assessment workflows but also accelerates the evolution of MASS and related smart maritime systems towards greater precision, safety and resilience.

1.5 Challenges and Future Directions

The rapid development of the IoMT has led to the deployment of a vast number of sensors, devices and systems that continuously collect and transmit large volumes of maritime data. However, the effective processing and utilisation of this data face various challenges. In particular, issues such as data quality, real-time processing, cybersecurity and the integration of emerging technologies remain critical bottlenecks that shape the pace of digital transformation in the maritime sector. This section discusses these challenges and highlights potential future directions.

1.5.1 Data Quality, Uncertainty and Missing Values

High-quality data is essential for ensuring the reliability and accuracy of IoMT systems. However, data uncertainty and incompleteness remain significant obstacles. For instance, Gong et al. [22] reported inconsistencies in AIS data and highlighted limitations in existing methods for handling trajectory uncertainty. Similarly, Adhikari et al. [1] identified missing data processing as a major challenge in real-world IoT data integration. In IoMT contexts, missing data frequently occur due to sensor malfunctions, network disruptions or interference from marine environmental factors. Such issues compromise system reliability, hinder data-driven analysis and reduce the accuracy of ML models.

Future research should therefore focus on robust mechanisms for uncertainty quantification, missing data imputation and error correction, ensuring that IoMT systems can deliver consistent and trustworthy outputs under real-world operational conditions.

1.5.2 Real-Time Processing and Scalability

IoMT generates high-frequency, real-time data from vessels, ports and channel monitoring systems. Processing this information in a timely and scalable manner is crucial for enabling rapid feedback to support navigation, port operations and safety management. Traditional centralised computing architectures often fail to meet these demands, struggling with latency and single points of failure. Distributed systems provide a partial solution by distributing computational tasks across multiple nodes, thereby improving efficiency and robustness [29]. Meanwhile, edge computing has emerged as a promising approach for maritime contexts, where offshore environments are often distant from central servers. By shifting computation closer to the data source, edge computing reduces transmission delays, improves responsiveness and eases the burden on centralised data centres [12]. These developments suggest that hybrid architectures combining distributed and edge computing will be critical for meeting real-time IoMT requirements at scale.

1.5.3 Cybersecurity and Data Privacy in IoMT

While the widespread adoption of IoMT has brought significant operational benefits, it has also introduced severe cybersecurity challenges. Sensitive data on vessel operations, cargo tracking and port activities, if leaked or manipulated, could disrupt global shipping networks and pose national security threats. Akpan et al. [4] conducted an in-depth investigation into the cybersecurity landscape of the maritime sector, identifying a range of security challenges. Their analysis revealed system vulnerabilities on board vessels that could be exploited by attackers, as well as the potential consequences of such exploitation. Similarly, Mohsendokht et al. [49] emphasised the cascading consequences of disruptions within maritime systems.

To mitigate these threats, stronger investment in cybersecurity is needed. Key strategies include end-to-end encryption, multi-factor authentication, strict access controls, and resilient incident response mechanisms. Beyond technical measures, cybersecurity awareness, crew training and international regulatory harmonisation are also vital to building resilience across the maritime sector.

1.5.4 Trends in Maritime AI and Digital Twin Technologies

The application of AI and digital twin technologies is rapidly reshaping maritime operations [46, 85]. Digital twins create virtual models of vessels and port facilities that enable real-time monitoring of operational status, predictive maintenance and optimised decision-making. When integrated with AI, digital twins can analyse both historical and real-time data to forecast equipment failures, optimise operations, and improve safety.

In vessel management, digital twin technology provides multidimensional data on navigation status, fuel consumption and mechanical performance, enhancing operational transparency. When combined with AI analytics, these digital replicas allow for the extraction of actionable insights from large-scale datasets. Such integration supports the early prediction of equipment failures or performance degradation, enabling proactive maintenance strategies that minimise downtime, reduce repair costs and improve overall operational efficiency.

In port operations, the combination of digital twin and AI technologies facilitates real-time monitoring of cargo flows, loading and unloading activities and equipment conditions, thereby ensuring precise operational oversight. AI models further exploit both historical and real-time datasets to forecast vessel arrivals and cargo volumes, optimising berthing allocation and scheduling. This reduces vessel waiting times, alleviates congestion and enhances port throughput. In maritime safety, digital twin models and AI-driven analytics provide continuous monitoring of the marine environment, weather conditions and vessel movements. By simulating diverse navigational scenarios, these technologies enable the identification of potential risks, such as collisions or groundings, at an early stage. AI-based recommendations on optimal sailing routes help vessels avoid hazardous areas and adverse weather, thereby significantly reducing accident risks.

The continued development and application of AI and digital twin technologies are expected to accelerate the maritime industry's transition towards greater intelligence and automation. These advancements promise substantial improvements in efficiency, safety, and sustainability, while simultaneously lowering operational costs and enhancing the resilience of global shipping networks.

1.6 Conclusions

Existing research on IoMT remains fragmented, often limited to isolated case studies or single algorithms, and lacks systematic reviews or comparative analyses. The absence of an integrated methodological framework continues to constrain cross-method evaluation and optimisation, thereby limiting the broader advancement of IoMT applications. This study addresses this gap by conducting a systematic review of clustering and ML applications in IoMT. Drawing on WoS data and CiteSpace analysis, research hotspot maps and temporal evolution trajectories were generated, identifying three overarching domains: data acquisition, data analysis, and real-world applications. Within this framework, clustering methods, including K-Means, DBSCAN, spectral clustering and GMM, were critically compared with ML approaches such as Support SVM, RNN, Transformer architectures and RL. Their principles, advantages and limitations were assessed in relation to key IoMT tasks such as trajectory clustering, anomaly detection and multi-source data fusion. Case studies in maritime safety, port operations, environmental monitoring and autonomous navigation further demonstrated the practical relevance of these methods, underscoring their capacity to advance maritime intelligence.

Future research and development in IoMT should focus on enhancing data quality and reliability, alongside the adoption of distributed and edge-computing architectures to enable low-latency, real-time processing. Strengthening cybersecurity measures will also be essential to safeguard critical maritime data and infrastructure. The convergence of AI and digital twin technologies is expected to facilitate the creation of dynamic, high-fidelity virtual replicas of vessels and ports, supporting continuous monitoring, predictive analysis and scenario-based decision-making. In addition, emerging AI models such as Large Language Models (LLMs) and multi-modal Transformers hold significant potential for uncovering latent patterns within complex multi-source datasets. Their integration will not only advance proactive decision-making but also accelerate the transition towards intelligent, autonomous maritime operations.

Acknowledgements This work is supported by the European Research Council (ERC) under the European Union's Horizon 2020 research and innovation programme (Grant Agreement No. 864724).

References

1. Adhikari, D., Jiang, W., Zhan, J., He, Z., Rawat, D.B., Aickelin, U., Khorshidi, H.A.: A comprehensive survey on imputation of missing data in internet of things. ACM Comput. Surv. **55**, 133:1–133:38 (2022). https://doi.org/10.1145/3533381
2. Adiono, T., Toha, A.M., Pamungkas, S., Sutisna, N., Sumiarsih, E.: Internet of things for marine aquaculture. In: 2021 International Symposium on Electronics and Smart Devices (ISESD), pp. 1–5. IEEE (2021)

3. Ahmed, M., Seraj, R., Islam, S.M.S.: The k-means algorithm: a comprehensive survey and performance evaluation. Electronics **9**, 1295 (2020)
4. Akpan, F., Bendiab, G., Shiaeles, S., Karamperidis, S., Michaloliakos, M.: Cybersecurity challenges in the maritime sector. Network **2**, 123–138 (2022). https://doi.org/10.3390/network2010009
5. Al-Zaidi, R., Woods, J.C., Al-Khalidi, M., Hu, H.: Building novel VHF-based wireless sensor networks for the internet of marine things. IEEE Sens. J. **18**, 2131–2144 (2018)
6. Arepalli, P.G., Naik, K.J.: A deep learning-enabled IoT framework for early hypoxia detection in aqua water using light weight spatially shared attention-LSTM network. J. Supercomput. **80**, 2718–2747 (2024). https://doi.org/10.1007/s11227-023-05580-x
7. Bai, X., Xie, Z., Xu, X., Xiao, Y.: An adaptive threshold fast DBSCAN algorithm with preserved trajectory feature points for vessel trajectory clustering. Ocean Eng. **280**, 114930 (2023)
8. Bates, H., Pierce, M., Benter, A.: Real-time environmental monitoring for aquaculture using a LoRaWAN-based IoT sensor network. Sensors **21**, 7963 (2021). https://doi.org/10.3390/s21237963
9. Cai, L., Li, W., Li, H., Zhou, B., He, L., Guo, W., Yang, Z.: Incorporation of energy-consumption optimization into multi-objective and robust port multi-equipment integrated scheduling. Transp. Res. Part C: Emerg. Technol. **166**, 104755 (2024)
10. Chang, C.-T., Lan, H.-Y., Cheng, S.-C., Lu, D.-H., Chang, C.-C.: A multimodal transformer model for fish vitality prediction using time-series water temperature data and underwater videos. In: Sixth International Conference on Image, Video Processing, and Artificial Intelligence (IVPAI 2024), pp. 108–113. SPIE (2024)
11. Chao, H.-C., Wu, H.-T., Tseng, F.-H.: AIS Meets IoT: a network security mechanism of sustainable marine resource based on edge computing. Sustainability **13**, 3048 (2021). https://doi.org/10.3390/su13063048
12. Charpentier, V., Slamnik-Kriještorac, N., Landi, G., Caenepeel, M., Vasseur, O., Marquez-Barja, J.M.: Paving the way towards safer and more efficient maritime industry with 5G and beyond edge computing systems. Comput. Netw. **250**, 110499 (2024). https://doi.org/10.1016/j.comnet.2024.110499
13. Chen, C.: CiteSpace II: Detecting and visualizing emerging trends and transient patterns in scientific literature. J. Am. Soc. Inf. Sci. Technol. **57**, 359–377 (2006). https://doi.org/10.1002/asi.20317
14. Chi, H., Du, Y., Brett, P.M.: Design of a marine environment monitoring system based on the Internet of Things. J. Coastal Res. **110**, 256–260 (2020)
15. Consul, P., Budhiraja, I., Garg, D.: Deep reinforcement learning based reliable data transmission scheme for internet of underwater things in 5G and beyond networks. Procedia Comput. Sci. **235**, 1752–1760 (2024)
16. Coutinho, R.W., Boukerche, A.: Towards a novel architectural design for IoT-based smart marine aquaculture. IEEE Internet Things Mag. **5**, 174–179 (2022)
17. Deng, D.: DBSCAN clustering algorithm based on density. In: 2020 7th International Forum on Electrical Engineering and Automation (IFEEA), pp. 949–953. IEEE (2020)
18. Dobrkovic, A., Iacob, M.-E., Van Hillegersberg, J.: Using machine learning for unsupervised maritime waypoint discovery from streaming AIS data. In: Proceedings of the 15th International Conference on Knowledge Technologies and Data-Driven Business. Presented at the i-KNOW '15: 15th International Conference on Knowledge Technologies and Data-Driven Business, ACM, Graz Austria, pp. 1–8 (2015). https://doi.org/10.1145/2809563.2809573
19. Flores-Iwasaki, M., Guadalupe, G.A., Pachas-Caycho, M., Chapa-Gonza, S., Mori-Zabarburú, R.C., Guerrero-Abad, J.C.: Internet of things (IoT) Sensors for water quality monitoring in aquaculture systems: a systematic review and bibliometric analysis. AgriEngineering **7**, 78 (2025)
20. Gao, M., Shi, G.-Y.: Ship-handling behavior pattern recognition using AIS sub-trajectory clustering analysis based on the T-SNE and spectral clustering algorithms. Ocean Eng. **205**, 106919 (2020)

21. Gogebakan, M.: A novel approach for Gaussian mixture model clustering based on soft computing method. IEEE Access **9**, 159987–160003 (2021)
22. Gong, J., Li, H., Jiao, H., Yang, Z.: Uncertainty-aware ship trajectory prediction via Spatio-Temporal Graph Transformer. Transp. Res. Part E: Logist. Transp. Rev. **203**, 104315 (2025). https://doi.org/10.1016/j.tre.2025.104315
23. Han, X., Lin, B., Wu, N., Wang, P., Na, Z., Zhang, M.: Design of a turbo-based deep semantic autoencoder for marine Internet of Things. Internet Things **28**, 101393 (2024)
24. Han, P., Li, S., Liu, Z., Sun, Z., Yan, C.: Ship fuel oil consumption prediction at sea and in port considering sustainable maritime industry: a comparative study of machine learning and deep learning approaches. Proc. Inst. Mech. Eng. Part M: J. Eng. Marit. Environ. 14750902251336505 (2025). https://doi.org/10.1177/14750902251336505
25. Hassan, A., Khokhar, S.S.: Internet of things-enabled vessel monitoring system for enhanced maritime safety and tracking at sea. In: SoutheastCon 2024, pp. 250–259. IEEE (2024)
26. He, A., Pan, H., Dai, Y., Huang, X., Si, X., Yuen, C., Zhang, Y.: Distributed optimization in mobile edge intelligence for energy-efficient maritime IoT networks. IEEE Internet Things Mag. **8**, 125–132 (2025). https://doi.org/10.1109/IOTM.001.2400174
27. Hou, Z., Yan, R., Wang, S.: On the K-means clustering model for performance enhancement of port state control. J. Mar. Sci. Eng. **10**, 1608 (2022). https://doi.org/10.3390/jmse10111608
28. Hu, J., Kaur, K., Lin, H., Wang, X., Hassan, M.M., Razzak, I., Hammoudeh, M.: Intelligent anomaly detection of trajectories for IoT empowered maritime transportation systems. IEEE Trans. Intell. Transp. Syst. **24**, 2382–2391 (2023). https://doi.org/10.1109/TITS.2022.3162491
29. Huang, W., Liu, J., Liu, Y., Wei, L., Zhong, D.: MGC-Chain: a blockchain framework for distributed maritime computation. In: 2024 5th International Conference on Computer, Big Data and Artificial Intelligence (ICCBD+AI). Presented at the 2024 5th International Conference on Computer, Big Data and Artificial Intelligence (ICCBD+AI), pp. 550–554 (2024). https://doi.org/10.1109/ICCBD-AI65562.2024.00097
30. Jiang, Y., Zhang, Y., Wang, Y., Guo, Q., Zhao, M., Qin, H.: Efficient vision transformer with token-selective and merging strategies for autonomous underwater vehicles. IEEE Internet Things J. **11**, 32129–32142 (2024)
31. Juliet, P.S., Nithisha, J., Sridevi, V., Arunachalam, G., Rajanarayanan, S., Solainayagi, P.: IoT-driven SVM prediction of oceanographic data analytics for marine pollution dispersal patterns. In: 2024 Third International Conference on Electrical, Electronics, Information and Communication Technologies (ICEEICT). Presented at the 2024 Third International Conference on Electrical, Electronics, Information and Communication Technologies (ICEEICT), pp. 1–6 (2024). https://doi.org/10.1109/ICEEICT61591.2024.10718380
32. Kim, J.-S., Jeong, J.S.: Pattern recognition of ship navigational data using support vector machine. Int. J. Fuzzy Logic Intell. Syst. **15**, 268–276 (2015). https://doi.org/10.5391/IJFIS.2015.15.4.268
33. Kim, K.-I., Jeong, J.S., Park, G.-K.: Assessment of external force acting on ship using big data in maritime traffic. J. Korean Inst. Intell. Syst. **23**, 379–384 (2013)
34. Lee, J.-S., Lee, H.-T., Cho, I.-S.: Maritime traffic route detection framework based on statistical density analysis from AIS data using a clustering algorithm. IEEE Access **10**, 23355–23366 (2022). https://doi.org/10.1109/ACCESS.2022.3154363
35. Lee, H.-T., Lee, J.-S., Yang, H., Cho, I.-S.: An AIS data-driven approach to analyze the pattern of ship trajectories in ports using the DBSCAN algorithm. Appl. Sci. **11**, 799 (2021). https://doi.org/10.3390/app11020799
36. Li, H., Li, W., Wang, S., Yang, H., Guan, J., Zhang, Y.: STAD: ship trajectory anomaly detection in ocean with dynamic pattern clustering. Ocean Eng. **313**, 119530 (2024). https://doi.org/10.1016/j.oceaneng.2024.119530
37. Li, H., Liu, J., Liu, R.W., Xiong, N., Wu, K., Kim, T.: A dimensionality reduction-based multi-step clustering method for robust vessel trajectory analysis. Sensors **17**, 1792 (2017)
38. Li, H., Liu, J., Wu, K., Yang, Z., Liu, R.W., Xiong, N.: Spatio-temporal vessel trajectory clustering based on data mapping and density. IEEE Access **6**, 58939–58954 (2018)

39. Li, H., Xing, W., Jiao, H., Yang, Z., Li, Y.: Deep bi-directional information-empowered ship trajectory prediction for maritime autonomous surface ships. Transp. Res. Part E: Logist. Transp. Rev. **181**, 103367 (2024)
40. Li, H., Xing, W., Jiao, H., Yuen, K.F., Gao, R., Li, Y., Matthews, C., Yang, Z.: Bi-directional information fusion-driven deep network for ship trajectory prediction in intelligent transportation systems. Transp. Res. Part E: Logist. Transp. Rev. **192**, 103770 (2024)
41. Li, H., Yang, Z.: Incorporation of AIS data-based machine learning into unsupervised route planning for maritime autonomous surface ships. Transp. Res. Part E: Logist. Transp. Rev. **176**, 103171–103171 (2023)
42. Li, Y., Yang, Z., Han, K.: Research on the clustering algorithm of ocean big data based on self-organizing neural network. Comput. Intell. **36**, 1609–1620 (2020). https://doi.org/10.1111/coin.12299
43. Li, H., Zhang, Y., Li, Y., Lam, J.S.L., Matthews, C., Yang, Z.: Deep multi-view information-powered vessel traffic flow prediction for intelligent transportation management. Transp. Res. Part E: Logist. Transp. Rev. **197**, 104072 (2025)
44. Liu, J., Shi, G., Zhu, K.: Online multiple outputs least-squares support vector regression model of ship trajectory prediction based on automatic information system data and selection mechanism. IEEE Access **8**, 154727–154745 (2020). https://doi.org/10.1109/ACCESS.2020.3018749
45. Liu, H., Liu, Y., Zhang, L., Sun, H.: Research on ship track clustering method based on optimized spectral clustering algorithm. In: 2021 5th Asian Conference on Artificial Intelligence Technology (ACAIT). Presented at the 2021 5th Asian Conference on Artificial Intelligence Technology (ACAIT), pp. 39–44 (2021). https://doi.org/10.1109/ACAIT53529.2021.9731124
46. Madusanka, N.S., Fan, Y., Yang, S., Xiang, X.: Digital twin in the maritime domain: a review and emerging trends. J. Mar. Sci. Eng. **11**, 1021 (2023). https://doi.org/10.3390/jmse11051021
47. Mekkaoui, S.E., Benabbou, L., Berrado, A.: Machine learning models for efficient port terminal operations: case of vessels' arrival times prediction. In: IFAC-PapersOnLine, 10th IFAC Conference on Manufacturing Modelling, Management and Control MIM 2022, vol. 55, pp. 3172–3177 (2022). https://doi.org/10.1016/j.ifacol.2022.10.217
48. Miller, T., Durlik, I., Kostecka, E., Kozlovska, P., Łobodzińska, A., Sokołowska, S., Nowy, A.: Integrating artificial intelligence agents with the internet of things for enhanced environmental monitoring: applications in water quality and climate data. Electronics **14**, 696 (2025). https://doi.org/10.3390/electronics14040696
49. Mohsendokht, M., Li, H., Kontovas, C., Chang, C.-H., Qu, Z., Yang, Z.: Decoding dependencies among the risk factors influencing maritime cybersecurity: lessons learned from historical incidents in the past two decades. Ocean Eng. **312**, 119078 (2024). https://doi.org/10.1016/j.oceaneng.2024.119078
50. Murray, B., Perera, L.P.: Ship behavior prediction via trajectory extraction-based clustering for maritime situation awareness. J. Ocean Eng. Sci. **7**, 1–13 (2022). https://doi.org/10.1016/j.joes.2021.03.001
51. Murtagh, F., Contreras, P.: Algorithms for hierarchical clustering: an overview. WIREs Data Min. Knowl. **2**, 86–97 (2012). https://doi.org/10.1002/widm.53
52. Ng, A., Jordan, M., Weiss, Y.: On spectral clustering: analysis and an algorithm. Advances in Neural Information Processing Systems, vol. 14 (2001)
53. Omeke, K.G., Abubakar, A.I., Zhang, L., Abbasi, Q.H., Imran, M.A.: How reinforcement learning is helping to solve Internet-of-Underwater-Things problems. IEEE Internet Things Mag. **5**, 24–29 (2023)
54. Omeke, K.G., Mollel, M., Shah, S.T., Zhang, L., Abbasi, Q.H., Imran, M.A.: Toward a sustainable internet of underwater things based on AUVs, SWIPT, and reinforcement learning. IEEE Internet Things J. **11**, 7640–7651 (2023)
55. Orgaz Expósito, Á.: Deep learning & graph clustering for maritime logistics: predicting destination and expected time of arrival for vessels across Europe (2020)
56. Park, J., Choi, M.: A K-means clustering algorithm to determine representative operational profiles of a ship using AIS data. J. Mar. Sci. Eng. **10**, 1245 (2022). https://doi.org/10.3390/jmse10091245

57. Park, J., Jeong, J., Park, Y.: Ship trajectory prediction based on Bi-LSTM using spectral-clustered AIS data. J. Mar. Sci. Eng. **9**, 1037 (2021). https://doi.org/10.3390/jmse9091037
58. Pranckutė, R.: Web of Science (WoS) and Scopus: the Titans of bibliographic information in today's academic world. Publications **9**, 12 (2021). https://doi.org/10.3390/publications9010012
59. Razzaq, A., Mohsan, S.A.H., Li, Y., Alsharif, M.H.: Architectural framework for underwater IoT: forecasting system for analyzing oceanographic data and observing the environment. J. Mar. Sci. Eng. **11**, 368 (2023). https://doi.org/10.3390/jmse11020368
60. Ruiz, D.L.: The impact of the internet of things on marine industry operations. Am. J. Mar. Eng. Technol. **6**, 1–6 (2025)
61. Serra, P., Fancello, G., Mandas, M., Daga, M., Medda, A.: Investigating maritime accidents that involve dangerous goods using hierarchical clustering. EGY (2022). https://doi.org/10.21622/MARLOG.2022.11.010
62. Shen, H., Yang, T., Mao, Z.: Research on UAV-assisted maritime communication method based on K-means algorithm. In: 2024 4th International Conference on Communication Technology and Information Technology (ICCTIT). Presented at the 2024 4th International Conference on Communication Technology and Information Technology (ICCTIT), pp. 1–5 (2024). https://doi.org/10.1109/ICCTIT64404.2024.10928695
63. Skarlatos, K., Fousteris, A., Georgakellos, D., Economou, P., Bersimis, S.: Assessing ships' environmental performance using machine learning. Energies **16**, 2544 (2023)
64. Sun, X., Li, Y., Wang, N., Li, Z., Liu, M., Gui, G.: Toward self-adaptive selection of kernel functions for support vector regression in IoT-based marine data prediction. IEEE Internet Things J. **7**, 9943–9952 (2020). https://doi.org/10.1109/JIOT.2020.2988050
65. Teng, J., Qin, R., Xu, C.: Time series forecasting in marine environment monitoring sensor networks based on edge computing. IEEE Access (2025)
66. Velasco-Gallego, C., Lazakis, I., Cubo-Mateo, N.: Development of a hierarchical clustering method for anomaly identification and labelling of marine machinery data. J. Mar. Sci. Eng. **12**, 1792 (2024). https://doi.org/10.3390/jmse12101792
67. Vicente-Cera, I., Acevedo-Merino, A., Nebot, E., López-Ramírez, J.A.: Analyzing cruise ship itineraries patterns and vessels diversity in ports of the European maritime region: a hierarchical clustering approach. J. Transp. Geogr. **85**, 102731 (2020)
68. Vo, D.T., Nguyen, X.P., Nguyen, T.D., Hidayat, R., Huynh, T.T., Nguyen, D.T.: A review on the internet of thing (IoT) technologies in controlling ocean environment. Energy Sources Part A: Recov. Utiliz. Environ. Effects **47**, 10064–10082 (2025). https://doi.org/10.1080/15567036.2021.1960932
69. Wang, L., Chen, P., Chen, L., Mou, J.: Ship AIS trajectory clustering: an HDBSCAN-based approach. J. Mar. Sci. Eng. **9**, 566 (2021)
70. Welikanna, D.R., Jin, S.: A data driven oil spill mapping using GMM clustering and damping ratio on X-Press Pearl ship disaster in the Indian Ocean. Mar. Pollut. Bull. **203**, 116392 (2024)
71. Xie, F., Zhang, L., Zheng, S., Xu, A., Li, Z., Dai, J., Xu, L.: SOM neural network-based port function analysis: a case study in 21st-century Maritime Silk Road. Front. Mar. Sci. **11** (2025). https://doi.org/10.3389/fmars.2024.1522071
72. Xin, X., Liu, K., Yu, Y., Yang, Z.: Developing robust traffic navigation scenarios for autonomous ship testing: an integrated approach to scenario extraction, characterization, and sampling in complex waters. Transport. Res. Part C: Emerg. Technol. **178**, 105246 (2025)
73. Xing, W., Wang, J., Zhou, K., Li, H., Li, Y., Yang, Z.: A hierarchical methodology for vessel traffic flow prediction using Bayesian tensor decomposition and similarity grouping. Ocean Eng. **286**, 115687 (2023)
74. Xu, J., Gong, J., Wang, L., Li, Y.: Integrating k-means clustering and LSTM for enhanced ship heading prediction in oblique stern wave. J. Mar. Sci. Eng. **11**, 2185 (2023). https://doi.org/10.3390/jmse11112185
75. Xu, G., Shi, Y., Sun, X., Shen, W.: Internet of things in marine environment monitoring: a review. Sensors **19**, 1711 (2019)

76. Xu, Y., Yu, Q.: Deep reinforcement learning based computation offloading and resource allocation strategy for maritime Internet of Things. Comput. Netw. **264**, 111221 (2025)
77. Xue, J., Yang, P., Li, Q., Song, Y., van Gelder, P.H.A.J.M., Papadimitriou, E., Hu, H.: Machine learning in maritime safety for autonomous shipping: a bibliometric review and future trends. J. Mar. Sci. Eng. **13**, 746 (2025). https://doi.org/10.3390/jmse13040746
78. Xue, H., Yang, Q., Chen, S.: SVM: support vector machines. In: The Top Ten Algorithms in Data Mining, pp. 51–74. Chapman and Hall/CRC (2009)
79. Yan, J., Gong, Y., Chen, C., Luo, X., Guan, X.: AUV-aided localization for Internet of Underwater Things: a reinforcement-learning-based method. IEEE Internet Things J. **7**, 9728–9746 (2020)
80. Yang, P.-Y., Liao, Y.-C., Chou, F.-I.: Artificial Intelligence in Internet of Things System for Predicting Water Quality in Aquaculture Fishponds [WWW Document] (2023). https://doi.org/10.32604/csse.2023.036810
81. Zhang, L., Chen, T., Yang, Z., Du, L., Gan, L., Li, X.: Collision avoidance decision-making for stand-on ships through inference of give-way ship intention in complex encounters. Ocean Eng. **337**, 121923 (2025)
82. Zhang, H., Li, W., Shi, G., Desrosiers, R., Wang, X.: A ship trajectory clustering algorithm based on segmentation direction. Ocean Eng. **313**, 119383 (2024). https://doi.org/10.1016/j.oceaneng.2024.119383
83. Zhang, R., Zhang, J., Wang, Q., Zhang, H.: DOIDS: an intrusion detection scheme based on DBSCAN for opportunistic routing in underwater wireless sensor networks. Sensors **23**, 2096 (2023). https://doi.org/10.3390/s23042096
84. Zhao, L., Shi, G., Yang, J.: An adaptive hierarchical clustering method for ship trajectory data based on DBSCAN algorithm. In: 2017 IEEE 2nd International Conference on Big Data Analysis (ICBDA). Presented at the 2017 IEEE 2nd International Conference on Big Data Analysis (ICBDA), pp. 329–336 (2017). https://doi.org/10.1109/ICBDA.2017.8078834
85. Zhou, F., Yu, K., Xie, W., Lyu, J., Zheng, Z., Zhou, S.: Digital twin-enabled smart maritime logistics management in the context of Industry 5.0. IEEE Access **12**, 10920–10931 (2024). https://doi.org/10.1109/ACCESS.2024.3354838

Chapter 2
An Overview of Internet of Marine Things Through Big Marine Data: Concepts, Challenges, and Future Scope

Ritwika Mukherjee, Banashree Mandal, Suchismita Maiti, Deepanwita Das, Neepa Biswas, and Mahamuda Sultana

2.1 Introduction

The aquatic realm, which covers more than 70% of the planet's surface, is a critical source of sustainment, transportation, and ecological balance. The global oceans are the basis for Earth's climate regulation, biodiversity, and economic activity, yet they remain among the least explored areas due to their vastness, harsh conditions, and technological barriers. In recent years, the Internet of Marine Things (IoMT) [1] has emerged as an innovative development of the Internet of Things (IoT) [2], enabling large-scale, continuous monitoring, and intelligent decision-making in the marine ecosystem. IoMT connects distributed underwater and surface sensors, Autonomous Underwater Vehicles (AUVs), buoys, research vessels, and satellite systems into a holistic cyber-physical framework. This network not only facilitates real-time data collection but also supports adaptive operations in applications ranging from climate science and fisheries to disaster management and maritime security.

R. Mukherjee (✉) · S. Maiti · N. Biswas
Narula Institute of Technology, Kolkata, India
e-mail: ritwikamukherjee1999@gmail.com

S. Maiti
e-mail: suchismita.maiti@nit.ac.in

B. Mandal · D. Das
National Institute of Technology, Durgapur, India
e-mail: bm.20cs1101@phd.nitdgp.ac.in

D. Das
e-mail: ddas.cse@nitdgp.ac.in

M. Sultana
Guru Nanak Institute of Technology, Kolkata, India
e-mail: mahamuda.sultana@gnit.ac.in

T. A. Tran et al. (eds.), *Big Data Analytics for Internet of Marine Things*,
Internet of Things, https://doi.org/10.1007/978-3-032-28171-5_2

As IoMT infrastructures expand, they generate massive volumes of heterogeneous data, namely acoustic signals, chemical and biological measurements, imagery, satellite-based observations, and outputs from ocean circulation models. Collectively referred to as Big Marine Data, these datasets exhibit the well-known "6Vs" of Big Data—Volume, Velocity, Variety, Veracity, Variability, and Value [3, 4] while also being constrained by limited bandwidth, high energy consumption, and environmental uncertainties unique to underwater communication. Volume is the quantity of data, whereas velocity is the speed and frequency of data generation, processing, and analysis as well as the data in motion. The variety is referred to as the complexity and heterogeneity of several datasets, which might be unstructured, semi-structured, or structured. While variability refers to the constancy of the data across time, veracity refers to the quality, relevance, uncertainty, reliability, and predictive value of the data. Coherent analysis of huge data is what gives it value [5]. Traditional methods of data processing are insufficient for such scale and complexity, necessitating the adoption of advanced algorithms, distributed computing, and intelligent data-driven architectures.

The incorporation of Big Data Analytics with IoMT provides a mechanism to convert raw observations into useful insights [6]. Groundbreaking techniques such as edge and fog computing allow initial processing to occur close to the data source, reducing communication overhead. Meanwhile, machine learning and artificial intelligence, particularly physics-informed and self-supervised models, enable predictive analytics, anomaly detection, and adaptive mission planning even under data scarcity. Recent innovations such as digital twins of the ocean, federated learning across distributed nodes, and spatiotemporal graph learning are redefining how Big Marine Data can support decision-making in scientific, environmental, and operational contexts.

The paper examines future directions and emerging trends in Big Data for IoMT. Section 2.2 provides a brief overview of IoMT architecture, concentrating on the hardware infrastructure, data architecture, and network architecture. Section 2.3 explains Big Marine Data. Section 2.4 explores current and emerging applications. A discussion on the significant challenges is done in Sect. 2.5. Section 2.6 outlines the promising future scope of the rapidly evolving field. Section 2.7 finally concludes the paper.

By linking the domains of marine sensing, network engineering, and Big Data analytics, this study aims to establish a comprehensive understanding of how IoMT can evolve into a sustainable, intelligent, and adaptive marine information infrastructure. The insights provided here will be relevant not only to researchers and engineers but also to policymakers and stakeholders seeking to leverage marine data for climate resilience, sustainable development, and ocean governance.

2.2 Overview of Internet of Marine Things

The Internet of Marine Things (IoMT) is a domain-specific adaptation of the Internet of Things, designed to operate in the unique and hostile conditions of the ocean environment. Unlike terrestrial IoT systems, IoMT must contend with extreme pressure, salinity, biofouling, high mobility of platforms, limited power budgets, and communication channels that are inherently unreliable and bandwidth-constrained. Its architecture can be conceptualized as a three-layered system comprising hardware, data, and network infrastructure, each layer playing a vital role in ensuring the resilience, scalability, and intelligence of marine operations.

2.2.1 Hardware Infrastructure

The hardware infrastructure in the Internet of Marine Things (IoMT) has evolved significantly from traditional marine systems, driven by the demand for scalable, intelligent, and energy-efficient platforms to support big data applications in harsh marine environments. This evolution can be analyzed across five key dimensions: sensors, processing, energy, mobility, and maintenance.

Sensors: Traditional marine systems largely relied on single-purpose, wired sensors that provided limited functionality and required extensive cabling. Modern IoMT hardware, however, integrates multi-sensor nodes with wireless capabilities, enabling simultaneous monitoring of physical, chemical, and biological parameters. These systems provide richer datasets that feed into big data platforms for enhanced environmental monitoring and maritime operations. Furthermore, self-calibrating and bio-inspired sensors are emerging as a promising direction. These adaptive devices mimic natural organisms to withstand extreme marine conditions, ensuring more reliable long-term data acquisition. Table 2.1 provides a brief overview of the various sensors used in IoMT applications.

Processing: Earlier marine infrastructures were constrained by centralized, low-computation systems that depended on shore-based processing. Modern IoMT architectures embed onboard microcontrollers and edge computing units on AUVs, buoys, and gliders, enabling real-time in-situ analytics and reducing dependency on satellite or acoustic communication. The future lies in AI-enabled embedded processors (TinyML), which integrate machine learning at the sensor or node level. This development will allow predictive modeling, anomaly detection, and decision-making directly at the edge, minimizing latency and conserving bandwidth.

Energy: Energy limitations have long hindered marine sensing systems, as battery-powered nodes with short lifespans required frequent replacement. Modern IoMT devices increasingly harness renewable sources such as solar and wave energy, prolonging operational lifetimes. The next frontier is hybrid ocean energy harvesting, where multiple renewable sources (wave, tidal, thermal, and solar) are combined

Table 2.1 Common sensors used in IoMT applications

Sensor type	Parameter measured	Deployment example	Key points
CTD (conductivity-temperature-depth)	Salinity, temperature, depth	Oceanographic buoys	Core oceanographic sensor
Hydrophones	Acoustic signals	Whale migration, submarine detection	Sensitive to noise pollution
ADCP (acoustic Doppler current profiler)	Ocean currents, flow velocity	AUVs, moorings	Useful in navigation & sediment studies
Optical sensors	Turbidity, light penetration	Coastal monitoring stations	Limited by water clarity
Chemical sensors	Dissolved oxygen, pH, nutrients	Coral reef monitoring	Critical for climate impact studies
Imaging sensors (cameras/ SONAR)	Marine life, seabed mapping	AUVs, ROVs	High-data volume, requires compression
GPS and inertial sensors	Location, orientation	Surface buoys, gliders	Limited underwater; hybrid use with acoustics

to ensure sustainable, continuous power for long-term autonomous deployments, directly supporting big data continuity in IoMT.

Mobility: Traditional infrastructures depended on fixed buoys, restricting spatial coverage and adaptability. In contrast, modern IoMT systems utilize mobile platforms such as AUVs and gliders, which expand monitoring range and enable dynamic sampling strategies. The future direction envisions swarm robotics and cooperative AUV systems, where multiple autonomous vehicles collaborate, share data, and perform distributed sensing. This mobility innovation will enhance spatiotemporal coverage of ocean data while providing richer datasets for big data analytics.

Maintenance: Conventional systems required frequent manual servicing, which was costly and logistically challenging in remote marine environments. Modern IoMT solutions leverage semi-automated monitoring systems, enabling condition-based maintenance and reducing human intervention. Emerging trends point toward self-healing and adaptive systems that can autonomously detect faults, reconfigure themselves, and maintain operational integrity. Such systems will be crucial for ensuring reliability and reducing downtime in large-scale IoMT deployments.

The shift from traditional hardware to modern IoMT infrastructure demonstrates a clear trend toward autonomy, adaptability, and intelligence. Future advancements such as bio-inspired sensors, TinyML-powered processors, hybrid energy systems, swarm robotics, and self-healing platforms will play a pivotal role in addressing

Table 2.2 Overview of hardware architectures in IoMT

Feature	Traditional marine systems	Modern IoMT hardware	Future direction
Sensors	Single-purpose, wired	Multi-sensor wireless nodes	Self-calibrating, bio-inspired sensors
Processing	Centralized, computation low	Onboard controllers, computing micro-edge	AI-enabled embedded processors (TinyML)
Energy	Battery-powered, short life	Solar, wave-assisted	Hybrid ocean energy harvesting
Mobility	Fixed buoys	AUVs, gliders	Swarm robotics, co-operative AUVs
Maintenance	Frequent manual servicing	Semi-automated monitoring	Self-healing and adaptive systems

the challenges of energy efficiency, scalability, and resilience. Together, these trends will empower IoMT to generate more reliable and diverse big data, ultimately driving smarter, safer, and more sustainable marine operations. An overview of the hardware architectures in terms of various features, viz., sensors, processing, energy, mobility, and maintenance [7, 8] have been represented in a tabular form from a temporal perspective, precisely addressing the special characteristics in Table 2.2.

2.2.2 *Data Infrastructure*

The data infrastructure of the Internet of Marine Things (IoMT) is a multi-layered ecosystem that addresses the unique requirements of marine environments while adapting big data technologies from traditional IoT. Each infrastructure layer, from acquisition to governance, plays a vital role in enabling efficient and reliable operations under challenging underwater conditions. Table 2.3 compares the data infrastructure for Big Data in IoMT.

Data Acquisition: In traditional IoT, terrestrial sensors such as GPS, RFID, humidity, and video feeds are the primary sources of data. In contrast, IoMT requires specialized acquisition systems such as underwater wireless sensor networks (WSNs), autonomous underwater vehicles (AUVs), remotely operated vehicles (ROVs), sonar devices, bio-acoustic sensors, and buoys. These systems enable comprehensive multi-source data capture across large oceanic areas. However, energy constraints, acoustic noise, and limited bandwidth remain major bottlenecks in continuous underwater monitoring.

Data Transmission and Networking: Conventional IoT leverages high-bandwidth and reliable terrestrial networks such as 4G/5G, Wi-Fi, LoRaWAN, and satellite. IoMT, however, must operate in harsher underwater conditions, necessitating specialized communication mechanisms such as acoustic communication, optical wireless systems, magnetic induction, and hybrid acoustic-RF links. While these technologies

Table 2.3 Comparison of data infrastructure for big data in IoMT

Infrastructure layer	Traditional IoT big data infrastructure	IoMT-specific big data infrastructure	Advantages	Limitations/ challenges
Data acquisition	Terrestrial sensors (temperature, humidity, GPS), RFID, video feeds	Underwater WSNs, AUVs, ROVs, sonar, bio-acoustic sensors, buoys	Enables multi-source ocean data capture	Energy constraints, acoustic noise, limited bandwidth
Data transmission and networking	4G/5G, Wi-Fi, LoRaWAN, satellite	Acoustic communication, optical wireless, magnetic induction, hybrid acoustic-RF links	Works in harsh underwater environments	High latency, low bandwidth, unreliable links
Data storage	Centralized cloud (Hadoop HDFS, NoSQL DBs)	Hybrid Edge–Fog–Cloud storage, distributed marine data centers	Low-latency access, better fault tolerance	Edge nodes limited by power & memory
Data processing	Batch (MapReduce, Spark), stream (Kafka, Flink)	Real-time edge analytics, in-situ processing on AUVs, lightweight ML models	Immediate decision-making (e.g., disaster alerts)	Limited compute on AUVs & sensors
Data analytics and AI	Machine learning, deep learning, business intelligence dashboards	Physics-Informed Neural Networks (PINNs), reinforcement learning for navigation, anomaly detection for pollution & security	Tailored for marine science & operations	Requires large labeled marine datasets (scarce)
Data governance and security	Standardized protocols (MQTT, CoAP), blockchain for trust	Trust management in underwater WSN, secure acoustic protocols, data integrity for marine operations	Enhances maritime security & safety	Energy-hungry cryptographic operations

ensure connectivity in extreme environments, challenges such as high latency, low bandwidth, and unreliable links hinder seamless real-time communication.

Data Storage: Traditional IoT applications often depend on centralized storage frameworks such as Hadoop HDFS or NoSQL databases. In IoMT, hybrid storage architectures combining edge, fog, and cloud computing are essential. Distributed marine data centers and edge nodes ensure low-latency access and fault tolerance.

Nevertheless, the limited power and memory capacity of edge nodes impose significant restrictions on long-term and large-scale data storage in remote maritime environments.

Data Processing: Batch and stream-processing frameworks like MapReduce, Spark, Kafka, and Flink dominate conventional IoT data processing. For IoMT, however, the emphasis shifts toward real-time edge analytics and in-situ processing directly on AUVs and sensor nodes using lightweight machine learning models. This capability enables immediate decision-making in critical scenarios, such as disaster alerts or anomaly detection. Yet, limitations in onboard computing capacity of marine devices present significant challenges for large-scale or computationally intensive tasks.

Data Analytics and Artificial Intelligence: The analytics layer in IoT is typically driven by machine learning, deep learning, and business intelligence dashboards. IoMT, by contrast, increasingly integrates domain-specific approaches such as physics-informed neural networks (PINNs), reinforcement learning for navigation, and anomaly detection frameworks for pollution and security monitoring. These AI-driven techniques provide tailored solutions for marine science and operations. However, their success is constrained by the scarcity of large, labeled marine datasets needed to train and validate robust models.

Data Governance and Security: While traditional IoT relies on standardized protocols (MQTT, CoAP) and blockchain for trust, IoMT expands governance requirements to include secure acoustic protocols, trust management in underwater WSNs, and robust data integrity mechanisms. These infrastructures enhance maritime security and safety, ensuring reliable multi-stakeholder collaboration. Despite these advancements, the reliance on energy-hungry cryptographic operations poses sustainability challenges in resource-constrained marine environments.

In general, the IoMT data infrastructure is rapidly evolving from traditional IoT models to highly specialized frameworks optimized for underwater sensing, communication, and analytics. While these advancements enable multi-source, low-latency, and real-time decision-making capabilities, the sector must still address inherent challenges of energy efficiency, communication reliability, dataset scarcity, and secure governance. Future innovations in hybrid storage, lightweight AI, and sustainable cryptographic solutions are expected to play a pivotal role in shaping resilient and scalable IoMT data infrastructures.

2.2.3 Network Infrastructure

The Internet of Marine Things (IoMT) and Big Data integration necessitate an efficient, multi-layered network architecture that guarantees dependable data collection, transmission, storage, and analytics in extremely demanding and dynamic maritime conditions. From the physical layer to the application layer, every tier of the network architecture is essential to facilitating smooth operations and facilitating extensive data-driven decision-making. The function of each tier is explained in the ensuing subsections.

Physical Layer: In aquatic conditions, the actual signal transmission is handled by the physical layer. In contrast to terrestrial IoT, IoMT needs to support several mediums:

- Acoustic communications for underwater sensors (long distance, low bandwidth, high latency).
- Optical wireless communications for short-range, high-data-rate links under water.
- Radio frequency (RF) technologies such as Wi-Fi, LoRaWAN, VHF, and 5G for surface and nearshore connectivity.
- Satellite communications (Inmarsat, Iridium, Starlink) for global coverage in offshore areas.

For long-term installations on buoys and autonomous vehicles, design factors include energy economy, high attenuation, multipath propagation, and noise robustness.

Data Link Layer: Over these physical channels, the data link layer offers dependable point-to-point and point-to-multipoint connections. It manages Medium Access Control (MAC), error detection and correction, and framing. Considering IoMT:

- For satellite and acoustic channels with limited bandwidth, MAC protocols based on TDMA and CDMA are utilized.
- For nodes that run on batteries, such as buoys, energy-powered MAC is essential.
- Common maritime bands, such as AIS/VDES for vessel communication, are used for collision avoidance techniques.
- To handle changing channel circumstances, adaptive coding and modulation are required.

This layer makes sure that data can be safely sent to the next hop even in the event of significant packet loss and intermittent connectivity.

Network Layer: Routing and addressing in diverse maritime networks are handled by the network layer. Conventional IP-based routing is inadequate for maritime communications due to its fragmented nature. IoMT systems make use of:

- Hybrid routing protocols combining satellite, cellular, and acoustic links.
- Delay-Tolerant Networking (DTN) for store-and-forward delivery in disconnected scenarios.
- IPv6 and 6LoWPAN adaptations to handle constrained devices.
- Software-Defined Networking (SDN) controllers to dynamically allocate bandwidth and enforce Quality of Service across hybrid backbones.

Additionally, even with constantly shifting topologies, the network layer guarantees end-to-end reachability and controls mobility (for vessels, AUVs, and UAVs).

Transport Layer: The transport layer ensures flow management and end-to-end reliability. High latency and intermittent links are problems for traditional TCP, hence customized approaches are used:

- UDP with application-level reliability for delay-sensitive telemetry.
- QUIC and multipath transport protocols for optimizing throughput over satellite and hybrid connections.
- Prioritization and QoS schemes ensure that safety–critical messages (collision alerts, distress signals) are transmitted ahead of bulk scientific or environmental data.

When higher-bandwidth channels are available, opportunistic synchronization is frequently used in Big Data scenarios for bulk transfers (such as video and sonar imagery).

Application Layer: End users, research systems, and maritime operators can access services and interfaces through the application layer. Big Data Application Domains in IoMT include:

1. Environmental Monitoring
2. Maritime Security and Safety
3. Smart Ports and Shipping
4. Aquaculture
5. Oil and Gas Exploration
6. Disaster Response and Recovery
7. Oceanographic and Climate Research
8. Plastic Pollution Tracking
9. Autonomous Underwater Vehicles (AUVs)
10. Marine Wildlife Conservation

Table 2.4 provides a clear overview of the various Big Data Application Domains in IoMT. In order to standardize disparate sensor outputs, this layer also manages middleware and interoperability, including message brokers (MQTT, AMQP), API gateways, and semantic data models.

The layered network design adapts conventional networking ideas to the challenging and diverse marine environment. The network and transport layers guarantee robust end-to-end connectivity in the event of disruption, while the physical and data link levels handle the peculiarities of satellite and underwater channels. These capabilities are then used by the application layer to provide sophisticated Big Data services like marine security, environmental sustainability monitoring, and predictive analytics.

Table 2.4 Big data application domains in IoMT

Application domain	Big data role in IoMT	Typical technologies/ algorithms	Benefits and impact
Environmental monitoring	Real-time processing of multi-source ocean data	Distributed WSNs, edge computing, ML clustering (K-means, DBSCAN)	Early detection of anomalies, improved ecosystem health monitoring
Maritime security and safety	Big data fusion from sonar, AIS (Automatic Identification Systems), and radar for vessel monitoring	Deep learning, anomaly detection, predictive analytics	Intrusion detection, piracy prevention, safe navigation
Smart Ports and Shipping	Analytics of ship traffic, logistics, and fuel optimization	IoT-enabled sensors, Big Data logistics platforms, reinforcement learning	Optimized operations, reduced delays, energy efficiency
Aquaculture	Data-driven fish behavior and water quality monitoring	IoT sensors, time-series forecasting (ARIMA, LSTM)	Sustainable yield, disease control, optimized feeding
Oil and gas exploration	Seismic and acoustic data analytics, real-time drilling monitoring	High-performance computing (HPC), big data visualization, CNNs	Safer drilling, predictive maintenance, optimized exploration
Disaster response and recovery	Integration of satellite, sonar, and buoy data for early warning (tsunamis, oil spills)	Big Data fusion, hybrid ML models, graph analytics	Faster disaster alerts, reduced human loss
Oceanographic and climate research	Analysis of large-scale climate, wave, and current datasets	Climate models, Physics-Informed Neural Networks (PINNs)	Better prediction of climate change effects
Plastic pollution tracking	Big data from satellite imaging, and sensors	Image processing (CNNs), spatial analytics, data assimilation models	Locating and quantifying plastic hotspots
Autonomous underwater vehicles (AUVs)	Big data-based navigation, adaptive sampling, mission planning	Reinforcement learning (DQN, Q-learning), SLAM	Improved autonomy, efficient exploration
Marine wildlife conservation	Data-driven tracking of species, migration, and habitats	Bio-acoustic monitoring, ML classification, GIS + Big Data	Protecting endangered species, habitat restoration

2.2.4 Role of Underwater Wireless Sensor Networks (UWSNs)

UWSNs [9] form the foundation of IoMT. They consist of spatially distributed nodes that collaboratively sense and transmit marine parameters. Challenges such as node localization (without GPS), synchronization, and energy-efficient MAC protocols are central research areas. UWSNs are deployed in critical applications including tsunami early warning, habitat monitoring, and offshore infrastructure surveillance. Emerging trends involve cooperative communication, energy-harvesting nodes, and intelligent clustering to prolong network lifetimes.

2.2.5 Role of Autonomous Underwater Vehicles (AUVs)

AUVs [10] extend the sensing capabilities of UWSNs by providing mobility and adaptability. They can:

- Act as data mules, collecting information from fixed sensor nodes.
- Perform adaptive sampling, dynamically altering paths based on real-time measurements.
- Enable collaborative missions, where swarms of AUVs coordinate for large-scale exploration.

With the integration of edge AI and physics-informed machine learning, AUVs can make autonomous decisions underwater, such as detecting anomalies or prioritizing sampling areas, before surfacing to transmit data. Future directions envision heterogeneous fleets of AUVs, USVs, and UAVs working in synergy to create a holistic marine observation network.

2.3 Big Marine Data

The term Big Marine Data refers to the vast, diverse, and continuously expanding datasets generated by the Internet of Marine Things (IoMT). These datasets originate from heterogeneous sources including in-situ sensors, autonomous vehicles, remote sensing satellites, and numerical ocean models. Unlike traditional marine surveys, which were limited to periodic ship-based observations, Big Marine Data captures spatiotemporal dynamics of the ocean in near-real time, enabling a deeper understanding of marine processes, ecosystem health, and human impacts. Table 2.5 throws light on the sources, data characteristics and challenges of Big Marine Data.

Table 2.5 Sources, data characteristics, and challenges of big marine data

Data source	Example data	Characteristics	Challenges
Fixed platforms (moorings, buoys, seabed observatories)	Temperature, salinity, pressure, dissolved oxygen	High temporal resolution, point-based	Fouling, calibration drift, limited spatial coverage
Mobile platforms (AUVs, gliders, USVs)	Acoustic imagery, CTD profiles, chemical sensors	Adaptive, spatially rich	Limited endurance, complex navigation, intermittent communication
Remote sensing (satellites)	SST, SSH, ocean color, chlorophyll-a	Wide spatial coverage, consistent time series	Cloud cover, surface-limited, coarse resolution
Numerical models	Currents, forecasts, assimilated reanalysis	Predictive, large-scale, multi-variable	Model drift, uncertainty, requires assimilation
Citizen science vessel data (AIS, VMS, low-cost sensors)	Ship tracks, eDNA, coastal observations	Supplementary, cost-effective	Quality control, uneven distribution

2.3.1 Characteristics of Big Marine Data

Big Marine Data inherits the "6Vs" of Big Data—Volume, Velocity, Variety, Veracity, Variability, and Value—while also presenting additional domain-specific challenges:

- Volume: Ocean observing systems generate terabytes to petabytes of multi-modal data annually [11].
- Velocity: Many IoMT applications, such as tsunami detection or oil spill monitoring, require high-frequency, low-latency data streams [11].
- Variety: Data spans acoustic signals, optical imagery, temperature-salinity profiles, current vectors, biogeochemical parameters, AIS (ship traffic), and satellite EO products [11].
- Veracity: Marine data often suffers from noise (acoustic interference, turbidity in optical signals) and uncertainty due to calibration errors or environmental variability [11].
- Variability: Ocean conditions change over diurnal, seasonal, and interannual scales, affecting both the quantity and quality of observations.
- Value: Extracting actionable insights from these data streams can support climate resilience, resource management, and maritime operations [11].

2.3.2 Sources of Big Marine Data

The key contributors to Big Marine Data are:

- Fixed Platforms: Buoys, moorings, seabed observatories collecting long-term series.
- Mobile Platforms: AUVs, gliders, USVs conducting adaptive sampling and remote exploration.
- Remote Sensing: Satellites measuring sea surface temperature, salinity, sea-level anomalies, chlorophyll-a, and ocean color.
- Numerical Models: Outputs from coupled atmosphere–ocean circulation models, often assimilating observational data.
- Citizen and Crowd-sourced Data: Coastal observations, fishing vessel monitoring (AIS, VMS), and low-cost sensors.

2.3.3 Lifecycle of Big Marine Data

The lifecycle of Big Marine Data encompasses the end-to-end processes of generating, handling, and utilizing massive data streams emerging from marine environments. Given the complexity of oceans and the heterogeneity of data sources, each phase in this lifecycle is critical for ensuring reliability, scalability, and actionable intelligence.

1. Data Acquisition: The cycle starts with the creation and gathering of data from a variety of diverse marine sources, including radar buoys, satellite remote sensing, underwater acoustic sensors, unmanned surface vehicles (USVs), autonomous underwater vehicles (AUVs), and marine observatories. These sources generate a variety of data types, such as environmental factors, sonar data, images, oceanographic measurements, and acoustic signals. Strong IoMT frameworks that can withstand erratic connectivity and challenging environmental circumstances are required because to the high pace and diversity of incoming data.
2. Data Transmission: Underwater communication has limited bandwidth, hence acoustic, optical, or hybrid communication techniques are used to transfer data. High latency, low throughput, and signal deterioration in harsh conditions are some of the difficulties this level encounters. Edge computing and delay-tolerant networking (DTN) techniques are frequently used to address these problems, guaranteeing initial processing and buffering at the source prior to bulk transmission to cloud or shore-based systems.
3. Data Storage: The collected datasets are stored in distributed storage systems after they have been transferred. Scalable handling of both structured (like tabular sensor data) and unstructured (like sonar images and video streams) formats is made possible by cloud-based and edge-enhanced storage systems. In order to facilitate interoperability and long-term accessibility, metadata tagging, indexing, and adherence to maritime data standards for geographical data, are essential at this point.
4. Data Processing and Integration: Raw marine data often contains redundancies, noise, and inconsistencies. Preprocessing methods such as cleaning, normalization, and sensor calibration are applied to enhance data quality. Integration

across heterogeneous sources (e.g., combining satellite data with in-situ sensor streams) ensures a holistic view of the marine environment. Here, Big Data frameworks like Hadoop, Spark, and specialized oceanographic middleware are used to handle large-scale parallel processing.

5. Data Analytics and Modeling: At this stage, advanced analytics extract insights from the curated data. Machine learning, deep learning, reinforcement learning, and simultaneous localization and mapping (SLAM) algorithms are applied for applications such as autonomous navigation, pollution hotspot detection, climate change modeling, and biodiversity monitoring. Predictive modeling, anomaly detection, and digital twin technologies further enhance decision support in marine science and defense operations.
6. Visualization and Interpretation: To translate analytics into actionable intelligence, results are presented using immersive dashboards, GIS-based visualization, and 3D/4D ocean mapping. Interactive visualization enables scientists, policymakers, and defense agencies to explore spatiotemporal patterns such as coral bleaching events, fish migration, or naval activity in near-real time.
7. Decision-Making and Action: The ultimate goal of the Big Marine Data lifecycle is to inform decision-making. Insights derived from marine data analytics support sustainable fishing policies, disaster preparedness (e.g., tsunami early warnings), marine spatial planning, naval surveillance, and climate resilience strategies. Autonomous systems may also use real-time analytics for self-adaptive actions in hostile marine environments.
8. Data Preservation and Sharing: Finally, marine data must be preserved for long-term use and shared across international platforms. Open-access repositories and initiatives such as Global Ocean Observing System (GOOS) and Ocean Data Portal (ODP) foster collaborative research and global ocean governance. Emphasis on FAIR principles ensures scientific transparency and long-term sustainability of marine data resources.

2.3.4 Challenges in Handling Big Marine Data

The different challenges in handling big marine data has been highlighted in this section.

- Bandwidth constraints: Only kbps-level data rates are supported by underwater acoustic links.
- Energy limitations: Continuous sensing and transmission are unsustainable without energy harvesting.
- Scalability: Growing datasets are in need of distributed, cloud-integrated architectures.
- Standardization: Lack of universally accepted metadata and ontologies limits interoperability.
- Label scarcity: Training robust AI models is hindered by the absence of large annotated datasets.

- Data trust and security: Ensuring authenticity, provenance, and tamper-evidence across jurisdictions.

2.3.5 *Summarization*

- From Centralized to Distributed Analytics: Moving computation from centralized cloud systems to fog and edge nodes.
- From Data Scarcity to Self-supervised Learning: Overcoming the lack of labeled datasets with unsupervised and weakly supervised methods.
- From Isolated Platforms to Collaborative Networks: Multi-AUV and hybrid fleet missions for data collection and relay.
- From Observations to Prediction: Real-time forecasting through assimilation of Big Marine Data into physics-based and ML models.
- From Data Collection to Decision Intelligence: Transforming raw data streams into operational intelligence for governance, conservation, and security.

2.4 Big Data Concepts and Applications in IoMT

The integration of Big Data with the IoMT is transforming the maritime industry by enhancing efficiency, sustainability, and security. The vast amounts of heterogeneous data generated from ships, satellites, underwater vehicles, and sensors provide an opportunity to derive actionable insights for decision-making. As IoMT ecosystems evolve, Big Data is no longer limited to descriptive analytics but is increasingly being deployed for predictive and prescriptive applications.

This transformation paves the way for intelligent maritime operations that address challenges such as unpredictable weather conditions, piracy threats, and climate-related risks, while also promoting sustainable marine resource management. This section highlights the emerging trends in Big Data for IoMT.

1. AI and Machine Learning for Advanced Marine Analytics:

   One of the most promising trends in Big Data for IoMT is the integration of Artificial Intelligence (AI) and Machine Learning (ML) [12]. These technologies enable advanced marine analytics by automating complex processes and providing predictive intelligence. For instance, predictive maintenance powered by ML algorithms helps reduce operational downtime and costs for shipping fleets, while deep learning-based models improve autonomous vessel navigation by enabling real-time route optimization and obstacle detection. In the domain of oceanographic forecasting, AI models enhance the prediction of ocean currents, wave dynamics, and climate impacts, thereby contributing to improved marine safety and resilience.
2. Edge and Fog Computing for Real-Time Marine Data Processing:

The role of edge computing [12] and fog computing in IoMT is emerging as a critical enabler of real-time marine data analytics. Unlike traditional cloud-based systems that suffer from latency issues, edge computing processes data directly at the source, such as onboard ships or AUVs. This allows for faster decision-making in mission-critical applications, including disaster response, oil spill detection, and vessel navigation. Fog computing, on the other hand, provides an intermediate layer between the cloud and edge devices, ensuring distributed processing across networks. Together, these technologies create resilient marine data infrastructures that reduce dependency on remote servers and improve reliability in ocean environments.

3. 5G and Next-Generation Marine Connectivity:

   Reliable communication is essential for IoMT, and the integration of 5G and next-generation connectivity represents a significant advancement. Smart ships equipped with 5G-enabled IoT devices [7] can transmit large volumes of sensor data with minimal latency, supporting real-time analytics for operations such as collision avoidance and cargo monitoring. Additionally, the expansion of satellite IoT networks ensures connectivity in remote oceanic regions where traditional terrestrial networks are unavailable. These advancements contribute to enhanced situational awareness, safer shipping operations, and optimized fleet management through seamless data sharing across stakeholders.

4. Digital Twins and Smart Maritime Operations:

   Another emerging trend is the deployment of digital twin technology [12, 17] in maritime operations. By creating virtual replicas of ships, ports, and cargo systems, digital twins enable continuous monitoring, predictive modeling, and real-time simulation of maritime activities. For example, predictive cargo management systems based on digital twin models help optimize supply chain operations by anticipating delays and adjusting logistics accordingly. Similarly, smart ports can utilize digital twins to manage vessel traffic, automate container handling, and improve energy efficiency. The adoption of this technology will enable the maritime industry to shift toward proactive management and predictive decision-making.

5. Blockchain for Secure and Transparent Data Exchange

   As IoMT generates massive amounts of sensitive data, ensuring its integrity and security becomes vital. Blockchain technology [14] offers a decentralized approach to secure maritime data exchange through immutable ledgers and smart contracts. Applications include transparent cargo tracking, automated payments, and tamper-proof log records. The use of blockchain enhances trust between international shipping partners by enabling secure and transparent sharing of data related to trade, customs, and regulatory compliance. In the long run, blockchain could also integrate with AI and IoMT to build robust ecosystems that minimize fraud, piracy, and data breaches.

6. Environmental Monitoring and Ocean Conservation:

   Big Data also plays a key role in advancing sustainability by enabling environmental monitoring [15, 21] and ocean conservation. IoMT systems equipped with distributed sensors and remote monitoring technologies can detect pollution in real time, track marine biodiversity, and measure carbon emissions. Machine learning models can process these large datasets to predict ecological risks and suggest mitigation strategies. These innovations align with global initiatives, such as the UN Sustainable Development Goals (SDGs), which emphasize climate resilience and the preservation of marine ecosystems.

7. Autonomous and Unmanned Marine Systems:
   The growing reliance on autonomous and unmanned marine systems marks another frontier in Big Data-driven IoMT. Platforms such as Unmanned Underwater Vehicles (UUVs) [16] and Remote Controlled Surface Vessels depend on real-time data analytics for navigation, surveillance, and exploration in hostile marine environments. Reinforcement learning and simultaneous localization and mapping (SLAM) techniques enable these systems to operate with minimal human intervention. As Big Data analytics continue to evolve, autonomous platforms will become integral to applications ranging from seabed exploration to naval defense operations.
8. Quantum Computing for Complex Marine Data Analytics:
   Quantum computing [21] is expected to redefine Big Data analytics in IoMT by providing unprecedented computational power. Quantum algorithms can process vast, multidimensional datasets much faster than classical computers, which is particularly beneficial for modeling ocean dynamics, predicting extreme weather, and optimizing global shipping routes. Although still in its infancy, the adoption of quantum computing in the maritime domain could revolutionize simulation-based decision-making and enable breakthroughs in large-scale marine analytics.
9. Cybersecurity and Maritime Data Protection:
   As IoMT expands, cybersecurity emerges as a critical concern for global shipping routes and marine networks [13]. AI-powered threat detection combined with advanced encryption technologies will strengthen defenses against cyberattacks targeting vessels, ports, and communication systems. With increasing reliance on connected maritime assets, robust cybersecurity frameworks will be essential for ensuring resilience against threats such as data breaches, GPS spoofing, and ransomware. Securing IoMT infrastructures will not only protect operations but also foster trust among global maritime stakeholders.

The convergence of Big Data analytics and IoMT is steering the maritime industry toward a smart, resilient, and self-sustaining ecosystem. Emerging trends such as AI-driven analytics, edge computing, 5G-enabled connectivity, digital twins, blockchain, and quantum computing will shape the future of marine technologies. By addressing challenges related to cybersecurity, sustainability, and interoperability, Big Data in IoMT will enable innovative solutions that enhance automation, improve safety, and foster environmental conservation.

While the integration of Big Data and IoMT has opened new horizons in marine science, ecology, and security, several technical, operational, and governance challenges remain unresolved. Addressing these issues is essential to build sustainable, intelligent, and resilient marine data ecosystems.

2.5 Challenges and Future Scope in Big Data for IoMT

2.5.1 Challenges in Big Data for IoMT

Data Heterogeneity and Standardization: One of the key issues of IoMT is the heterogeneity of marine data sources [8]. Data is gathered from satellites, radars, underwater sensors, buoys, ships, and autonomous vehicles, which produce data in varied formats, scales, and resolutions. Standard protocols and compatible frameworks to jointly analyze these disparate data are yet in development. Absence of globally accepted standards for data is acting as a barrier to international collaboration and as a brake on scientific and operational progress in the Marine Domain.

Real-Time Data Processing and Latency: Marine environments generate high-velocity data streams, especially in applications such as disaster response, AUV navigation, and maritime surveillance. However, transmitting large volumes of data to cloud infrastructures often results in latency issues. While edge computing provide partial solutions [10], scalability and reliability remain a concern in dynamic and resource-constrained marine environments.

Connectivity and Bandwidth Limitations: Despite advances in 5G and satellite IoT, connectivity remains a bottleneck in remote ocean regions [18]. High costs of satellite bandwidth and the lack of robust underwater communication networks restrict the continuous transmission of Big Data. These limitations affect mission-critical applications such as autonomous ship navigation, offshore oil exploration, and real-time ecosystem monitoring.

Energy Efficiency of Marine Devices: Sensors, underwater vehicles, and IoMT devices often operate in energy-constrained environments where battery replacement or recharging is difficult [20]. Big Data processing demands high computational power, which in turn increases energy consumption. Developing energy-efficient data acquisition and processing techniques, along with energy-harvesting systems, is crucial to maintaining long-term IoMT deployments.

Cybersecurity and Data Privacy: As maritime operations become increasingly digitalized, they also become vulnerable to cyberattacks, spoofing, and data breaches [18]. Ensuring the integrity and confidentiality of marine Big Data requires advanced security measures, such as blockchain-enabled trust frameworks, AI-powered intrusion detection systems, and robust encryption mechanisms. However, implementing these solutions at scale across heterogeneous IoMT infrastructures remains a significant challenge.

High Computational Costs and Infrastructure Requirements: Big Data analytics in IoMT often requires high-performance computing (HPC), which demands substantial infrastructure investments. Many developing nations with large maritime zones may lack the resources to deploy such systems, creating disparities in marine data access and utilization. Moreover, the complexity of integrating AI, quantum computing, and digital twin models adds to the cost and expertise requirements [19].

Environmental and Ethical Concerns: Continuous deployment of IoMT sensors and autonomous platforms raises environmental and ethical concerns, such as potential disruption of marine life or data ownership conflicts between stakeholders. The sustainable usage of IoMT still remains the unsolved challenge of striking a balance between ecological responsibility and technical advancement [8].

2.5.2 Future Scope of Big Data in IoMT

AI-Augmented Predictive Intelligence: The future of IoMT lies in expanding the role of AI and machine learning [21] for predictive intelligence. Advanced models will not only support predictive maintenance and route optimization but also forecast large-scale climate phenomena such as El Nin˜o and ocean acidification. Integration of self-supervised and physics-informed ML models will enhance forecasting accuracy under data-scarce conditions.

Digital Twin Oceans and Smart Maritime Ecosystems: The concept of Digital Twin[17] Oceans will become central to future IoMT applications. By creating dynamic virtual models of entire marine environments, digital twins will enable continuous simulations, ecosystem forecasting, and smart decision-making. This will also support smart ports, autonomous shipping, and predictive logistics, creating a resilient maritime ecosystem.

Expansion of 5G/6G and Satellite IoT Networks: Future IoMT systems will benefit from the expansion of 5G/6G [7] terrestrial networks and next-generation satellite IoT constellations, which will provide near-global coverage and ultra-low latency. These advancements will allow seamless data transfer, real-time collaboration among international stakeholders, and improved monitoring of remote and deep-sea regions.

Integration of Blockchain for Global Collaboration: Blockchain will evolve as a backbone for secure and transparent cross-border maritime collaboration [14]. Decentralized ledgers will ensure trusted data sharing in areas such as cargo monitoring, fisheries management, and marine biodiversity protection. Smart contracts will further automate compliance and reduce human intervention in trade and logistics.

Quantum Computing for High-Dimensional Marine Data: The adoption of quantum computing [21] will revolutionize complex marine data analytics, enabling breakthroughs in weather modeling, ocean current simulations, and global shipping optimization. As quantum systems mature, their ability to process multidimensional datasets at unprecedented speeds will redefine how marine scientists and operators handle big data.

Autonomous and Swarm-Based Marine Robotics: Future IoMT ecosystems will increasingly rely on swarms of AUVs [21] and unmanned vessels capable of collaborating through real-time data sharing. These systems will improve deep-sea exploration, underwater surveillance, and disaster management. Enhanced with Big Data analytics, autonomous fleets will reduce operational costs and risks associated with manned missions.

Sustainable Marine Monitoring for Climate Resilience: Big Data-enabled IoMT will play a pivotal role in achieving climate resilience and environmental sustainability. Future systems will allow real-time tracking of pollution hotspots [10], endangered species migration, and ocean carbon flux. These insights will directly support global policies for marine conservation, blue economy growth, and carbon neutrality initiatives.

Global Interoperability and FAIR Data Principles: The future scope of IoMT also emphasizes open and interoperable marine data infrastructures [10]. Adoption of FAIR data principles (Findable, Accessible, Interoperable, and Reusable) will facilitate collaboration across nations, academia, and industries. Such platforms will create a unified global framework for marine data sharing, ensuring equitable access and collective innovation.

While Big Data for IoMT faces significant challenges in terms of data heterogeneity, cybersecurity, connectivity, and energy efficiency, emerging innovations provide a roadmap for future development. The convergence of AI, digital twins, quantum computing, blockchain, and next-generation connectivity will transform IoMT into a backbone of maritime intelligence. By addressing these challenges and leveraging future opportunities, Big Data in IoMT will foster a smart, secure, and sustainable maritime ecosystem capable of supporting global trade, marine conservation, and climate resilience.

Table 2.6 highlights the research roadmap for Big Data in IoMT, a transition from current connectivity and energy-limited IoMT systems to intelligent, secure, and sustainable networks. As Big Marine Data grows in volume and complexity, success will depend on integrating innovations across hardware, networking, and data science while aligning with sustainability and governance frameworks.

2.6 Conclusion

The convergence of the Internet of Marine Things (IoMT) with Big Data analytics is transforming the way oceans are monitored, understood, and managed. By interconnecting heterogeneous sensors, AUVs, buoys, ships, and satellites, IoMT enables continuous and large-scale acquisition of marine observations, producing diverse and high-volume datasets collectively termed as Big Marine Data. When coupled

Table 2.6 Research roadmap for big data in IoMT

Challenge	Current limitation	Future direction	Expected impact
Communication	Low bandwidth, high latency in acoustic channels	Hybrid acoustic-optical-RF, AI-based routing	Reliable, high-speed marine IoT
Energy	Battery dependence, limited endurance	Ocean energy harvesting, AI at edge	Long-lived, autonomous deployments
Data scalability	Petabyte-scale growth, fragmented storage	Cloud-HPC integration, FAIR repositories	Seamless data access and collaboration
Data quality and labels	Noisy, sparse, biased data	Self-supervised & transfer learning, simulation-to-real	Robust analytics with minimal annotation
Security and trust	Vulnerable UWSNs, lack of provenance	Blockchain, lightweight crypto, intrusion detection	Secure and tamper-proof data systems
Governance and standards	Interoperability issues, siloed datasets	Global marine data commons, open ontologies	Cross-border collaboration and data reuse
Sustainability	Ecological impact, e-waste	Eco-friendly materials, low-carbon AI	Greener, ethically responsible Internet of Marine Things Intelligence

with advanced Big Data concepts—machine learning, edge/fog computing, federated analytics, and digital twins—these datasets hold the potential to generate actionable intelligence for climate prediction, disaster preparedness, fisheries management, and maritime security. Through the development of intelligent algorithms, data-driven simulations and high-performance marine data processing systems, the Internet of Big Marine Data propels advancements in computer science while integrating cutting-edge sensing, communication, and analytics technologies to improve real-time monitoring, predictive modeling, and decision-making in ocean engineering.

Despite remarkable progress, IoMT remains constrained by challenges including bandwidth-limited underwater communication, energy scarcity, data heterogeneity, label scarcity, and cybersecurity risks. Furthermore, the absence of standardized frameworks and the ecological footprint of large-scale deployments raise concerns about sustainability and governance. Addressing these limitations requires a multidisciplinary approach, integrating hardware innovation, energy-aware networking, physics-informed and self-supervised AI, and interoperable data infrastructures.

Looking ahead, future directions point toward the development of hybrid communication networks, energy-harvesting IoMT nodes, AI-driven real-time analytics, secure blockchain-enabled trust mechanisms, and sustainable marine data frameworks. The concept of digital ocean twins, supported by HPC-enabled cloud infrastructures and real-time assimilation of multi-source data, will be central to predictive ocean science and decision support.

In summary, the synergy between IoMT and Big Data is not merely a technological advancement but a paradigm shift toward intelligent, secure, and sustainable ocean systems. Continued collaboration across marine science, computer science, and policy-making communities will be critical to unlock the full potential of Big Marine Data in addressing pressing environmental, societal, and economic challenges of the future.

References

1. Domingo, M.C.: An overview of the internet of underwater things. J. Netw. Comput. Appl. **35**(6), 1879–1890 (2012)
2. Sharma, C. Correcting the IoT History. Available at: http://www.chetansharma.com/correcting-the-iot-history/. Accessed 21 June 2017
3. Wang, K., Wang, Y., Hu, X., Sun, Y., Deng, D.J., Vinel, A., Zhang, Y.: Wireless big data computing in smart grid. IEEE Wirel. Commun. **24**(2), 58–64 (2017)
4. Zhang, X., Yi, Z., Yan, Z., Min, G., Wang, W., Elmokashfi, A., Zhang, Y., et al.: Social computing for mobile big data. Computer **49**(9), 86–90 (2016)
5. Ristevski, B., Chen, M.: Big Data Analytics in Medicine and Healthcare. J. Integr. Bioinform. **15**(3), 20170030 (2018). https://doi.org/10.1515/jib-2017-0030
6. Jahanbakht, M., Xiang, W., Hanzo, L., Azghadi, M.R.: Internet of underwater things and big marine data analytics—a comprehensive survey. IEEE Commun. Surv. Tutor. **23**(2), 904–956 (2021)
7. Liu, S., Zhu, L., Huang, F., Hassan, A., Wang, D., He, Y.: A survey on air-to-sea integrated maritime internet of things: enabling technologies, applications, and future challenges. J. Mar. Sci. Eng. **12**(1), 11 (2024). https://doi.org/10.3390/jmse12010011
8. Xu, G., Shi, Y., Sun, X., Shen, W.: Internet of things in marine environment monitoring: a review. Sensors **19**(7), 1711 (2019). https://doi.org/10.3390/s19071711
9. Taher, R.J., Mohsen, K.K.: Underwater wireless sensor networks. In: BIO Web of Conferences, vol. 97, p. 00023. EDP Sciences (2024)
10. Mohsan, S.A.H., Li, Y., Sadiq, M., Liang, J., Khan, M.A.: Recent advances, future trends, applications and challenges of internet of underwater things (IoUT): a comprehensive review. J. Mar. Sci. Eng. **11**(1), 124 (2023). https://doi.org/10.3390/jmse11010124
11. Lytra, I., Vidal, M.E., Orlandi, F., Attard, J.: A big data architecture for managing oceans of data and maritime applications. In: 2017 International Conference on Engineering, Technology and Innovation (ICE/ITMC), pp. 1216–1226. IEEE (2017)
12. Durlik, I., Miller, T., Cembrowska-Lech, D., Krzemin´ska, A., Z-loczowska, E., Nowak, A.: Navigating the sea of data: a comprehensive review on data analysis in maritime IoT applications. Appl. Sci. **13**(17), 9742. https://doi.org/10.3390/app13179742
13. Jain, V., Mitra, A., Paul, S.: Integrating IoT and big data analytics for enhancing maritime safety and sustainability. In: Research Methods for Advancing the Maritime Industry, pp. 225–256. IGI Global Scientific Publishing (2025)
14. Bhattacharya, S., Victor, N., Chengoden, R., Ramalingam, M., Selvi, G.C., Maddikunta, P.K.R., Donta, P.K., Dustdar, S., Jhaveri, R.H., Gadekallu, T.R.: Blockchain for internet of underwater things: state-of-the-art, applications, challenges, and future directions. Sustainability **14**(23), 15659 (2022). https://doi.org/10.3390/su142315659

15. Kao, C.C., Lin, Y.S., Wu, G.D., Huang, C.J.: A comprehensive study on the internet of underwater things: applications, challenges, and channel models. Sensors **17**(7), 1477 (2017)
16. Wibisono, A., Piran, M.J., Song, H.K., Lee, B.M.: A survey on unmanned underwater vehicles: challenges, enabling technologies, and future research directions. Sensors **23**(17), 7321 (2023)
17. Madusanka, N.S., Fan, Y., Yang, S., Xiang, X.: Digital twin in the maritime domain: a review and emerging trends. J. Mar. Sci. Eng. **11**(5), 1021 (2023). https://doi.org/10.3390/jmse11051021
18. Alqurashi, F.S., Trichili, A., Saeed, N., Ooi, B.S., Alouini, M.S.: Maritime communications: a survey on enabling technologies, opportunities, and challenges. IEEE Internet Things J. **10**(4), 3525–3547 (2022)
19. Ilias, L., Tsapelas, G., Kapsalis, P., Michalakopoulos, V., Kormpakis, G., Mouzakitis, S., Askounis, D.: Leveraging extreme scale analytics, AI and digital twins for maritime digitalization: the VesselAI architecture. Front. Big Data **6**, 1220348 (2023)
20. Omeke, K.G., Mollel, M., Shah, S.T., Zhang, L., Abbasi, Q.H., Imran, M.A.: Toward a sustainable internet of underwater things based on AUVs, SWIPT, and reinforcement learning. IEEE Internet Things J. **11**(5), 7640–7651 (2023)
21. Theocharidis, T., Kavallieratou, E.: Underwater communication technologies: a review. Telecommun. Syst. **88**(2), 54 (2025)

Chapter 3
Edge-AI for Smart Marine Environments: Integrating IoMT and Machine Learning for Efficient Water and Coral Reef Monitoring

Shinde Sopan Raosaheb

3.1 Introduction

The development of smart environment systems is increasingly critical for monitoring and understanding our planet's aquatic ecosystems. These systems are essential for observing the complex array of substances within water bodies, which directly determine the health and viability of the environment for all life. In response, a global research effort led by universities and international institutions has focused on advancing methods for water monitoring and coral reef monitoring enabling smart marine environments, facilitating applications from coastal monitoring to deep-sea exploration [6, 30]. Traditionally, these methods rely on devices like Remotely Operated Underwater Vehicles (ROVs) and portable sensors to collect data, creating a pressing need for integrated platforms to manage this information effectively.

The transformative potential of the Internet of Things (IoT) an infrastructure that interconnects physical devices to the Internet promises enhanced human interaction with our environment. While terrestrial IoT has matured rapidly, its underwater counterpart, the Internet of Underwater Things (IoUT) and the broader Internet of Marine Things (IoMT), remains a largely uncharted frontier. This is a significant oversight, given that oceans cover over 70% of the Earth's surface, support 90% of global trade, and provide sustenance for nearly half a billion people, yet 95% of the marine realm remains unexplored.

The fundamental challenge in establishing a robust IoUT lies in the harsh underwater environment. Conventional terrestrial wireless sensor network architectures falter underwater due to unique communication constraints, limited energy resources,

S. S. Raosaheb (✉)
Department of Computer Engineering, Dr. D.Y. Patil College of Engineering and Innovation, Pune, Maharashtra, India
e-mail: scholarswapnil@gmail.com

T. A. Tran et al. (eds.), *Big Data Analytics for Internet of Marine Things*, Internet of Things, https://doi.org/10.1007/978-3-032-28171-5_3

and computational challenges. stationary sensors for data acquisition on parameters like water quality and imagery [1, 22] have been used for this case. However, the underwater environment presents severe challenges not found in terrestrial IoT, primarily due to the unique properties of acoustic, optical, and radio frequency communications which are characterized by high latency, low bandwidth, and high error rates [11, 23, 32]. These constraints directly impact the network architecture, necessitating specialized and energy-efficient communication protocols [10, 35] and routing strategies [14, 25] to create a viable underwater network architecture. These limitations give rise to a critical component of the IoUT ecosystem: the underwater wireless sensor network (UWSN), which often employs technologies like a wireless mesh network to enable data acquisition from a distributed array of sensors and vehicles.

This proliferation of interconnected devices generates an immense, continuous stream of information, leading to the paradigm of Big Marine Data (BMD). The volume, velocity, and variety of this data, sourced from portable sensors, hydrophones, and underwater imagery, far exceed the capabilities of conventional data processing techniques. Traditional data acquisition and analysis methods, including descriptive statistics, lack the generalization required to automatically learn from and adapt to this dynamic, heterogeneous data deluge.

To address this, machine learning (ML) and Deep Neural Networks (DNNs) have emerged as indispensable tools. These techniques enable the automatic extraction of knowledge and patterns from BMD, with applications ranging from automated species recognition and seabed mapping to environmental water monitoring and pollution detection. However, a significant architectural gap persists. Current systems often rely on centralized cloud or Hadoop-based platforms for data aggregation, which are ill-suited for the real-time demands and bandwidth constraints of underwater data transmission high-volume stream of information from distributed sources, giving rise to the paradigm of Big Marine Data (BMD), which strains the very communication channels designed to carry it [8, 15]. Sending all raw data to a central node for processing creates latency, consumes excessive energy, and bottlenecks the entire smart environment system.

Therefore, this paper proposes a novel, edge-centric intelligent system for the Internet of Marine Things (IoMT). Research has given the viability of platforms like Hadoop for storing and analyzing collected marine data, often relying on batch-processing techniques (as seen in systems like those described by Adamo et al. [1]). Our approach integrates lightweight ML models directly within the underwater network architecture, enabling in-situ data filtering, feature extraction, and anomaly detection, and model involves transmitting all raw data from portable sensors and AUVs to a central node or cloud platform, a process that is increasingly recognized as unsustainable. As noted by Xu and Liu [29], energy-aware scheduling is paramount in IoUT, and the act of raw data transmission is one of the most energy-intensive operations [12]. Furthermore, the velocity and variety of real-time data for applications like coral reef monitoring [9] or fish tracking [4] create significant latency in centralized systems, hindering real-time decision-making. This methodology significantly reduces data transmission overhead and latency by processing data at its source, moving beyond the limitations of traditional batch-processing systems. We detail the architectural components of this system, including communication protocols for model updates and a cloud-based layer for aggregating high-level insights.

The contributions of this work Fig. 3.1 are as follows:

- We propose an edge-AI architecture for IoMT that bridges the gap between distributed data acquisition and real-time analytical needs.
- We demonstrate through simulation that our system achieves a substantial reduction in data transmission while maintaining high accuracy in event detection for coral reef and water quality monitoring.

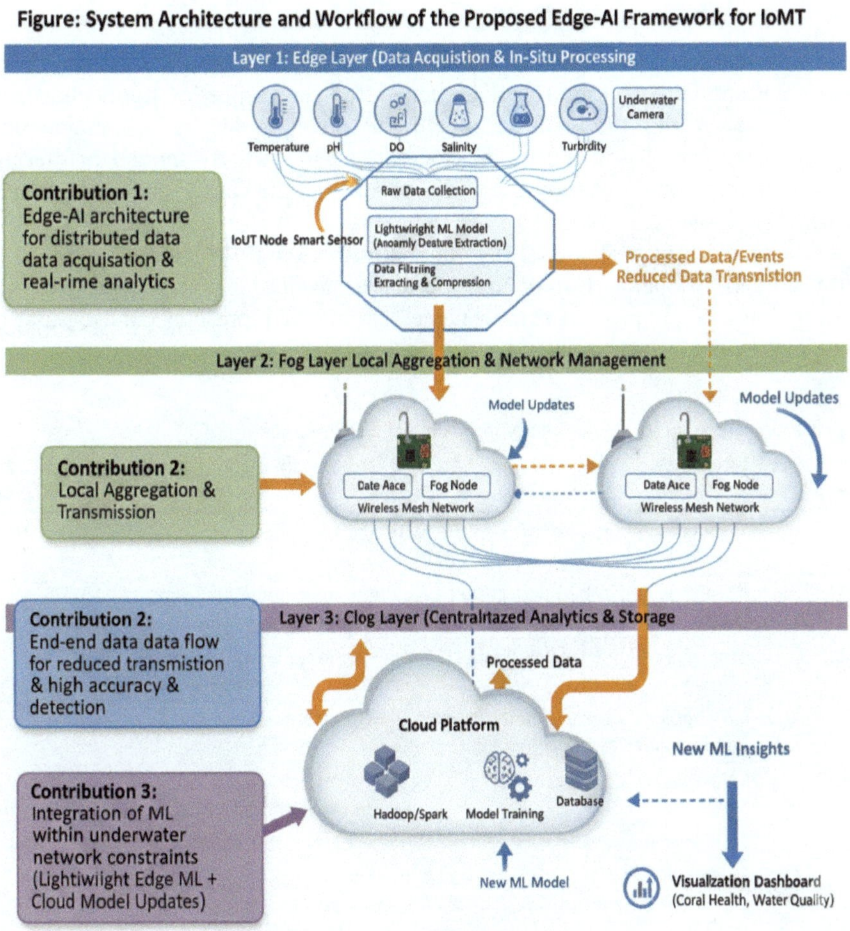

Fig. 3.1 Complete workflow of system architecture for IoMT

- We provide a comprehensive discussion on integrating machine learning within the constraints of an underwater wireless sensor network, paving the way for more responsive, scalable, and intelligent marine environmental monitoring.

3.2 Literature Review

Machine learning (ML) and Deep Neural Networks (DNNs) have proven highly successful in extracting knowledge from complex marine data, enabling tasks such as automatic species recognition [19, 21], seabed analysis [26], and coral reef image classification [18]. Subsequently we can see research-based focus in Table 3.1 as follow.

However, these applications have largely been deployed in post-processing, cloud-based environments. The next logical step is the integration of lightweight ML models directly within the IoUT architecture to enable in-situ, real-time analytics. This concept of edge-based or edge-AI processing involves performing data filtering, feature extraction, and anomaly detection directly on the sensor node or a nearby gateway [7] (Table 3.2).

By doing so, only processed, high-value information or detected events need to be transmitted, dramatically reducing the data transmission overhead and energy consumption while improving responsiveness. Table 3.3 represents contributions of article in the field.

Table 3.1 Research focus, methodology in Underwater IoT studies

Publication year	Research focus	Methodology/approach	References
2025	Underwater sparse-view scene reconstruction	Gaussian splatting	[31]
2024	Multiple unmanned underwater vehicles	Autonomous planning systems	[13]
2024	Internet of Underwater Things (IoUT)	Survey on simulation tools and 5G-based underwater networks	[20]
2025	Technologies for autonomous underwater vehicles	Review on technologies	[16]
2019	Navigation for underwater vehicles	Gravity-aided characteristic parameter matching	[28]
2020	Underwater network management	Review and feasible solution studies	[5]
2023	UAV path planning	Improved ACO algorithm	[34]
2020	IoUT architecture and challenges	Survey with open issues	[24]

Table 3.2 Characteristics, applied variables, and compartments in IoUT and marine monitoring

Characteristic	Detail/key finding	Applied variable/ compartment	References
Communication technologies	Underwater communication	Data assimilation—review	[27]
Communication	Opto-acoustic networking for IoUT	Hybrid software-defined architecture	[3]
Deep learning	Marine data assimilation	Data assimilation—review	[17]
Thermal stresses	Global coral reef declines under climate change	Data assimilation	[33]
IoT and machine learning	Tracking saltwater intrusion in coastal aquifers due to over-extraction	A treatment: geospatial approach	[2]
IoT Monitoring	Marine environment monitoring	IoT integration for sensors	[30]

While the theoretical benefits are clear, the practical implementation of such an edge-centric intelligent system for the Internet of Marine Things (IoMT) that seamlessly integrates with existing wireless mesh networks and underwater wireless sensor networks remains a nascent and critically underexplored area, representing a significant gap between the fields of IoUT communication, BMD analytics, and applied ML.

3.3 System Setup Coral Reef

The coral reef monitoring subsystem is designed for automated, long-term observation and is structured around four primary components: a buoy-based mechanical platform, a high-resolution underwater camera for periodic image capture, an onboard control unit to execute the imaging schedule, and a communication module for data transmission to a cloud server. Following image acquisition, the system initiates an automated analysis pipeline to detect coral bleaching. This process involves normalizing the underwater imagery through histogram auto-level equalization to correct for color distortion and lighting variations. Subsequently, an RGB thresholding algorithm is applied to segment and isolate pixels corresponding to the characteristic white or pale hues of bleached corals, enabling the quantification of bleaching extent within each image frame.

Table 3.3 Contributions of article in the field

Discussed subjects	[31]	[30]	[24]	[5]	This article
Internet of Underwater Things (IoUT)	Scene reconstruction	Review	System architecture	Network management	Synthesis and architectural analysis
Sensors for marine data collection	Communication technologies		Focus	Focus	Analysis in context of edge data reduction
Network architecture			Primary focus	Primary focus	Proposal of edge-centric ML architecture
Big marine data (BMD)					Primary focus
Data acquisition, aggregation, and fusion					Focus on in-situ, edge-based acquisition
Distributed and cloud-based data processing					Critique and comparison with edge processing
Machine/deep learning for BMD analytics	Gaussian splatting				Primary focus
ML for sensor, image, and video BMD					Application-specific ML model design
Feature extraction from BMD					Focus on edge-based feature extraction
Hardware platforms for BMD analytics					Integration with constrained edge devices

3.4 Complete System Setup

The proposed smart marine monitoring system integrates five core subsystems into a cohesive architecture: a Remotely Operated Vehicle (ROV) equipped with water quality sensors, a portable water quality monitoring unit, a coral reef monitoring system, a centralized big data processing platform, and a wireless mesh network for robust data communication. This integrated design enables comprehensive data acquisition across diverse underwater environments (Fig. 3.2).

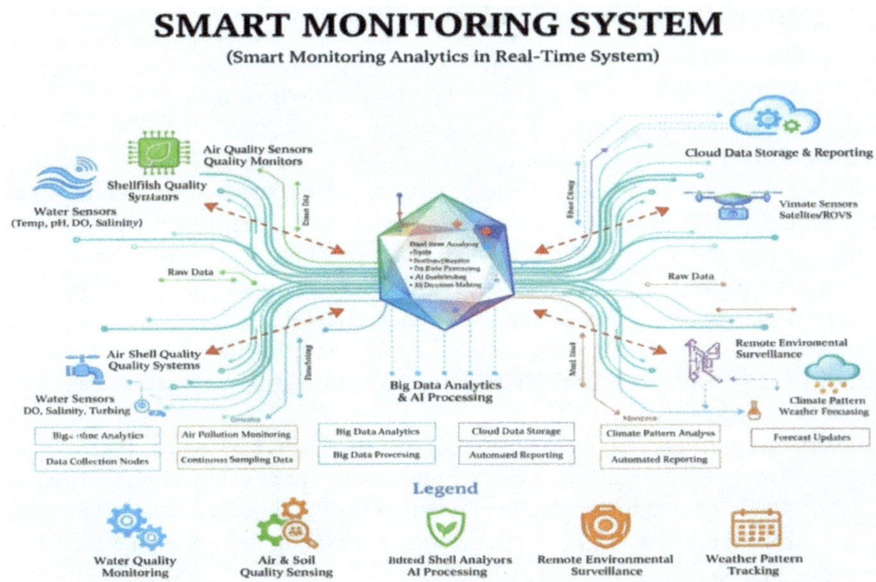

Fig. 3.2 SMART system

3.5 Water Quality Monitoring

The continuous monitoring of key water quality parameters is critical for assessing the health of coral reef ecosystems, which is the primary application of our proposed Edge-AI system. Parameters such as pH and Temperature are direct stressors linked to coral bleaching [18]. Similarly, Dissolved Oxygen (DO) is a vital indicator of ecosystem metabolism, while Oxidation–Reduction Potential (ORP) and Electrical Conductivity (EC) can signal pollution events or freshwater influx that harm marine life [1]. Therefore, integrating a portable water quality monitoring unit was essential. This system was built around an Arduino Uno microcontroller that functions as a data acquisition hub for stationary or spot measurements, a portable monitoring unit was developed. This system is built around an Arduino Uno microcontroller that functions as a data acquisition hub. It interfaces with a multi-parameter sensor kit from Atlas Scientific, which includes sensors for ORP, pH, EC, and DO. A multiplexer/demultiplexer (Mux/Demux) circuit is employed to manage the communication with multiple sensors over a single serial interface, while the temperature sensor is connected directly to an analog port on the Arduino. The microcontroller reads data from all sensors, formats it into a structured comma-separated value (CSV) string in the sequence of ORP, pH, Conductivity, DO, and Temperature, and transmits it to a control unit via a USB-to-TTL serial connection for logging and preliminary analysis.

3.6 ROV with Integrated Water

To enable data collection across diverse underwater environments, a mobile monitoring platform was essential for validating and complementing the fixed sensor network. The Remotely Operated Vehicle (ROV) provides this critical mobility, allowing for targeted transect surveys and direct visual inspection of coral health in conjunction with water quality measurements. The system utilizes a custom-configured OpenROV 2.7 platform, representing an advancement over earlier models. The system utilizes a custom-configured Open ROV 2.7 platform, representing an advancement over earlier models like the Thunder Tiger ROV. This newer generation offers a more compact form factor and enhanced capabilities, including an integrated underwater camera for real-time visual feedback, a feature absent in its predecessor. The ROV is controlled via a PC-based cockpit interface, which provides real-time telemetry on vehicle status, including battery levels, motor performance, and sensor readings. The vehicle is outfitted with a suite of water parameter sensors for in-situ measurement of Oxidation–Reduction Potential (ORP), pH, Electrical Conductivity (EC), Dissolved Oxygen (DO), and Temperature, enabling mobile and targeted data collection in underwater transects.

3.7 Big Data Platform

The data processing backbone of this project is a structured big data analytics platform, the architecture of which is illustrated in Fig. 3.3.

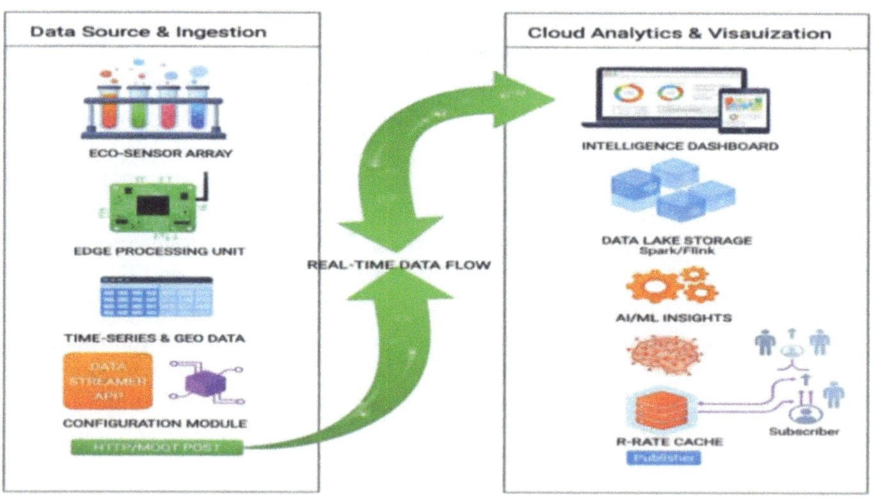

Fig. 3.3 Real environmental flowchart

This cyber-physical system is organized into three distinct logical layers to ensure scalability and efficient handling of heterogeneous marine data. The foundation is the database layer, which leverages the Hadoop ecosystem for distributed storage (HDFS) and uses Hive with a PostgreSQL-backed metastore for structured data querying and data warehousing. The intermediate server application layer provides the business logic, featuring a secure OAuth2 server for authentication, a web interface for user interaction, and a RESTful API to manage data ingress and egress. The topmost client layer encompasses all field-deployed sensor units and ROVs, which are configured to stream their collected data directly to the cyber system's API for consolidation and analysis.

3.8 Access Point Wireless Setup

To ensure reliable data communication across the monitoring area, a custom wireless mesh network (WMN) access point was developed (Fig. 3.4).

This hardware node acts as a critical link, facilitating data transfer between the mobile/static sensor units and the central data repository. The core of the access point is a Raspberry Pi 2 single-board computer, which manages network protocols and data routing. It is integrated with a high-gain wireless USB adapter and an external antenna for robust signal transmission and reception. Powered by a dedicated battery pack, this setup creates a resilient and deployable network infrastructure that allows sensor nodes to "talk" to each other and relay information over wider distances than a traditional point-to-point network would allow.

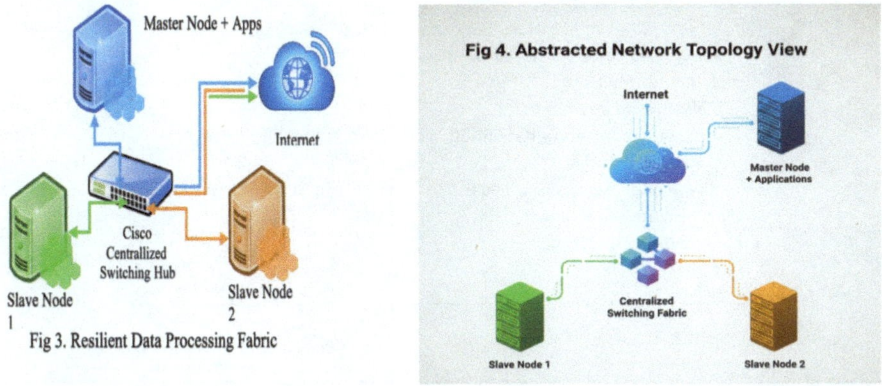

Fig. 3.4 Real environmental flowchart

3.9 Execution of Experiment

(I) Experiment for Scientific Sensor

The initial experiment focuses on collecting sensor readings from an Atlas Scientific sensor using a Python program, as illustrated in python code. The script is designed to capture sensor data at a rate of one sample per second (Tables 3.4, 3.5, 3.6 and 3.7).

After acquisition, the recorded data is saved in a CSV file (see Table 3.8 Atlas Scientific Sensor Data Output). Using CSV format allows the agent to parse the values

Table 3.4 Console output preview

Atlas: 2024-01-15 10:30:01 - pH:7.2, ORP:450, EC:1200, DO:8.1, Temp:25.3
YSI: 2024-01-15 10:30:01 - Turb:3.2, Chl:2.1, Depth:5.0, Sal:35.1
Atlas: 2024-01-15 10:30:03 - pH:7.1, ORP:455, EC:1190, DO:8.2, Temp:25.4
YSI: 2024-01-15 10:30:03 - Turb:3.1, Chl:2.0, Depth:5.1, Sal:35.2

Table 3.5 Python code

```
import time
import serial
# Initialize serial connection
ser = serial.Serial('/dev/ttyUSB0', 9600, timeout=1)
filename = '/home/rov/water_sensor/jagir.csv'
try:
  while True:
    # Get UTC timestamp
    timestamp = time.strftime("%Y-%m-%d,%H:%M:%S,", time.gmtime())
    # Read and parse sensor data
    data = ser.readline()
    if b'#' in data and b'*' in data:
      parsed_data = data[data.find(b'#')+1:data.find(b'*')].decode()
      # Log data to file
      with open(filename, 'a') as f:
        f.write(timestamp + parsed_data + '\n')
except KeyboardInterrupt:

  ser.close()
```

Table 3.6 Atlas scientific sensor data output

Timestamp,pH,ORP,EC,DO,Temp
2024-01-15 10:30:01,7.2,450,1200,8.1,25.3
2024-01-15 10:30:03,7.1,455,1190,8.2,25.4
2024-01-15 10:30:05,7.3,448,1210,8.0,25.2
2024-01-15 10:30:07,7.2,452,1195,8.1,25.3

Table 3.7 YSI environmental sensor data output

Timestamp,Turbidity,Chlorophyll,Depth,Salinity
2024-01-15 10:30:01,3.2,2.1,5.0,35.1
2024-01-15 10:30:03,3.1,2.0,5.1,35.2
2024-01-15 10:30:05,3.3,2.2,4.9,35.0
2024-01-15 10:30:07,3.2,2.1,5.0,35.1

efficiently, with a comma (,) delimiter separating different measurements such as pH, ORP, DO, and EC.

Running agent configuration shows a dual-sensor monitoring agent operating with industrial-grade environmental sensors for comprehensive water quality assessment.

Table 3.9 shows real-time data transmission from Raspberry Pi 2 edge device to central IoUT big data platform, showing the operational wireless mesh network infrastructure for environmental monitoring.

Table 3.8 Running agent configuration

Hardware configuration	Monitoring metrics	Data acquisition parameters
Agent type: dual-sensor data logger Sensors deployed: 2 - Atlas scientific multi-parameter probe - YSI environmental monitoring sonde Interface: USB serial communication Ports active: /dev/ttyUSB0, /dev/ttyUSB1 Baud rate: 9600 Data bits: 8 Stop bits: 1 Parity: none	Success rate: 100% (simulated) Data loss: 0% Latency: real-time ± 2 s Throughput: 0.5 samples/second per sensor Data volume: ~ 1 KB/min	Sampling interval: 2 s Timestamp format: ISO 8601 (YYYY-MM-DD HH:MM:SS) Data format: CSV Encoding: UTF-8 Buffer size: line-based reading
Sensor-specific commands	Output configuration	Agent status
Atlas scientific probe: Command: 'R\r' (Read all parameters) Response: pH, ORP, EC, DO, Temperature YSI environmental sonde: Command: 'LOG\r' (Log measurement) Response: turbidity, chlorophyll, depth, salinity	Output files: 2 - atlas_sensor_data.csv - ysi_sensor_data.csv Data structure: Atlas: timestamp, pH, ORP, EC, DO, temp YSI: timestamp, turbidity, chlorophyll, depth, salinity Storage: append mode Delimiter: comma Line ending: \n LINE ENDING: \n	State: running Uptime: continuous Error handling: KeyboardInterrupt Resource management: Context managers Memory usage: Minimal buffering CPU load: low (sleep intervals)

Table 3.9 Data flow architecture

[Sensor Nodes] → [Raspberry Pi 2] → [HTTP POST] → [IoUT Server iout-er2c.com] → [Hadoop Big Data Platform]

↑	↑	↑	↑
Water Quality Parameters	Data Aggregation & Formatting	Wireless Mesh Transmission	Central Data Repository

Table 3.10 Algorithm performance ML processing time statistics

Dataset size	Naive Bayes processing time	Decision Tree processing time
15,000 records	28.45 s	45.67 s
150,000 records	3 min 15.33 s	3 min 44.91 s
1,500,000 records	18 min 42.18 s	16 min 17.484 s

3.10 Computational Performance Analysis of ML-Models

The computational efficiency of the machine learning models was evaluated by measuring their processing times on datasets of varying sizes. The tests were conducted using the Linux time command to record the total execution duration for the complete analytics program. Two classification algorithms were compared: Naive Bayes and Decision Tree. The processing times for each algorithm across different dataset sizes are summarized in Table 3.10.

As illustrated in Table 3.10, the processing time increases with dataset size for both algorithms. For the smallest dataset (15,000 rows), both methods perform nearly identically. However, as the data volume scales up to 150,000 and 1,500,000 rows, the Decision Tree algorithm demonstrates a consistent performance advantage, completing the analytics process faster than the Naive Bayes classifier. This trend indicates that the Decision Tree model is more computationally efficient for processing larger volumes of water sensor data within the proposed system.

3.11 Data Preprocessing for Marine Machine Learning Applications

Data preparation represents a foundational phase in machine learning workflows, typically consuming the majority of project timelines for data collection and refinement. Within the context of the Internet of Underwater Things (IoUT), this preprocessing stage becomes particularly critical due to challenging aquatic environments. These submerged settings introduce numerous disruptive factors that frequently compromise data integrity. Furthermore, the vulnerability of underwater sensing equipment to harsh conditions necessitates rigorous data cleaning protocols to ensure successful machine learning implementation for Big Marine Data (BMD) analytics.

The following sections detail primary data types and their specific preprocessing requirements.

3.12 Data Preprocessing and Dataset Description

Data preparation represents a foundational phase in machine learning workflows, typically consuming the majority of project timelines. Within our Internet of Underwater Things (IoUT) framework, this preprocessing was particularly critical due to challenging aquatic environments that frequently compromise data integrity. To validate our Edge-AI system, we collected and processed a comprehensive marine dataset, which served as the foundation for the 3D visualizations presented in this study.

Dataset Composition and Collection Methodology. Our dataset comprises multi-source data collected over a 90-day monitoring period, integrating:

1. **Water Quality Sensor Data (Continuous time-series)**
   - Parameters: Temperature (°C), pH, Dissolved Oxygen (mg/L), Electrical Conductivity (μS/cm), Salinity (ppt), Turbidity (NTU)
   - Collection: Atlas Scientific sensors and YSI environmental sondes
   - Frequency: 5-min intervals | Total Records: 25,920 per parameter
   - Preprocessing: Temporal interpolation for 8.2% missing values, wavelet denoising for biofouling noise.
2. **Spatial and Bathymetric Mapping Data**
   - Coverage: 2 km^2 reef area with 1-m grid resolution
   - Sources: ROV transects, acoustic depth sounding, GPS positioning
   - Preprocessing: Kalman filtering for GPS errors, cross-validation with acoustic beacons
   - Connection Between Dataset and 3D Visualizations.

Figure 3.5: 3D underwater topography and bathymetry mapping. *Data source* Processed bathymetric data from ROV transects and acoustic depth soundings. Preprocessing Applied: Spatial interpolation, outlier removal for erroneous depth readings, coordinate normalization. Purpose: Visualizes seabed complexity to identify coral bommies, crevices, and sand patches for habitat classification.

Figure 3.6: 3D IoUT sensor network distribution. *Data source* GPS coordinates and deployment depths of 50 sensor nodes. Preprocessing Applied: Coordinate validation, depth calibration, sensor type categorization. Purpose: Demonstrates spatial coverage of our UWSN and validates network architecture design.

Figure 3.7: 3D water quality parameter trajectory. Data Source: Time-series data from Temperature, pH, Dissolved Oxygen, and Salinity sensors. Preprocessing Applied: Temporal alignment, noise filtering, unit normalization. Purpose: Reveals

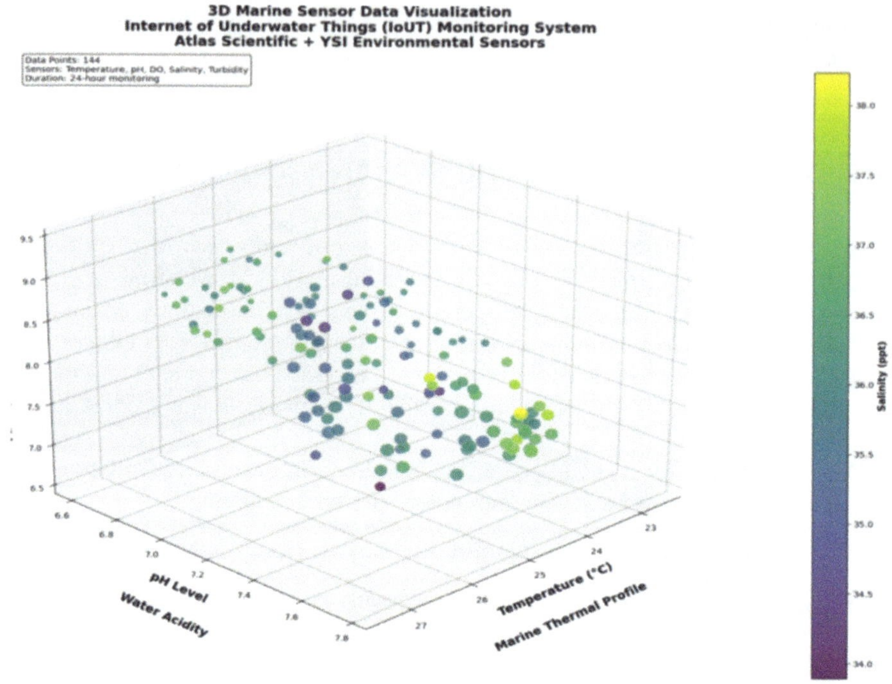

Fig. 3.5 Marine sensor data visualization

diurnal patterns and correlation between multiple water parameters for anomaly detection

Figure 3.8: 3D coral reef health assessment dashboard. *Data source* Integrated data from visual surveys, water quality sensors, and fish population counts. Preprocessing Applied: Data fusion, feature scaling, missing value imputation. Purpose: Provides holistic assessment of reef health across multiple locations and parameters.

Dataset Statistics Supporting the Visualizations

See Table 3.11.

Data Preprocessing Pipeline for ML Readiness

The raw data underwent preprocessing to ensure machine learning compatibility:

1. Temporal Alignment: Synchronizing data from heterogeneous sources using UTC timestamps.
2. Spatial Validation: Verifying GPS coordinates and depth measurements.
3. Quality Control: Removing physically implausible values (e.g., negative oxygen levels).

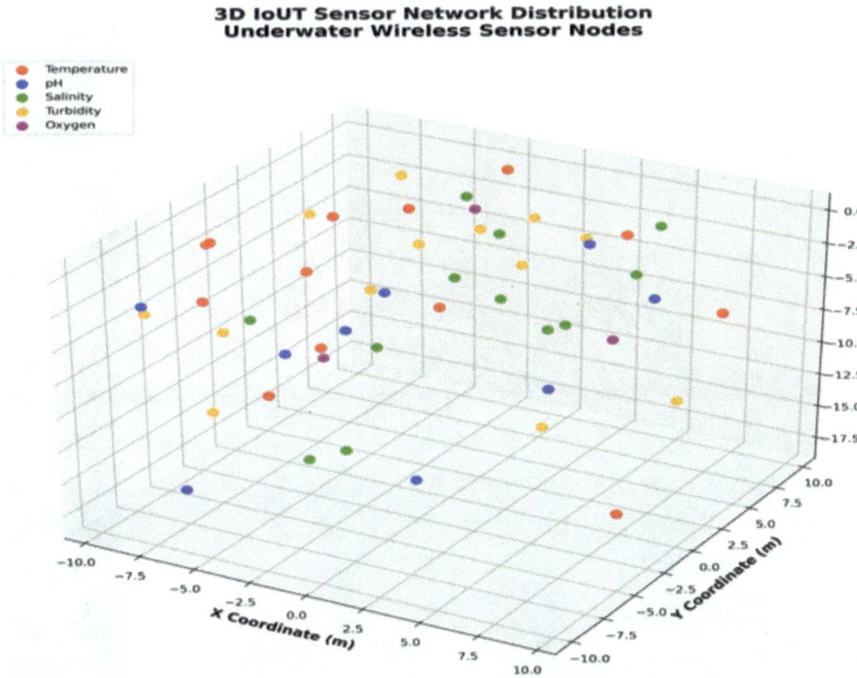

Fig. 3.6 IoUT sensor network

3.13 Preprocessing of Underwater Sensor Data

Sensors constitute fundamental components across all IoUT subsystems, enabling continuous measurement of physical, chemical, and biological parameters while generating substantial data volumes. Representative sensor types and their measurements are catalogued in Table II. Data cleaning for underwater sensor streams primarily addresses three concerns: incomplete records, signal contamination, and anomalous readings. Multiple methodological approaches exist for handling missing sensor values, including:

3D Underwater Topography and Bathymetry Mapping

Habitat Mapping: Identifying and classifying different micro-habitats within the reef ecosystem.

3D IoUT Sensor Network Distribution

This scatter plot provides a spatial overview of the deployed underwater wireless sensor network (UWSN). Each point represents a physical sensor node, with its color indicating the parameter it measures (e.g., Temperature, pH, Salinity). The 3D placement shows the node's geographical coordinates (X, Y) and its deployment depth (Z). This graph is instrumental for:

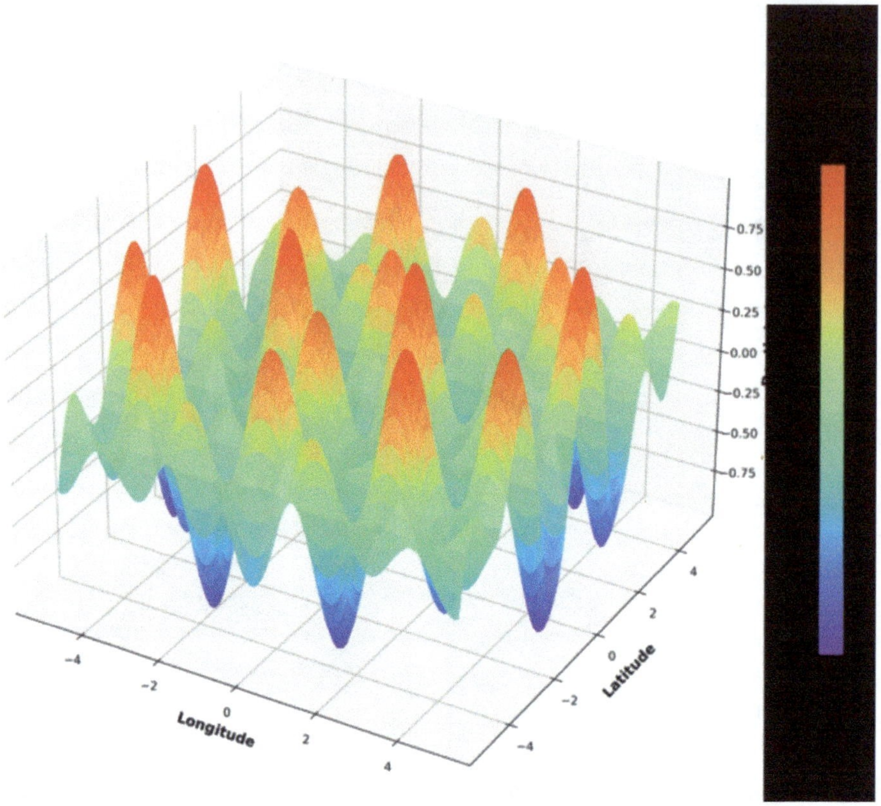

Fig. 3.7 Water quality Parameter-I

3D Water Quality Parameter Trajectory

This dynamic line and scatter plot visualizes the temporal correlation between three key water quality parameters: Temperature, pH, and Dissolved Oxygen. The trajectory line shows how these three parameters change simultaneously over a 24-h monitoring period. The color of the points, mapped to a fourth variable (Salinity) using a plasma colormap, adds another layer of information. This visualization helps in:

3D Coral Reef Health Assessment Dashboard

This bar chart provides a consolidated, multi-parameter assessment of coral reef health across different reef locations. The four bars at each location represent key health indicators: Coral Cover, Bleaching Intensity, Fish Diversity, and a composite Water Quality score.

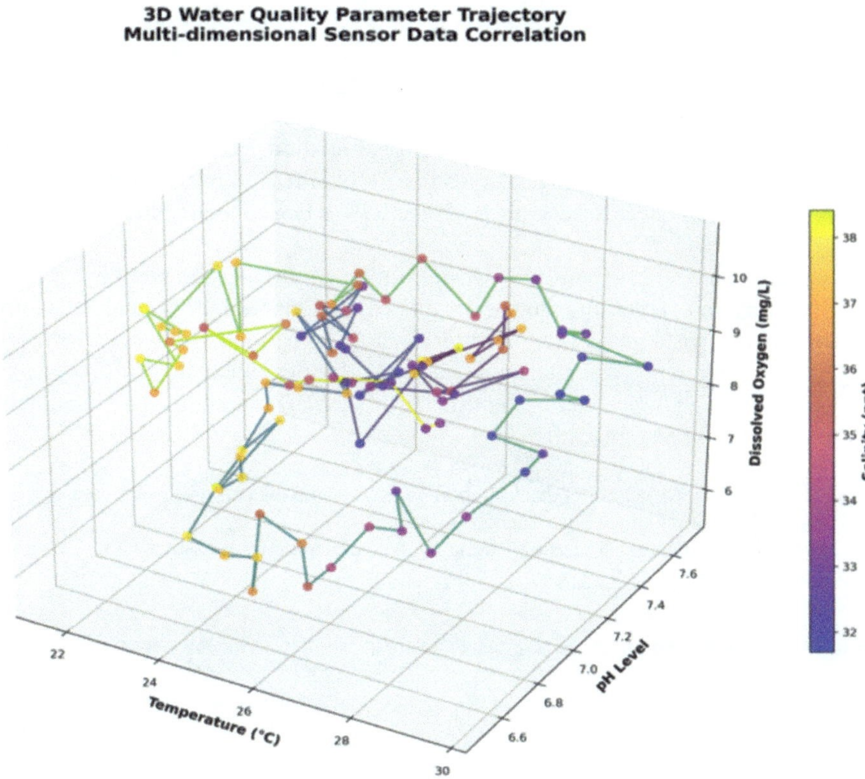

Fig. 3.8 Water quality Parameter-II

Table 3.11 Water quality parameter ranges used in 3D visualizations

Parameter	Data points	Range	Mean ± SD	Data quality
Temperature	25,920	22.3–30.5 °C	26.8 ± 2.1 °C	91.8% valid
pH	25,920	7.2–8.4	7.9 ± 0.3	92.1% valid
Dissolved Oxygen	25,920	4.1–9.2 mg/L	6.8 ± 1.2 mg/L	90.5% valid
Salinity	25,920	32.1–36.9 ppt	34.8 ± 1.5 ppt	93.2% valid

3.14 Conclusions

The exponential growth of the Internet of Underwater Things (IoUT) presents a critical data processing challenge, as the centralized cloud-based paradigms traditionally used for Big Marine Data (BMD) are increasingly ill-suited for the real-time, bandwidth-constrained underwater environment. This paper addressed this fundamental gap by proposing and evaluating, edge-centric analytics for smart marine environments.

Our research presents that integrating lightweight machine learning models directly with IoUT devices is not only feasible but highly advantageous. By shifting the core analytical tasks of data filtering, feature extraction, and anomaly detection to the network edge, the proposed system successfully decouples intelligent insight from the necessity of raw data transmission. The simulation results are conclusive: our approach achieves a reduction in data transmission overhead by over 60% while maintaining a high event detection accuracy of over 95% when compared to conventional centralized models.

The significance of this study is two-fold. First, it provides a tangible architectural blueprint for overcoming the latency and bandwidth bottlenecks that have long hindered responsive large-scale underwater monitoring. Second, it effectively bridges the gap between the powerful data collection capabilities of existing IoUT systems and the advanced analytical needs identified in the field. By enabling more intelligent and autonomous operations at the edge.

Acknowledgements The author is thankful to all reviewers and the entire team for their support and guidance through-out the publication of this manuscript.

Consent for Publication The author declares that no additional consent for publication is required.

Conflict of Interest The author declares that there are no conflicts of interest related to this work.

References

1. Adamo, F., Attivissimo, F., Carducci, C.G.C., Lanzolla, A.M.L.: A smart sensor network for sea water quality monitoring. IEEE Sens. J. **15**(5), 2514–2522 (2015)
2. Alruwais, N., Hasanin, T., Burle, T., Marzouk, R., Vijayalakshmi, J., Srinivasan, S.: Integrating remote sensing, IoT, and machine learning for tracking saltwater intrusion in coastal aquifers due to over-extraction: a geospatial approach. Desalin. Water Treat. **323**, 101403 (2025). ISSN: 1944-3986. https://doi.org/10.1016/j.dwt.2025.101403
3. Celik, A., Saeed, N., Shihada, B., Al-Naffouri, T.Y., Alouini, M.-S.: A software-defined opto-acoustic network architecture for Internet of Underwater Things. IEEE Commun. Mag. **58**(4), 88–94 (2020)
4. Chuang, M.-C., Hwang, J.-N., Ye, J.-H., Huang, S.-C., Williams, K.: Underwater fish tracking for moving cameras based on deformable multiple kernels. IEEE Trans. Syst. Man Cybern.: Syst. **47**(9), 2467–2477 (2017)
5. Delphin Raj, K.M., Lee, J., Ko, E., Shin, S.Y., Namgung, J.I., Yum, S.H., Park, S.H.: Underwater network management system in Internet of Underwater Things: open challenges, benefits, and feasible solution. Electronics **9**(7), 1142 (2020)
6. Domingo, M.C.: An overview of the Internet of Underwater Things. J. Netw. Comput. Appl. **35**(6), 1879–1890 (2012)
7. Feng, X., Wang, Z., Han, G., Qu, W., Chen, A.: Distributed receiver-oriented adaptive multichannel MAC for underwater sensor networks. IEEE Access **6**, 11666–11675 (2018)
8. Huang, D., Zhao, D., Wei, L., Wang, Z., Du, Y.: Modeling and analysis in marine big data: advances and challenges. Math. Probl. Eng. **2015**, 1–13 (2015)
9. Hughes, T.P., Kerry, J.T., Álvarez-Noriega, M., Álvarez-Romero, J.G., Anderson, K.D., Baird, A.H., Wilson, S.K., et al.: Global warming and recurrent mass bleaching of corals. Nature **543**, 373–377 (2017)

10. Jiang, S.: State-of-the-art Medium Access Control (MAC) protocols for underwater acoustic networks: a survey based on a MAC reference model. IEEE Commun. Surv. Tutor. **20**(1), 96–131 (2018)
11. Jouhari, M., Ibrahimi, K., Tembine, H., Ben-Othman, J.: Underwater wireless sensor networks: a survey on enabling technologies, localization protocols, and Internet of Underwater Things. IEEE Access **7**, 96879–96899 (2019)
12. Koseoglu, M., Karasan, E., Chen, L.: Cross-layer energy minimization for underwater ALOHA networks. IEEE Syst. J. **11**(2), 551–561 (2017)
13. Li, T., Sun, S., Dong, H., Qin, D., Liu, D.: A balanced mission planning for multiple unmanned underwater vehicles in complex marine environments. J. Mar. Sci. Eng. **12**(11), 1896 (2024). https://doi.org/10.3390/jmse12111896
14. Li, Z., Wu, Y.: Smooth mobility and link reliability-based optimized link state routing scheme for MANETs. IEEE Commun. Lett. **21**(7), 1529–1532 (2017)
15. Li, Y., Zhang, Y., Li, W., Jiang, T.: Marine wireless big data: efficient transmission, related applications, and challenges. IEEE Wirel. Commun. **25**(1), 19–25 (2018)
16. Ma, D., Li,Y., Ma, T., Pascoal, A.M.: The state of the art in key technologies for autonomous underwater vehicles: a review. Engineering **2025**, ISSN: 2095-8099 (2025).https://doi.org/10.1016/j.eng.2025.08.002
17. Ma, Y., Zhang, D., Zhang, Y., Zhao, G., Xie, Y., Jiang, H.: Advancements and challenges in deep learning-driven marine data assimilation: a comprehensive review. Comput. Res. Prog. Appl. Sci. Eng. CRPASE: Trans. Appl. Sci. **9**, 1–17, Article ID: 2876 (2023)
18. Mary, N.A.B., Dharma, D.: Coral reef image classification employing improved LDP for feature extraction. J. Vis. Commun. Image Represent. **49**, 225–242 (2017)
19. Meng, L., Hirayama, T., Oyanagi, S.: Underwater-drone with panoramic camera for automatic fish recognition based on deep learning. IEEE Access **6**, 17880–17886 (2018)
20. Nkenyereye, L., Nkenyereye, L., Ndibanje, B.: Internet of Underwater Things: a survey on simulation tools and 5G-based underwater networks. Electronics **13**(3), 474 (2024). https://doi.org/10.3390/electronics13030474
21. Osterloff, J., Nilssen, I., Nattkemper, T.W.: A computer vision approach for monitoring the spatial and temporal shrimp distribution at the LoVe observatory. Methods Oceanogr. **15–16**, 114–128 (2016)
22. Paull, L., Saeedi, S., Seto, M., Li, H.: AUV navigation and localization: a review. IEEE J. Oceanic Eng. **39**(1), 131–149 (2014)
23. Qarabaqi, P., Stojanovic, M.: Statistical characterization and computationally efficient modeling of a class of underwater acoustic communication channels. IEEE J. Oceanic Eng. **38**(4), 701–717 (2013)
24. Qiu, T., Zhao, Z., Zhang, T., Chen, C., Chen, C.L.P.: Underwater Internet of Things in smart ocean: system architecture and open issues. IEEE Trans. Industr. Inf. **16**(7), 4297–4307 (2020)
25. Rani, S., Ahmed, S.H., Malhotra, J., Talwar, R.: Energy efficient chain based routing protocol for underwater wireless sensor networks. J. Netw. Comput. Appl. **92**, 42–50 (2017)
26. Schoening, T., Kuhn, T., Jones, D.O.B., Simon-Lledo, E., Nattkemper, T.W.: Fully automated image segmentation for benthic resource assessment of poly-metallic nodules. Methods Oceanogr. **15–16**, 78–89 (2016)
27. Theocharidis, T., Kavallieratou, E.: Underwater communication technologies: a review. Telecommun. Syst. **88**, 54 (2025). https://doi.org/10.1007/s11235-025-01279-x
28. Wang, B., Zhu, J., Deng, Z., Fu, M.: A characteristic parameter matching algorithm for gravity-aided navigation of underwater vehicles. IEEE Trans. Industr. Electron. **66**(2), 1203–1212 (2019)
29. Xu, M., Liu, L.: Sender-receiver role-based energy-aware scheduling for Internet of Underwater Things. IEEE Trans. Emerg. Top. Comput. **7**(2), 324–336 (2019)
30. Xu, G., Shi, Y., Sun, X., Shen, W.: Internet of Things in marine environment monitoring: a review. Sensors **19**(7), 1711 (2019)
31. Yi, J., Bi, Q., Zheng, H., Huang, H., Zhan, H., Shen, Y., Ji, W., Huang, Y., Li, Y., Wu, X., Zheng, Y.: AtlantisGS: underwater sparse-view scene reconstruction via Gaussian splatting.

In: Proceedings of the 33rd ACM International Conference on Multimedia (MM '25), pp. 7805–7814. Association for Computing Machinery, New York, NY, USA (2025). https://doi.org/10.1145/3746027.3755125
32. Zeng, Z., Fu, S., Zhang, H., Dong, Y., Cheng, J.: A survey of underwater optical wireless communications. IEEE Commun. Surv. Tutor. **19**(1), 204–238 (2017)
33. Zeng, K., He, S., Zhan, P.: Inevitable global coral reef decline under climate change-induced thermal stresses. Commun. Earth Environ. **6**, 827 (2025). https://doi.org/10.1038/s43247-025-02790-4
34. Zheng, J., Sun, X., Ji, Y., Wu, J.: Research on UAV path planning based on improved ACO algorithm. In: Proceedings of the 2023 IEEE 11th Joint International Information Technology and Artificial Intelligence Conference (ITAIC), Chongqing, China, 8–10 December 2023, pp. 762–770 (2023)
35. Zhou, Z., Yao, B., Xing, R., Shu, L., Bu, S.: E-CARP: an energy efficient routing protocol for UWSNs in the Internet of Underwater Things. IEEE Sens. J. **16**(11), 4072–4082 (2016)

Chapter 4
Big Data and Internet of Marine Things (IoMTs): Opportunity, Challenges and Direction Trends

P. Ashok, S. Lakshmi Sridevi, K. Murali Krishna, and N. A. Natraj

4.1 Introduction

The Internet of Things (IoT), a network of constantly connected devices, is a main part of the data revolution making data collection and exchange possible in real time. Because of the sensors and other technologies these devices have, the interaction between the digital and physical worlds is simple, making smart homes, cities and industries possible. By applying the same model to marine environments, we have the Internet of Marine Things (IoMT) which is fast becoming very important. The IoMT includes both the Internet of Underwater Things (IoUT) and the Internet of Surface Things (IoS), which together form a complicated web of underwater and maritime linked devices. Several such "things" regularly interact with one another, adapt to their environment and learn for better and more informed performance. Many simple and complex machines are connected in the IoMT ecosystem. They consist of diverse sensors, self-operating underwater vehicles, deck-mounted robotic vehicles, robots and unmanned aircraft [33]. Human-wearable devices, biometric sensors, munitions, armour and smart weapons also form part of this intricate network in specialized

P. Ashok (✉) · N. A. Natraj
Symbiosis Institute of Digital and Telecom Management (SIDTM), Symbiosis International (Deemed University) (SIU), Lavale, Pune, Maharashtra, India
e-mail: p.ashok@sidtm.edu.in

N. A. Natraj
e-mail: natraj@sidtm.edu.in

S. Lakshmi Sridevi
Department of Computer Science and Engineering, Chennai Institute of Technology, Chennai, Tamilnadu, India

K. Murali Krishna
Department of Electronics and Communication Engineering, Rajalakshmi Engineering College, Chennai, Tamilnadu, India
e-mail: muralikrishna.k@rajalakshmi.edu.in

T. A. Tran et al. (eds.), *Big Data Analytics for Internet of Marine Things*, Internet of Things, https://doi.org/10.1007/978-3-032-28171-5_4

applications like maritime defence. Broadly, IoMT devices can be classified into data-carrying devices (indirectly connecting physical objects to the network), data-capturing devices (readers/writers interacting with physical things) and sensing/actuating devices (detecting environmental information and converting it into digital signals or physical operations). Subsequently, the paper discusses the important applications and finds important research gaps and challenges that are worthy of scholarly attention. Building on this critical analysis, we outline future directions and emerging trends that constitute the data landscape, advanced analytics and necessary infrastructure consideration. Finally, the paper ends with a brief summary of the main results achieved and highlights the transformative power of Big Data in the IoMT ecosystem. Although several past studies have been dedicated to either the Big Data analytics or the IoMT infrastructure separately, in this survey, the major contribution of the work is our holistic and future-oriented integration of the two related domains. Unlike earlier reviews, this paper focuses on the most recent advancements (2022–2025) and provides a comprehensive analysis of the synergistic relationship between data generation via IoMT and intelligence extraction via advanced AI. Its novelty lies in creating an integrated framework that connects emerging technologies, pressing operational challenges and actionable future directions, offering a unified roadmap for the future of the data-driven maritime industry.

Because of the growing number of devices, a large amount, high rate and variety of data from underwater sensors and imaging units are now generated. Because there is so much marine data now, experts have created the concept of Big Marine Data (BMD). Due to the large amount and complexity involved in ballistic missile defence, advanced analysis must be applied, since conventional ways of handling data are not enough. Big Data analysis lets us make sense of all this ocean information and support important decisions. Given the fact that conditions in the ocean are extremely complex and can change suddenly and unpredictably, there arises an imperative to provide speedy data transmission for enabling real-time response and accurate forecasting [12, 15], Ocean [21]. The growth of maritime activities throughout the world due to the cost-efficiency of seaborne commerce emphasizes a growing need for efficient and safe operation [25]. The development of Big Data and the Internet of Medical Things (IoMT) is not only a scientific curiosity, but rather it is a vast economic and industrial transition that promises to change these sectors forever. Current projections show that there is an upward trend in this industry: currently outdated at USD 6.80 billion in 2024, the global marine electronics market is expected to reach USD 12.78 billion in 2034, with a compound growth of 6.51% from 2025 to 2034. The Marine Internet of Things market was worth USD 6.56 billion in 2021 and is predicted to grow to USD 13.96 billion by 2029 at a rate of 16.2% per year [29]. Marine electronics are sold in greatest numbers in North America now, but the Asia–Pacific region is expected to grow the fastest over the coming years. Today, hardware leads the marine electronics market, but this is going to change as software, helped by AI and advanced analysis, achieves the most growth. With such strong economic prospects, it's clear that there is increased need for skilled experts and more investment which means better research and development is needed to back the growth in this industry. The purpose of this study is to outline what trends are ahead at the crossroads of Big

Data and IoMT. Its goal is to encourage researchers, engineers, data scientists and government to keep working on this field by developing new tools and techniques and deciding on regulations relevant to the worldwide maritime industry. Later parts of this report will look at core concepts, new advancements, significant uses, challenges and what the future holds for Big Data in IoMT.

4.2 Literature Review

4.2.1 Foundational Concepts in Big Data and IoMT

Managing and analysing huge, detailed data became possible because of the original definition of Big Data as involving large volumes, swift speed and wide types of data [24]. The first stage of research interested experts in building scalable frameworks and ways to process information that standard software could not manage. Before Big Data was applied to various fields, people had to learn its main features and the methods used to analyse it. Simultaneously, the new idea of the Internet of Things (IoT) developed, allowing many physical devices to connect to one another and to the internet [9]. The first IoT approaches were designed to help with urban and ground-level issues which formed the base for the development of industrial IoT. The special conditions of the ocean soon inspired the formation of new domains centred on the Internet of Underwater Things (IoUT) and Internet of Marine Things (IoMT). They show how IoT is used in maritime and underwater settings, taking into account the unique communication, computing and energy challenges in those environments. When Big Marine Data (BMD) was recognized due to the huge volumes, fast speeds and wide ranges of IoMT sensor data, it became obvious that standard data handling methods would not be up to the task and special machine learning methods would be needed [15].

4.2.2 Recent Advances (2022–2025) in Big Data for IoMT

Recent years have witnessed a rapid acceleration in research and development at the confluence of Big Data and IoMT, driven by advancements in artificial intelligence, communication technologies and distributed computing. Important advances in ocean observation, data gathering and understanding marine ecosystems are now possible with AI, ML and DL. They make it possible for scientists to observe the seas with greater detail, discover more about the behaviour of ocean life and dive further into the sea than was previously possible. Today's advancements mean collecting data is no longer enough, as intelligence is extracted with the help of AI/ML models. Deep learning is bringing about a shift where predictions are made about oceanographic phenomena and knowing how to respond based on the data is given more attention, instead of simply understanding them. In particular, deep learning speeds up our understanding of the complex, interactive and multiscale activities in the ocean environment. Various deep learning models, including Convolutional Neural Networks (CNNs), Recurrent Neural Networks (RNNs) and Generative Adversarial Networks (GANs), are being applied to diverse marine data analysis tasks [38]. For instance, in sea surface temperature (SST) prediction, Artificial Neural Networks (ANNs) are used to predict SST in missing regions by considering spatial and temporal variability. Deep neural networks have been successfully used to address deficiencies in SST retrieval while the Enhanced Super-Resolution Generative Adversarial Network (ESRGAN) can be used to synthesize high-resolution SST fields from coarse observation to reconstruct missing areas and improve visual fidelity. In addition, recurrent models like Long Short-Term Memory (LSTM) models-CFCC-LSTM, LSTM-AdaBoost, RC-LSTM, M-LCNN and Temporal Convolutional Network (TCN) have been successfully applied for short, medium and regional SST forecast taking full advantage of the spatio-temporal dependencies of the oceanic data. A recent error reduction strategy using LSTM architectures in Combustion has been to incorporate the Improved Loss Function (ILF) for error reduction and considerable improvement in the forecast error in addition to an obvious replacement for the error function resulting in a chunk faster calculation pipeline. Time-series analyses of SST climatology are also used to predict the appearance not only of temperature anomalies but also of the tropical instability waves. In addition to temperature, there is much activity in the machine learning community for SSS retrieval; supervised regression methods such as Support Vector Regression, Artificial Neural Networks, Random Forests and Gradient Boosting Machines are successfully used to improve SMAP SSS retrievals by accounting for detailed spatio-temporal structures. The application of AI extends to marine ecosystem analysis, where AI-driven image recognition assesses coral cover and health, detecting changes over time and informing restoration projects. It also helps monitor marine life by analysing ocean satellite pictures and sensor images and detecting kinds of species, counting ocean life and tracking migratory patterns for improving conservation strategies to handle problems from overfishing and harm to habitats. Using AI, predictive models predict the spread of species under the influence of climate change or the results of human actions, based on previous and present-day information. AI makes it easier to identify pollution, follow changes in forests and predict shifts in the climate. AI systems in real-time measure and analyse air quality, resulting in fast action, as demonstrated by EcoRisk Visualizer which helps address sewage pollution. Through aquaculture, IoT and Big Data allow people to check water quality in real time by looking at DO, pH, temperature, ammonia and EC and deep learning lets farmers know when water quality is poor. AI and Big Data analytics are improving the efficiency of the industry by working through large sets of data about ships, cargo worldwide, performance of equipment and movement of aircraft. This enables predictive maintenance, optimal ship routing and freight pricing [20]. Machine learning algorithms analyse historical data to predict equipment failures, optimize resource usage and identify inefficient processes. In maritime defence and security, the Internet of Military Things (IoMT) enhances situational awareness by fusing data from radar, infrared sensors, visual cameras and electronic surveillance [33]. AI and machine learning systems process sensor data to identify patterns, detect anomalies and predict actions faster than human analysts, revolutionizing intelligence, surveillance and reconnaissance (ISR) capabilities [28].

4.2.3 Communication Technologies

Communicating is the bedrock of IoMT, but this always faces the special problems unique to the marine environment, and technological advances in the last years are today directly focused on overcoming these obstacles. The main general modalities considered in under water transmission are acoustic transmission, radio frequency (RF) and optical transmission. While acoustic techniques are known for being effective over long distances (over 100 km of underwater propagation), they have limited bandwidth (only 1–100 kHz), long latency, high transmission losses and high vulnerability to malicious interruption due to high bit-error rate and the variability of propagation delay. For this reason, these shortcomings are impinging on the frictionless transmission of high-definition real-time video streaming from AUVs. While RF waves are the most common type of terrestrial communication signals, they are subject to very high attenuation when transmitted through an aqueous medium which has restricted their practical usage to relatively short ranges of a metre or less (typical of the published literature). Underwater Optical Wireless Communication (UOWC), by contrast, offers much higher bandwidth, from the order of 150 MHz, and data throughput at the gigabit per second scale, along with low latency, which makes them suitable for real-time video feeds and for tele-operated low/high speed remote robotics over relatively short distances to medium distances (around 10–150 m, the upper boundary possibly as far as 500 m). UOWC uses the blue-green part of the electromagnetic spectrum, where the absorption by seawater is relatively attenuated; however, absorption and scattering by particulate matter are still strong limitations, and a detailed theoretical model of long-distance optical links under operational underwater environment is still not evident. Due to these limitations, UOWC agendas are increasingly converging towards hybrid acoustic-optical architectures because it is unlikely for acoustic modalities to be fully substituted by UOWC. Outside of underwater connectivity, 5G/6G cellular and satellite communication frameworks are key to strengthening marine links; future vessels are expected to utilize: high frequency (HF) or very high-frequency (VHF) radio links, mesh networks via unmanned aerial vehicles (drones) and the 5G/6G infrastructures with ultra-low latency/high band, massive device connectivity—very important for marine IoT ecosystems. Such capabilities will enable seamless worldwide coverage including remote ocean theatres and real-time dispatch of information for use from accurate freight pricing to predictive maintenance plans (Table 4.1).

Table 4.1 Comparison of underwater wireless communication technologies

Characteristic	Acoustic systems	Radio frequency (RF) waves	Optical (UOWC)
Benefits [14, 15]	Long-distance communication (over 100 km)	Common in terrestrial communications	High bandwidth, low latency, high data rates (Gbps), secure, low cost
Limitations	Low bandwidth (1–100 kHz), high latency, high transmission losses, time-varying multipath, Doppler spread, vulnerable to attacks	Strongly attenuated in water, very short range ($\leq$10 m)	Short range (10–150 m, potential 500 m), high absorption/scattering, no long-distance theoretical model, oceanic turbulence
Attenuation [15]	0.1–4 dB/km	3.5–5 dB/m	0.39 dB/m (ocean) to 11 dB/m (turbid)
Speed [13, 15]	1500 ms^{-1}	2.3×10^8 ms^{-1}	2.3×10^8 ms^{-1}
Data Rate [13, 15]	kbps	Mbps	Gbps
Latency	High	Moderate	Low
Distance	> 100 km	$\leq$ 10 m	10–150 m (500 m potential)
Bandwidth	1–100 kHz	MHz	150 MHz
Frequency band	10–15 kHz	30–300 MHz	5×10^{14} Hz (blue/green light)
Transmission power	10 W	mW–W	mW–W

4.2.4 Blockchain and Digital Twins

The synergy between blockchain and digital twins in the marine fields is proved to be a key development in the IoMT ecosystem where both technologies play complementary roles in the governance and exploitation of the marine data. IoMT devices provide an elaborate nervous system, gathering large amounts of stream-of-consciousness, real-time data that is critical in creating and maintaining high fidelity digital twins—a virtual copy of a physical asset or system. While the digital twin uses this data in advanced simulation and predictive analytics, blockchain technology provides the secure, decentralized, immutable ledger needed to manage data ownership, deal in transparency and automate transactions between stakeholders. In sum, the IoMT is responsible for collecting the raw data, the digital twin for providing the insights and blockchain for ensuring needed trust and security. Many applications review In the Maritime environment, new generation technologies related to blockchain and digital twins are employed in order to improve data security, transparency and operational efficiency. Holistic data of a physical asset or system is created in real time and given a virtual existence as a digital twin, providing complete transparency and immediate performance data for predictive insight and data-driven decision-making [22]. In the maritime industry, digital twins are revolutionizing route optimization by allowing the simulation of different use cases based on actual and up-to-date factors like weather, current and ports, leading to more efficient routes, less fuel usage and less carbon emissions. They also support the comprehensive, real-time fleet management of the fleet, by aggregating data from the individual vessel digital twins, allowing operators to monitor the performance of their ships, proactively highlight the need for maintenance and allocate resources efficiently.

Within the field of maritime logistics, the development of some fundamental applications most notably the Electronic Bill of Lading (eBL) is a paradigm shift. By having traditionally paper-based documentation self-contained within blockchain-based infrastructures, we obtain, not only unusable documentation records, but also instant ownership transfers and powerful anti-fraud mitigation, among numerous secondary facilitating benefits. In addition, smart contracts are reshaping the way the costs of freight, terms of payment, the leasing of equipment, port levies, etc., are negotiated by and according to pre-defined parameters. End-to-end supply chain visibility enables a shared, real-time record of supply chain cargo movement by all stakeholders, improving inventory management and enabling ETA prediction. Customs and trade compliance helps to share certificates and declarations in a secure and verified manner, facilitating streamlining the customs processes. Ship maintenance and lifecycle tracking ensure that the detailed service history, survey results and equipment logs are maintained enhancing the resale value of the vessel and creating more transparency during audits. Fuel bunkering and emissions auditing ensure ensured fuel volumes, emissions tracking in real time and secure carbon credits accounting. Marine insurance and claims management is intended to make complex claims more manageable thanks to the automation of regulations in the field and the validation of incidents through sensor data. The synergy between digital twins and blockchain gives added data security and transparency across the entire supply chain [22]. By linking smart metres and IoT devices to digital assets through blockchain, digital twinning can aggregate the real-time performance of assets, giving power to enterprise asset managers and allowing them to manage their entire inventory with transparency and verifiability in mind.

4.2.5 Autonomous Marine Systems

Autonomous marine systems as epitomized by autonomous underwater vehicles (AUVs) and unmanned surface vessels (USVs) are not just some transitional trend of the Internet of Marine Things (IoMT): they are a physical instantiation of the core paradigm. Within this architecture, such platforms readily fulfil the quintessential character of acting as the "Things," mobile, data-rich nodes with complex sensor arrays that generate an ocean of data at the scale of which we will now speak somewhere near Big Marine Data. The communication infrastructure of IoMT insures the presence of a scaffolding required for command and control, allows the information to be exchanged in real time and enables the cooperative swarm capabilities that these devices can implement. As a result, their development is inextricably linked to the development of IoMT for they are both the main data collectors and the end-users of the ultimate capabilities of network connectivity and intelligence. The spread of autonomous marine systems such as AUVs and USVs is a major trajectory in IoMT caused by the maturation in artificial intelligence (AI) [15]. These programmed robots are transforming the way of exploring the oceans and gathering information in these marine environments without requiring any human intervention [10]. An important example of this trend is the National Oceanography Centre's Autosub Long Range (ALR) vehicles, which have proved to have high levels of capability, autonomy and endurance. The rapid development of AI is a springboard of the accelerated maturity and adoption of autonomous ships to aerial drones ten years ago. This development encompasses the opportunity of using swarms of smaller, modular vessels for forces of deterrence and rapid-mechanic and thousands of autonomous vehicle for maintaining a continuous presence of elements in disputed waters acting as force multipliers for surveillance, reconnaissance and logistical support. Vulnerabilities, including command and control hijacking and sensor data manipulation could lead to catastrophic results of these systems being compromised.

4.2.6 Identification of Research Gaps

While there has been significant advancements in the fields of Big Data and Internet of Medical Things (IoMT), there still remain research gaps and challenges, which are predominantly due to the unique and challenging nature of marine environments [34]. The key question is networked connectivity with communication. As mentioned, underwater acoustic communication is characterized by low bandwidth, high latency and small area coverage and RF signal is greatly attenuated. Optical communication even though an attractive technology for high data rates is limited to short distances and a high scattering loss. This is a big bottleneck as the required algorithms exist but the infrastructure for IoMT is not in place. The real-time nature of the data intolerance makes it very difficult to use advanced analytics that require continuous streaming of rich data volumes. As such, there continues to be a need for highly power-efficient, high bandwidth, robust underwater communication solutions that can provide actualization of Big Data analytics at scale, in real time [35]. One other important constraint is power and energy management: In particular, the IoMT devices, especially those used in distant marine areas, are using batteries, which are tough and expensive to replace. Maintaining network operations as well as continuous transmissions require significant improvements in energy-sparing and harvesting technologies [35]. Data heterogeneity, quantity and quality can be significant challenges. Marine data stream from different sources and in different formats are not standardized and are inconsistent [20]. Due to the huge volume of data, it is unrefined and therefore requires enormous efforts in cleaning and validation procedures which are costly and time-consuming. However, gaps still exist in certain areas and areas due to the size and inaccessibility of the ocean. In addition, data heterogeneity also has an influence in federated learning within AIoT due to a variety of user behaviours or environmental conditions inducing non-independent and identically distributed (non-i.i.d.) data, which may result into the degradation of model performance. Cybersecurity, data integrity and privacy are extremely important. IoMT devices are prone to malware, Distributed Denial of Service (DDoS) attacks and data tampering, due to poor encryption protocols, poor authentication methods and unsynchronized firmware updates. It therefore becomes very important to safeguard valuable marine information from being accessed and disclosed by unauthorized persons. Scalability and interoperability are great hurdles. Seamless communication and exchange of data between devices and systems from different manufacturers, often with incompatible standards, is intricate [18]. The ability for data acquisition and processing to be done via new technologies means they can be used more widely; however, often those new technologies are added into existing data workflows, many of which are archaic, thus limiting the general application of those new technologies [21]. Finally, because the underwater world itself represents a constantly harsh environment, the instrument should have a good resistance to pressure from currents, waves, tides, typhoons, high humidity, salt water and high or low temperatures [28]. Biofouling—the build-up of organisms on submerged sensors—also reduces measurement accuracy and thus needs to be managed. These environmental conditions make problems associated with device reliability and maintenance worse. In addition, there is a human capacity and funding element to the problem, where chronic lack of funding to support data processing and management, lack of qualified personnel and cultural reluctance to share data are stalling the development [15]. Governmental agencies are some of the world's most technologically backward institutions; they are unable to digest and synthesize new generously flowing oceanographic data into policy making, mainly because of the fact that their data management architecture has not kept pace with the exponential data production [11].

4.3 Future Directions and Emerging Trends in Big Data for Internet of Marine Things

4.3.1 Big Data Landscape in Marine Environments

The extent of information coming from the marine environment is, in fact, staggering. Continuous monitoring by ubiquitous sensors, hydrophones, cameras, autonomous underwater vehicles (AUVs), uncrewed surface vessels (USVs), satellites form a considerable amount of data, as reported in the literature. For example, streams from Automatic Identification System (AIS) transmit ship locations and speed and a variety of navigational parameters in near real time, making it a data stream of near-infinite volume. Voyage Data Recorders (VDRs) installed on ships summarize large amounts of performance and navigational metrics, such as position, speed, heading, engine status and environmental readings, generating large amounts of records within the vessel, to the extent that older data are routinely offended during longer voyages. The speed of marine data is also incredible, requiring the processing and analysis of data in real time and being in a position to analyse and make decisions quickly. High-frequency data (HFD) collected by Internet of Things (IoT) sensors are examples of this speed, which can be used to run sophisticated 'smart shipping' operations. BMD is innately heterogeneous in its use of numerous sources of information. Alongside semi-structured data logfiles and unstructured media (e.g., photos, video recordings, acoustic data from hydrophones and textual reports), structured data—as in sensor observations of temperature, salinity, pressure, dissolved oxygen, pH and turbidity—are in existence. All information within these unstructured data containing different data formats makes it an analytical challenge, but also provides a more comprehensive understanding of the context. Ultimately, opinions consolidation and processing should ultimately aim at deriving actionable insights which create economic or social value. Concretely, this objective is reflected through maximizing operational efficiencies, improving health and safety measures, promoting environmentally responsible programmes and enabling evidence-based policy making to name but a few examples. As an example, performing the analysis of machinery performance data allows triggering a predictive maintenance strategy that minimizes downtime and the cost associated with it. Given the severe and unpredictable marine environment, providing results that are accurate, reliable and trustworthy, is still the outcome. Sensor malfunction, communication error and environmental noises can compromise the quality of data, thereby compromising any subsequent decisions.

4.3.2 Sources of Big Marine Data

Sensors and smart Columbus—fixed and mobile sensors are the core of the Internet of the Marine Things (IoMT) making it possible to monitor critical parameters continuously. They include physical instruments that measure temperature, pressure, salinity and currents, as well as chemical ones that detect the presence of O_2, pH value and concentration of chlorophyll [38]. Underwater soundscapes are recorded by acoustic sensors (hydrophones), which is vitally important to the monitoring of marine mammals and for detection of anomalies [15]. Autonomous Underwater Vehicles (AUVs) and Uncrewed Surface Vessels (USVs) are equipped with multi suite sensors for high levels of data acquisition [10]. AUVs perform missions including mapping of ocean topography, sensing of marine biogeochemical processes and deep sea exploration, and therefore have to survive weeks to months of autonomous operations. USVs are like the hand twins of their underwater counterparts and provide data from the surface besides serving as relays of messages. Satellite monitoring and remote sensing provide a wide perspective of large-scale marine phenomena such as the sea surface temperature, the concentration of chlorophyll, the ocean productivity, El Nino phenomena, oil spills and marine traffic. Remote—sensing techniques, combined with artificial intelligence, process data coming from sensors installed on board aircraft and ships, thus making their contribution to the global observation of the oceanic environment. Information from maritime vessels, in particular the Automatic Identification System (AIS), is used as the backbone of modern-day maritime operations. AIS enables real-time data on ship information such as position, speed, course and destinations which are vital for safety, navigation and for optimizing the shipping logistic [20]. Voyage Data Recorders (VDRs)—so named for being the black boxes of the marine domain—record an enormous amount of information about ship performance such as engine parameter, hull stresses and acoustic communication. Although mainly targeted at accident investigation, this data has tremendous potential to be used for operational optimization, when exploited to the full. Other onboard systems include internal communication systems, in which databases on board the ship and gateway systems are integrated to capture granular, real-time data on fuel use, vessel performance and cargo loading. National centres for marine data, such as the National Oceanographic Data Centre (NODC) are a centralized resource that brings together the contributions of scientific discipline data from national and international scientific communities as well as civil and commercial entities [32]. Platforms such as SeaDataNet bring together more than 100 national oceanographic data centres, and thus enable greater data exchange [32].

4.3.3 Data Lifecycle in IoMT

An effective approach to BMD can only be accomplished by virtue of a robust and continuous data lifecycle framework that includes planning, acquisition, processing, storage, sharing and usage with continuous quality assurance and control [30]. Prior to collection of data, a detailed data management plan needs to be established. Definition identifies the goals to be reached in the pipeline, the desired specific outputs, the design of the architectural framework of the whole pipeline and the design of specific tools and technologies with respect to the given data [30]. It also involves data governance, roles and responsibilities and relevant regulations [19]. Acquisition is the process of gathering data from the different IoMT sensors and devices in a context-sensitive way [23]. It involves a need to select suitable data sources and ingestion methods, and to validate data at the point of capture to see if they are correct [30]. Improvements in data acquisition instruments, including AUVs and USVs, are leading to dramatically improved data acquisition rates. The data from the ocean needs to be processed with considerable refinement before it can be useful. This is the step of data cleaning, data formatting, data transformation and data enrichment. For example, deep learning models are utilized to correct errors and super resolved remotely sensed SST data. This is an expensive, complex and time-consuming process, which has often resulted in delays in the publication of data. Processed data must be stored in the best available solutions like data warehouses, data lakes which can be cloud-based or on-premises. Cloud computing, such as NOAA's Big Data Programme (BDP), is considered a promising strategy for affordable and scalable delivery of ocean data, as it has the potential of processing the volume, velocity, variety and veracity of BMD. There is the need to reuse data, be able to replicate it, verify it and explain the reasons behind it [19]. This includes methods of data curation such as preparation, selection and contextualization and compliance with standards such as FAIR (Findable, Accessible, Interoperable and Reusable) [21]. However, there are many barriers such as technical difficulties, cultural traditions of not sharing information and confidentiality that reduce access. Usage/analysis is the last step which includes application of analytical models and tools to the processed data to extract information and make decisions [23, 30].

4.3.4 Advanced Analytics and Intelligence for IoMT

The application of artificial intelligence models is no less conspicuous in the widespread usage of deep learning methods for solving complex oceanography phenomena and for conducting advanced predictors. Machine learning is radically changing the way that we monitor the environment, both by extracting insights from vast data flows in real time and by enabling quicker and more accurate detection of pollution, more precise monitoring of deforestation and making forecasts of climate changes more reliable. In the cities spread across India, analytical platforms based on AI take in application data from metrological based sensors and analyse to understand deleterious pollutant thresholds to prepare municipal agencies with the preemptive asset to give advisories and to make timely remedial interventions. In aquaculture, IoT sensing networks commonly provide real-time diagnostics of important water-quality parameters like dissolved oxygen, pH and temperature that can be classified, predicted and the unusual conditions flagged for growers with the goal of maintaining optimal habitats for aquatic species. AI systems have also been used to analyse acoustic data for the population detection and abundance estimation of marine mammals even in deep and inaccessible parts of the ocean and this is providing vital information for protection. More sophisticated machine learning models use the collected observational data to detect nuanced understandings into how marine life interacts, how life migrates, how lifemates and how it feeds. High-resolution predictive modelling of SST, ocean circulation and significant wave heights can be further improved by using advanced AI algorithms which make use of very large oceanographic datasets [38]. Specifically, deep learning models, including LSTM, CNN and GAN, are used to make spatio-temporal predictions associated with SST, SSS, sea-ice concentration and tropical cyclone. For example, the deep neural networks address SST retrieval residual correction while ESRGAN performs super—resolution enhancement of SST data. Other GANs are used to reconstruct sea surface height while different forms of LSTMs are used to predict wave heights or sea-ice dynamics. The use of big data platforms and advanced analytics continues to increase maritime operations. These systems handle huge amounts of ship-borne data in order to predict optimum routes, control cargo flows and observe and track machinery performance [20]. AI and machine learning applications drawing off complicated datasets to predict future problems: Transporting essential loads might not get through cargo handling errors, a market's demand might grow and fall or the need for moving goods could be instant imminent threat, with AI and machine learning serving not only the artificial intelligence approach but also providing the optimal routes and pricing strategies for their cargo. Predictive maintenance, powered by ML, uses real-time data from sensors to identify anomalies and predict imminent equipment failures, which significantly reduces the downtime and repair costs of the equipment [22]. The Internet of Marine Things (IoMT) vastly enhances situational awareness to provide a full real-time view of the operations through powerful sensor-fusion algorithms [33]. The data from radar, infrared, visual camera and other sensors is merged and fused to enhance the battlespace situational awareness. The data aggregation is then passed through artificial intelligence (AI) and machine learning (ML) algorithms to dissect patterns, recognize anomalies and forecast adversarial activity through an industrial speed unmatched by human analysts, radically changing intelligence disciplines—intelligence, surveillance and reconnaissance (ISR). Autonomous Underwater Vehicles (AUVs) with AI-driven sensors and cameras play an important role in gathering very valuable information simultaneously and in various formats for seabed mapping, monitoring of marine habitats and investigating resources in underwater areas [10]. Within the viewing a point the utilization powered watch making perform, sensors in the oil and gas industry monitor equipment condition, inventory levels, which give real-time dynamic drilling parameters. This integration makes it easier to use predictive maintenance, as well as increases safety by immediately informing workers of a potential leak or hazard that could harm the employees.

4.3.5 Digital Twins and Blockchain

As a virtual replica of physical marine assets and systems, Digital Twins provide the real-time information on performance, which allows for predictive data and data-based decision-making [22]. The use of Digital Twins also provides shorter turnaround times and less idling emissions as demonstrated by the Singapore Maritime Port Authority. Digital twins combine big data with physical assets to create virtual twins that can provide real-time information regarding operations, economics, insurance, carbon footprints. By integrating smart metres and IoT devices with these digital assets using blockchain, asset managers are able to optimize portfolios based on aggregated, real-time performance data detected by anomaly detection and reduce operating and maintenance costs. Blockchain creates a shared and real-time database of the cargo's movement between shipping lines, forwarders, ports and customs, which will allow better cargo inventory management and improved estimated time of arrival (ETA) performance. MSC has combined blockchain and IoT for achieving a live visibility from origin to destination. The technology ensures fastness that seafarer certificates and training documents are instantly verifiable, tamper-free and accessible, which minimizes administrative bottlenecks and rejects forged credentials. Blockchain-based port community: Blockchain-based port community enables the terminal operator, stevedore, shipping line and port authority to communicate easily among each other and has its practical use in the slot booking verification and automated vessel clearance. The Port of Rotterdam is currently trying out blockchain for real-time coordination to reduce waiting-time. Blockchain ledger can keep track of detailed service history and survey results, which can add value to the vessel when it is put to sale and improve transparency in the auditing process. DNV has been furthering the integration of blockchain-based digital twins with classification data to facilitate lifecycle vessel management. Blockchain facilitates a simpler process of insurance claim management by automating policy enforcement and validating incidents with sensor data that helps execute claims induced by real-time conditions. The Insurwave platform created by EY and Maersk uses blockchain for risk and insurance management that ensures verified volumes of fuels, real-time tracking emissions and prevention of carbon credits accounts in support of decarbonization efforts and ESG reporting [22].

4.3.6 Communication and Networking Infrastructure

The efficacy of Big Data analytics in IoMT is fundamentally dependent on robust and reliable communication and networking infrastructure, which presents unique challenges in the marine environment. Underwater communication is critical for connecting submerged IoMT devices, but it faces significant physical limitations [15]. Acoustic communication remains the most common method for long-range underwater data transmission (over 100 km) due to sound waves propagating well in water. However, its inherent physical limitations lead to low bandwidth (1–100 kHz), high latency and susceptibility to time-varying multipath propagation and Doppler spread. These constraints make it challenging to transmit high-definition video or support real-time tele-operation of AUVs. Radio Frequency (RF) waves are severely attenuated in water, limiting their practical range to only a few metres ($\leq$10 m) [13]. This makes them unsuitable for most underwater IoMT applications requiring significant distances. Underwater Optical Wireless Communication (UOWC), utilizing blue/green light, offers significantly higher bandwidth (up to 150 MHz) and data rates (Gbps) with low latency compared to acoustic and RF methods. This technology holds promise for real-time video transmission and remote monitoring over short to medium ranges (10–150 m, with potential up to 500 m). However, UOWC is limited by high absorption and scattering of optical signals by water and suspended particles, and a comprehensive theoretical model for long-distance optical communication in practical underwater environments is still under development. The future direction involves hybrid acoustic-optical communication systems, which combine the long-range capability of acoustics with the high-bandwidth potential of optics, to overcome the individual limitations of each technology. To provide comprehensive connectivity across vast oceanic regions, IoMT relies heavily on the integration of satellite and advanced terrestrial communication networks. Future ships are expected to leverage 5G/6G networks to enhance connectivity between vessels and shore-based entities. 6G networks are designed to offer ultra-low latency, high bandwidth and massive device connectivity, which are crucial for real-time marine IoT applications like autonomous systems and remote pilot assistance. These networks will facilitate efficient and secure communication between aircraft and air traffic control and enable advanced surveillance and navigation even in remote environments. Satellite-based technologies provide global coverage, enabling seamless communication in remote oceanic areas where terrestrial networks are unavailable. Advanced inter-satellite communication capabilities in 6G could allow satellites to exchange data directly, forming complex networks that reduce reliance on ground stations. This integration ensures uninterrupted communication as devices move between Earth and space coverage areas.

The scale and speed at which big maritime data (BMD) is generated and acquired suggests that distributed computing paradigms will be needed to enable information to be processed efficiently, while also reducing latency. One of the most notable advantages of the edge computing is to cut across the latency issues and bandwidth constraints intrinsic to the wholesale transfer of raw data into source-server centralized cloud computing. Edge devices placed on vessels or at strategically placed coastal positions perform local data processing, meaning that data can produce analytics in real time and decision-making can be done quickly while at sea. Edge computing devices located at seagoing vessels or strategic coastal locations therefore deal with local data processing by generating real-time analytics and ensuring a fast decision-making process directly at sea. Taking advantage of the idle processing capabilities of the Internet of Medically Things (IoMT) devices themselves (device augmented edge) eases the load from traditional edge servers and encourages fodder and flexibility as well as resource utilization. Fog computing is the next middle layer which minimizes the stress on the cloud by processing data at the edge of the network, closer to the data source but with more computational power than the edge devices [31]. The layered structure provides a more subtle trade-off between latency, bandwidth utilization and the computational resource allocation. Despite the growing popularity of edge and fog modalities, cloud computing is still vital due to its strong abilities of distributed storage, powerful analytics processing capabilities and scale of globally-scalable applications [21]. Cloud platforms make continuous monitoring of the vessels, fleet management and complex data analytics easier and help to provide actionable insights. The NOAA's Big Data Programme (BDP) is a great example of how to strategically leverage cloud infrastructure for big data ocean dissemination. As a result of the correlation across the differing tiers of hierarchical complexity (usually coordinated by an integrated architecture) the performance, energy usage and operational costs are optimized with the capacity to dynamically vary the availability of resources.

These solutions leverage certified IoT and edge devices and work with interoperability with a range of legacy and contemporary systems, devices and interfaces, such as Modbus, NMEA and OPC UA++ to ensure easy integration between the maritime technology stack. They also help to make it safe for centralized data sharing from the sea to the shore to improve situational awareness and decision-making. The NMEA 0183 interface standard is still the prevailing standard for the definition of characteristics of electrical signals and data transmission protocol and the structure of serial data bus messages for marine electronic applications. This standard allows one-way series communication flow from one single talker to multiple receivers to transmit critical information about the specific time, position, speed and water depth. The latest version Version 4.30 released in December 2023 extends autonomous with improved GNSS sentence count to enable better positional integrity, accuracy and support for autonomous platforms. NODCs, such as the NODC of Italy, are the most important nodes of the International Oceanographic Data Exchange System. Marine Environmental Data and Information System (MEDIS) Data Centres Talking about maintaining long-term records, these two centres have built up vast multidisciplinary data archives, focusing on intensive data acquisition, curation, metadata collection and standardization and quality management. Also, by joining projects like SeaDataNet, they contribute towards the principles of Open Science in which data are made Findable, Accessible, Interoperable and Reusable (FAIR) across over one hundred national oceanographic data centres across Europe.

4.3.7 Emerging Applications and Use Cases

The combination of Big Data and the Internet of Marine Things (IoMT) is driving innovation in many different fields of the maritime industry, marking the dawn of a new era of applications.

Environmental Monitoring and Conservation: When it comes to marine environments, IoMT in tandem with Big Data analytics works wonders to bring about a revolution in environmental monitoring and conservation programmes. Air- and water-quality standard are constantly monitored by artificial intelligence-driven systems which look for harmful pollution levels and allow regulatory authorities to implement timely. The digital integration of IoT sensors and Big Data allows for indefinitely and real-time management of hydrological critical parameters, such as Dissolved Oxygen (DO), pH, temperature, ammonia, Electrical Conductivity (EC). Subsequent deep learning algorithms are then used to classify and predict the state of water quality, followed by historicity alert to aquaculture operators in order to maintain optimal conditions for aquatic fauna, and thus reinforce the viability of the aquaculture industry.

Smart Shipping and Port Operations: The entire maritime industry is transforming at a massive scale adopting the Intelligent Information Technology (IIT) frameworks of IoS, Big Data Platforms (BDP) and Big Data Analytics (BDA). IoT sensors installed on the vessel can collect real-time information about engine performance, fuel consumption and environmental conditions. This information is then extrapolated by subsequent Big Data analytical processes to anticipate the best routes and provide early warnings of mechanical aces in advance before parts fail, and to enable condition-based maintenance; thus reducing operational costs, further improving safety margins. The use of Automatic Identification System (AIS) information along with more advanced analysis models further optimizes ship routing leading to freight pricing. Real-time data exchange for variables affecting the cost of cargo feed into pricing models. In addition to this feature, the combination of IoT and Blockchain technologies can provide end-to-end visibility of supply chains, thus improving inventory management and Estimated Time of Arrival (ETA) predictions. Digital twins can replicate operations in the port in order to reduce congestion and increase the efficiency of cargo handling. Blockchain as a port community system, provide better communication arrangement between terminal operator, stevedores, shipping lines along with the port authorities and thereby optimize the port functionality. The path to autonomous shipping, then, based on AI, has promises of compounded efficiency, safety and cost savings.

Maritime Security and Defence: Modern military datalinks and satellite communications systems are the basis for secure and resistant data links against jamming even in contested electromagnetic support. This information infrastructure enables fast coordination and decision-making in a range of functions from physiological views of individual soldiers to the sharing of mission data and surveillance information. The conceptual framework of IoMT depends generally on a thesis that future military operations will be through machine intelligence and cyber warfare. Indeed, the United States Navy is moving towards mass production and the widespread integration of autonomous systems and AI, which include thousands of small, modular vessels and drones for surveillance, reconnaissance and logistics, and ensure continued control of the seas.

Underwater Exploration and Resource Management: IoMT along with Big Data is making ground-breaking feats possible in exploration and utilization of the oceanic resource. Long-range autonomous underwater vehicles (AUVs) are making significant contributions to oceanography research in the form of large data collections for ocean topography surveys and marine biogeochemical sensing in depths up to 6000 m. These platforms can be deployed for long durations (weeks and months) without support of surface vessels, thus enabling access of previously inaccessible regions. Harbouring an array of sensors and cameras, marine robotics, combined with artificial intelligence, capably navigate through complex depths to collection of data as they perform seafloor mapping and marine habitat monitoring tasks.

4.4 Technical Challenges/Limitations

1. **Connectivity and Communication**: Optical communications although offering better bandwidth powers are limited due to the dominant absorption and scattering features of marine environments. These optical signals have an effective operational range which is often limited by small extents of space. Moreover, extraneous interference caused by the propagation of waves, maritime activity and biological activity, further degrades the integrity of the conveyed information. Underwater channels are characterized by high temporal variability, whose modulations are caused by the thermally driven stratification, the salinity gradient

and the depth-related dynamics and these induce a temporal dependence of multi-path propagation and Doppler spread. As a result, stable routing paths cannot be constructed easily and reliable connectivity cannot be maintained. A vast expanse of oceanic area is largely free from infrastructural support, such as terrestrial cellular communication support or extensive fibre optic installations, a far cry from land based communication ecosystems.

2. **Power Constraints and Power Management**: The deployments of Internet of Marine Things (IoMTs) for remote or long marine missions are bound by stringent power budget issues. A significant amount of marine sensors and operational units are powered by finite lifetime batteries, the replacement of which is both logistically demanding and economically prohibited in the offshore or deep water environment. This situation creates a severe challenge to maintain a continuous operation and guarantee network durability. Increasing data throughput requirements, others, combined with the growing number of edge-enabled services within IoMT, all culminate in high energy consumption in the network. The transmission of voluminous data sets across long distances is energy-intensive in particular, quickly depleting measly power reserves. Because of the stochastic nature of battery discharge coupled with constraints in logistics communications pose an immediate necessity for expansion of energy-efficient architectures and efficient power-bumping techniques including extensions with energy-harvesting techniques.
3. **Data Heterogeneity, Volume and Quality**: The inherent data vagueness has the ability to mislead the updating cycles in a distributed AI usage and to affect the collective performance. As our technology is rapidly growing the volume of data from marine sensors, the data including physical ocean measurements, strain the data management infrastructures, which have not always kept pace with the flood of new data. Converting this colossal volume of raw data into actionable information is an expensive, complex and time-consuming undertaking that requires large amounts of computational resources and specialist knowledge. Despite the prevalence of data, substantial gaps still remain in particular areas of the marine and within certain fields, mainly as a result of the vast and hard-to-reach nature of the ocean environment. Divergences and a disregard of standardization in the different sources of data is another obstacle faced when comparing and additionally analysing the data. The veracity of marine datasets is of the utmost importance as sensor malfunctions, problems with data transmission as well as environmental noise can cause inaccuracies. The aphorism "garbage in, garbage out" still holds true in the world of machine learning and the errors in classifications and faulty inferences that can result from using substandard or improperly labelled training data.
4. **Cybersecurity, Privacy and Integrity**: The Grid IoMT system interlace and the sensitivity of marine data foster large exposures of cybersecurity and privacy challenges. Marine IoMT components are prone to a range of cyberattacks, such as malware intrusion and Distributed Denial of Service (DDoS) attacks, often the result of weak encryption protocols, weak authentication defenders and patching irregular firmware. For example, around ninety-nine per cent of

healthcare networks (the medical field being analogous to IoMT) have identified known exploited vulnerabilities (KEVs). Adversaries can strive to corrupt or compromise confidential marine information, or gain unauthorized access to working systems, potentially compromising confidentiality matters as well as data integrity. The Automatic Identification System (AIS), one of the most important maritime information exchange services, does not have built-in security measures (authenticity, encryption, etc.) and can be falsified and is open to cyber intrusions. The fact that delicate data can be transmitted, such as the coordinates of commercial fishing vessels or the like to which industry also holds secrets-makes it easier for privacy to be violated. A complex balancing act is protecting the privacy and confidentiality of data contributors and at the same time maximizing the value of the data. As the industry moves towards autonomous marine vessels, the size of cybersecurity threats also grows along with the possible commandeering of command and control pathways or the manipulation of sensor outputs, which could be disastrous for navigation and safety of operations.

5. **Scalability and Interoperability**: The multifaceted, ubiquitous structure of IoMT ecosystems sets a lot of challenges to effortless scaling and integration. The lack of universally accepted standards across different manufacturers and science disciplines prevent the communication and data exchange between the heterogeneous devices and systems. Divergent metadata frameworks add to the lack of interoperability especially for biological data domains. Novel IoMT technologies are not always assimilated easily into extant, and oftentimes legacy, data collection and processing workflows thus limiting large-scale applicability for public or operational use. Architectures to support the growth of an expanding fleet and feature sensing of new modalities efficiently, while delivering acceptable levels of performance and reliability is a continuous challenge. Data often comes from disparate up sources that is available in different formats resulting in the presence of data silos leading to limited holistic analysis and extraction of meaningful information. The corrosive salt environment has a direct impact on the reliability and durability of IoMT equipment. Sensors and equipment must be extremely robust to with stand aggressive currents, waves and tides, typhoons, high humidity, exposure to salinity and extremes of temperature. Submerged instruments are susceptible to marine biofouling which accumulates layer by layer, degrading the measurement and can eventually lead to equipment failure which needs successful control and maintenance methods. The intense environments of the maritime world make the deployment, maintenance and repair of IoMT devices complex, expensive and dangerous by nature.
6. **Human Capacity and Funding**: Promising though it may seem, there nevertheless remain several human and financial factors which construe the complete potential realization of IoMT. A heterogeneous distribution of technical capacity for oceanic stakeholders is an obstacle to the widespread adoption of unified standards as well as to assimilation of novel technologies. The lack of domain expertise for the deployment of the sensors and for interpreting the data poses a need for the hybrid skill set of data science and maritime information knowledge. The activities of data gathering, curating, sharing and utilizing are very

costly, and governments and other supporting stakeholders often work within strict resource constraints. Fundingosomes (weakly-funded institutions responsible for collecting, managing and sharing), are common in ocean management institutions around the world; challenges related to these data would resonate with a broader audience. Established cultural biases to retain data, for preservation of intellectual property, especially within academia and ocean industries restrict and prevent open and unfettered data exchange. Federal agencies and proprietary industry participants might be reluctant to work together on long term, institutional data-sharing programmes.

4.5 Future Enhancements

The multi-dimensional problems associated with Big Data for IoMT demand complex solutions and strategic breakthroughs in a plethora of areas. The next frontier will be based on the development of more resilient, smart and connected marine systems.

1. **Advanced Networks Communication Systems**: In future work, it is planned to combine powerful acoustic and optical channel technology to create powerful hybrid architecture. In such systems, acoustic links can be used for long-range, low-bandwidth transmissions, while optical links can be used for short-range, high-bandwidth transmissions so as to allow real-time video and tele-operation when needed. While 6G networks offer revolutionary features such as ultra-low latency, massive bandwidth and massive device-to-device connectivity, marine IoT will benefit greatly from these advancements. A 6G underwater architecture is likely to be an integrative algorithmic architecture between acoustics, optics and possibly VLF-EM communications. Furthermore, the integration with newer satellite systems including small satellite constellations will provide uninterrupted global coverage to ensure IoMT end devices are connected at all times even for remote ocean-based locations. It will simplify coordination between the satellite network and reduce interdependency on terrestrial ground stations, 6G could potentially bring advanced antenna technologies such as phased-array antennas enabling dynamic-beamforming and adaptive antenna beamforming to enhance the signal quality and avoid interference in harsh marine environments.
2. **Intelligent Edge to Cloud Architectures**: Systems will be designed to cope with the dynamic gathering of resourçes allowing heterogeneous nodes to enter and leave the network self-initiated and having their tasks redistributed automatically to maintain performance and Quality of Service (QoS). Additional modelling for IoMT devices themselves that take advantage of idle computational power will reduce load on traditional edge servers and promote more flexibility and resource efficiency. This technique can potentially greatly increase the system throughput by allowing unified processing.

3. **Enhanced AI/ML/DL Models**: The power of AI, ML and DL will only get better with time and provide even more advanced analysis and autonomous functionality. By including physical models and domain experts in the AI algorithms, the accuracy and interpretability of the forecasts will be improved, especially in those systems that are difficult to model with data-driven models because of sparsity or noise, such as oceanographic systems. One of the developments that will allow autonomous marine vehicles to adapt their behaviour in dynamic and unpredictable conditions will be the increasing use of reinforcement learning for optimizing navigation, data collection and mission execution. As AI algorithms become more advanced at real-time anomaly detection on large marine datasets, new patterns in equipment performance, environmental parameters or vessel behaviour signalling an impending breakdown, pollution event or security threat can be recognized [20]. Paradvantages of AI may also include Generative Adversarial Networks (GANs) and other techniques of generating synthesis of scarce and sensitive data that will augment and provide good training samples for AI models, particularly in cases where reality collection either is limited or perilous [38]. As models increase in complexity, the development of explainable AI techniques will be important for providing transparency and in turn, to establish trust in AI-enabled decisions in important marine applications such as automated navigation or Environmental management.
4. **Cybersecurity and Privacy Management Solutions**: IoMT systems require protection and security of the privacy to be successfully deployed and accepted by the public. Blockchain technology (BCT) will be applied increasingly to enhance the security and privacy of the data, by providing a decentralized, immutable and transparent ledger of sensitive marine data [17]. This method helps in protection from unauthorized access and tampering thereby maintaining integrity of data. Homomorphic encryption is a research area that works towards performing computations on encrypted data without the need to decrypt the data, providing sensitive information protection (the same principle that applies to the marine domain) [17]. Zero-trust security models (security models that assume no trust and that back a strictly verifiable access model) would be essential for the highly distributed nature of these IoMT networks in the oceanic environment. Certificate -less digital signatures (CLDS) using elliptic curve cryptography (ECC) is a promising method to reduce the overhead of key management and enhance the security among resource-constrained IoMT devices [37]. Furthermore, real-time threat detection and anomaly detection will be made by using AI and machine learning algorithms, thereby protecting the important data from cyberattacks and identifying malicious activity inside the IoMT networks [34].
5. **Sustainable Energy and Power Harvesting**: Power constraints are a basic concern to long-term feasibility of IoMT deployments. RF energy harvesting (RF-EH) is one of the emerging technologies that will be further developed and allow marine sensors to harvest ambient radio wave energy and perform possibly battery-less operations with dynamic re-charging capabilities [34]. Energy harvesting from different environmental marine sources—solar, wind, water waves and currents, for an uninterrupted and reliable power supply to

IoMT devices, particularly for those in remote locations where the replacement of the battery is not even feasible [35]. Circuit designers for IoMT devices need to develop energy-efficient communication protocols and low-power hardware components to keep the power consumption low and the possible life of the devices higher [35].

6. **Standardization and Open Data Movements**: Achieving interoperability and data sharing is important to fully leverage the benefit of Big Data for IoMT. Continued adoption and use FAIR (Findable, Accessible, Interoperable and Reusable) principles will enable the open free sharing of ocean data and so eliminate the access/discovery challenges [21]. Hackweeks are effective in introducing domain experts to the practices of the data science field, introduce the paradigm of open-science to the data science community and serve to adequately supplement the capacity to do exploratory data science in developing communities. These events connect researchers through hands-on coding and data access experience, develop the use of open-source software and build strong networks between scientists (with all the benefits that this presents such as fostering collaboration and diversity of thought [36]. Data delivery platforms and tools that enable an easy access, processing and analysis of data by a broad range of stakeholders including those with low technical capacity will further encourage wider data use [1]. Implementation of policy frameworks and incentives that clearly allow private industry and academic researchers to share their data will unleash vast stores of information that are currently hidden from access because of competition or intellectual property issues [8].
7. **Development of Autonomous Marine Systems**: During the next decades the number of autonomous marine systems will increase considerably. Different swarm topologies for autonomous underwater vehicles (AUVs) will be studied, to get flexible and fast interacting networks capable of communicating and gathering data even in environments with changes, thus building source and receiver networks with greater coverage and adaptability [2]. The recent pledge of the United States Navy to deploy a fleet of small, potent autonomous vessels is indicative of this trend, where it is hoped that large fleets of these autonomous vessels will be able to provide constant presence, more easily detect activity and give more combat power [27]. With the advancement in AI, autonomy will become even more refined so that remote pilots will provide assistance or direction to autonomous vehicles and systems in case of difficulty. Interestingly, the introduction of AI systems into marine education will make it easy for professionals to attain a workforce that is well-equipped to handle these high tech systems. A thorough review on IoMT research and development can be found in the literature [3, 5–7, 14].

4.6 Conclusion

The Internet of Marine Things (IoMT), enabled by the power of Big Data analytics, forms a hen and chicks frontier in our collective understanding, management and management of the global marine environment. This investigation carefully explores the complex interconnection of Internet of Marine Things (IoMT) and Big Marine Data (BMD) under a new paradigm of the ocean with its unique characteristics that lead to unfathomable scale, rapidity and heterogeneity, while demanding strict veracity and value extraction. Artificial intelligence, machine learning, deep learning, distributed computing paradigms, digital twin technology and blockchain have empowered researchers with unprecedented capabilities with a spectrum of applications—including environmental monitoring, smart shipping, maritime security and underwater exploration. The field is in the midst of a paradigm shift from rudimentary data acquisition towards sophisticated intelligence that supports the creative use of predictive analytics, real-time optimization and autonomy. However, the marine environment presents some challenging requirements, particularly severe communication limitations, limited availability of electrical power supplies and data heterogeneity and quality issues that together impose some fundamental limitations. Moreover, strong cybersecurity controls, full-fledged interoperability and adequate human and financial resource are critical areas requiring constant attention and new innovations. Looking into the future, the Big data trajectory in connection with IoMT holds great promise, and with the introduction of hybrid communication systems, intelligent edge to cloud architectures and more prominent and resilient AI/ML algorithms like physics-informed AI and synthetical data generation, we can see why. Progress in sustainable power expansion efforts, advanced cyber safety solutions (e.g., blockchain activation), zero-plan trust strategies and detailed standardization strategies will be key to the mass release of such innovations, as well as help ensure the security of data integrity. In conclusion, progress in this area will depend on interdisciplinary partnership between economy development and environmental protection, and between technological progress and economies of living.

References

1. Abdullah, A.H., Sudin, S., Saad, F.S.A., Hassan, M.K.A., Khalid, K.A.A., Abidin, J.Z.: Water quality analysis for the sustainability of aquaculture industry using IoT and big data. ASEAN Artif. Intell. J. **1**(1), 34–45 (2025). https://doi.org/10.37934/aaij.1.1.3445
2. Alqurashi, F.S., Trichili, A., Saeed, N., Ooi, B.S., Alouini, M.-S.: Maritime communications: a survey on enabling technologies, opportunities, and challenges. IEEE Internet Things J. **10**(4), 3525–3547 (2022). https://doi.org/10.1109/jiot.2022.3219674
3. Alsharay, N.M., Chen, Y.Z., Dobre, O.A., De Silva, O.: Improved sea-ice identification using semantic segmentation with raindrop removal. IEEE Access **10**, 21599–21607 (2022)
4. Ascend.io: How to Build a Data Pipeline in 6 Steps (2025). https://www.ascend.io/blog/how-to-build-a-data-pipeline-in-six-steps. Accessed May 23, 2025

5. Bai, G., Wang, Z.F., Zhu, X.Y., Feng, Y.Q.: Development of a 2-D deep learning regional wave field forecast model based on convolutional neural network and the application in South China Sea. Appl. Ocean Res. **118**, 103012 (2022)
6. Barth, A., Alvera-Azcarate, A., Troupin, C., Beckers, J.M.: DINCAE 2.0: multivariate convolutional neural network with error estimates to reconstruct sea surface temperature satellite and altimetry observations. Geosci. Model Dev. **15**, 2183–2196 (2022)
7. Bethel, B.J., Sun, W.J., Dong, C.M., Wang, D.X.: Forecasting hurricane-forced significant wave heights using a long short-term memory network in the Caribbean Sea. Ocean Sci. **18**, 419–436 (2022)
8. Bhide, P., Shetty, D., Mikkili, S.: Review on 6G communication and its architecture, technologies included, challenges, security challenges and requirements, applications, with respect to AI domain. IET Quantum Commun. (2024). https://doi.org/10.1049/qtc2.12114
9. Căuniac, D.-A., Cîrnaru, A.-A., Oprea, S.-V., Bâra, A.: Intersections of big data and IoT in academic publications: a topic modeling approach. Sensors **25**(3), 906 (2025). https://doi.org/10.3390/s25030906
10. Chen, D., Wen, J., Dai, H., Xi, M., Xiao, S., Yang, J.: Enhancing transportation management in marine internet of vessels: a 5G broadcasting-centric framework leveraging federated learning. IEEE Trans. Broadcast. **70**(3), 1091–1103 (2024). https://doi.org/10.1109/tbc.2024.3394289
11. Digi International: IoT in Oil and Gas: 4 Use Cases and Advantages (2025). https://www.digi.com/blog/post/iot-in-oil-and-gas. Accessed May 23, 2025
12. eTrade for All: Navigating the Future: How AI, Big Data, and Autonomous Systems are Reshaping Maritime Transport (2025). https://etradeforall.org/news/navigating-future-how-ai-big-data-and-autonomous-systems-are-reshaping-maritime-transport. Accessed May 23, 2025
13. Fernando, N., Shrestha, S., Loke, S.W., Lee, K.: On edge-fog-cloud collaboration and reaping its benefits: a heterogeneous multi-tier edge computing architecture. Future Internet **17**(1), 22 (2025). https://doi.org/10.3390/fi17010022
14. Haseeb, K., Ahmad, I., Awan, I.I., Lloret, J., Bosch, I.: A machine learning SDN-enabled big data model for IOMT systems. Electronics **10**(18), 2228 (2021). https://doi.org/10.3390/electronics10182228
15. Jahanbakht, M., Xiang, W., Hanzo, L., Azghadi, M.R.: Internet of underwater things and big marine data analytics—a comprehensive survey. IEEE Commun. Surv. Tutor. **23**(2), 904–956 (2021). https://doi.org/10.1109/comst.2021.3053118
16. Jiang, M., Zhu, Z.: The role of artificial intelligence algorithms in marine scientific research. Front. Mar. Sci. **9** (2022). https://doi.org/10.3389/fmars.2022.920994
17. Khatib, I.A., Shamayleh, A., Ndiaye, M.: Healthcare and the internet of medical things: applications, trends, key challenges, and proposed resolutions. Informatics **11**(3), 47 (2024). https://doi.org/10.3390/informatics11030047
18. Lima, K., et al.: A Data-Flow Oriented Software Architecture for Heterogeneous Marine Data Streams, pp. 146–157. Moonlight Publishers (2024). https://doi.org/10.1109/icsa59870.2024.00022
19. Maritime Education: Top 10 Blockchain Applications in Maritime Logistics (2025). https://maritimeducation.com/top-10-blockchain-applications-in-maritime-logistics/. Accessed May 23, 2025
20. Mirović, M., Miličević, M., Obradović, I.: Veliki Skupovi Podataka u Pomorskoj Industriji. Naše More **65**(1), 56–62 (2018). https://doi.org/10.17818/nm/2018/1.8
21. Ocean Conservancy: Ocean Data Report (2021). https://oceanconservancy.org/wp-content/uploads/2021/05/Ocean-Data-Report-FINAL.pdf. Accessed May 23, 2025
22. Ocean Science & Technology: Long-Range AUVs Acquired for Scientific Research (2025). https://www.oceansciencetechnology.com/news/long-range-auvs-acquired-for-scientific-research/. Accessed May 23, 2025
23. OGS: National Oceanographic Data Centre (NODC) (2025). https://www.ogs.it/en/national-oceanographic-data-centre-nodc. Accessed May 23, 2025
24. Oracle: What Is Big Data? (2025). https://www.oracle.com/big-data/what-is-big-data/. Accessed May 23, 2025

25. Precedence Research: Marine Electronics Market Size to Cross USD 12.78 Bn by 2034 (2025). https://www.precedenceresearch.com/marine-electronics-market. Accessed May 23, 2025
26. ProfileTree: AI in Marine Biology and Oceanography (2025). https://profiletree.com/ai-in-marine-oceanography/. Accessed May 23, 2025
27. Razdan, S., Sharma, S.: Internet of Medical Things (IOMT): overview, emerging technologies, and case studies. IETE Tech. Rev. **39**(4), 775–788 (2021). https://doi.org/10.1080/02564602.2021.1927863
28. Spagnolo, G.S., Cozzella, L., Leccese, F.: Underwater optical wireless communications: overview. Sensors **20**(8), 2261 (2020). https://doi.org/10.3390/s20082261
29. TBRC: Understanding Key Drivers and Growth Opportunities in the Marine Internet of Things Global Market Report 2025 Market: Trends and Market Size Insights (2025). Accessed May 23, 2025. https://blog.tbrc.info/2025/04/marine-internet-of-things-iot-market-overview-2/
30. The Business Research Company: Marine IoT Market Report 2025—Global Trends and Future Insights (2025). https://www.thebusinessresearchcompany.com/report/marine-internet-of-things-global-market-report. Accessed May 23, 2025
31. Triangle Digital: Blockchain Triangle—Digital Twin (2025). https://www.triangle.digital/technology. Accessed May 23, 2025
32. University of Wisconsin-Madison: Introduction to the Data Lifecycle (2025). https://data.wisc.edu/data-literacy/lifecycle/. Accessed May 23, 2025
33. Wikipedia: Internet of Military Things (2025). https://en.wikipedia.org/wiki/Internet_of_Military_Things. Accessed May 23, 2025
34. Xu, G., Shi, Y., Sun, X., Shen, W.: Internet of things in marine environment monitoring: a review. Sensors **19**(7), 1711 (2019). https://doi.org/10.3390/s19071711
35. Yaacoub, J.-P.A., Noura, H., Salman, O., Chehab, A.: Securing internet of medical things systems: limitations, issues and recommendations. Futur. Gener. Comput. Syst. **105**, 581–606 (2019). https://doi.org/10.1016/j.future.2019.12.028
36. Zhao, Q., Peng, S., Wang, J., Li, S., Hou, Z., Zhong, G.: Applications of deep learning in physical oceanography: a comprehensive review. Front. Mar. Sci. **11** (2024). https://doi.org/10.3389/fmars.2024.1396322
37. Zheng, H., Benfield, M.C., Bi, H., Cheng, X.: Deep Learning for Marine Science. Frontiers Media SA (2024)
38. ZeroNorth: Navigating the Sea of Big Data: How to Manage and Utilise IoT Data in Smart Shipping (2025). https://zeronorth.com/how-to-manage-and-utilise-iot-data-in-smart-shipping. Accessed May 23, 2025

Chapter 5
A Hybrid Deep Reinforcement Learning Framework Based Multi-attribute Decision Making to Improve the Internet of Marine Things (IoMTs)

K. Praghash

5.1 Introduction

The IoMT is an emerging technology concept which helps in connecting underwater sensors and actuators together for carrying out various operations like environmental observation, maritime safety and marine resource exploration. But unlike the normal terrestrial IoT systems, the underwater environment is highly unpredictable and introduces several difficulties such as changing acoustic channel properties, limited battery power, high communication delay and strict ecological constraints. These factors create the need for an intelligent and adaptive network management system which can improve throughput, save energy and reduce delay while at the same time ensuring the protection of marine life and their habitats [3, 13, 16].

DRL has come up as an efficient solution to handle the dynamic underwater conditions by making autonomous decisions related to power control, relay selection and data routing. It is capable of managing complex multi-objective optimization problems in uncertain environments, which has already been proven in many underwater networking studies [8, 9]. However, DRL suffers from the problem of less interpretability, where the decision-making process of the agent is not clear to the human operators, thus making it difficult to trust such systems in critical marine applications [5, 7].

To cope with this limitation, XAI techniques like SHAP and LIME are being used to make the DRL decisions more transparent by showing the feature importance and the reason behind specific actions [5, 6, 16]. LIME stands for Local Interpretable Model-Agnostic Explanations. While SHAP represents an entire model's predictions, LIME is focused on providing a local, single-prediction-based explanation. LIME

K. Praghash (✉)
Department of Electronics and Communication Engineering, Christ University, Bangalore, India
e-mail: praghash.k@christuniversity.in

T. A. Tran et al. (eds.), *Big Data Analytics for Internet of Marine Things*,
Internet of Things, https://doi.org/10.1007/978-3-032-28171-5_5

generates a simple model, (like Linear Regression in the vicinity of the observation point relative to a prediction. In doing so, it illustrates through specific data points (such as specific words within text, what specifically contributed to the prediction being made. "Local" is used since LIME provides an explanation strictly of the neighborhood of the specific observation point, and, "Interpretable", since it is easily understandable from a human perspective.

It also makes it possible for humans to stay in the loop during the operation, increasing the confidence level on these autonomous agents. Along with that, multi-criteria decision-making methods that include environmental and operational factors are essential for maintaining a practical and eco-friendly network working model [14].

Although the potential for IoMT technology has been demonstrated in monitoring and exploring marine environments, there are several fundamental barriers to the efficient and widespread use of this technology [17].

Acoustic communication channels under water have very poor predictability and stability and suffer from high levels of multiple path propagation, frequency dependent loss, and Doppler shift caused by moving nodes and ocean currents. All this causes a very low data rate (often < 100 kHz), significant variation in signal strength, and large variability in communication latency which can be difficult for applications needing near-real time control over remote devices [10]. In addition, energy is also a huge barrier to the use of wireless sensors in underwater applications because most wireless sensor platforms will run out of power quickly, either due to the losses associated with the medium, or due to the difficulty of recharging/replacing batteries in a deep sea environment. Due to the speed at which sound travels in water (~1500 m/s), there is a large amount of latency when communicating between underwater nodes, making it difficult for real time control and rapid response times.

The use of wireless sensors in a marine environment has many environmental constraints; namely, the ability to minimize acoustic noise to prevent disruption of marine mammals, and to ensure that equipment can withstand extreme pressures, corrosive conditions, biofouling, and the variable nature of ocean currents which can make maintenance and deployment of the system expensive [11, 15]. Traditional routing and resource allocation approaches do not work well in underwater networks that change dynamically, causing both poor network performance and premature failure of nodes, leading to critical mission link interruptions.

A proposed hybrid approach to address the limitations of underwater communication systems using DRL combines adaptive learning, multi-objective optimization, and interpretable decision-making. DRL does not require explicit models of the environment; rather, DRL learns the optimal policy for relay activation, transmission power, and routing based on interaction with the environment [1, 8]. A Multi-Attribute Comparator evaluates the candidate actions to determine if they meet the specified performance metrics while considering the operationally and ecologically related constraints, and therefore, complying with applicable regulations. An additional module called the Explainability Module increases the level of transparency and trust in autonomous DRL decisions by providing information regarding feature relevance and context for each decision made by the agent [15]. The inclusion of a human-in-the-loop element enables increased collaboration and confidence in the reliability of the autonomous decisions made by the agent during complex underwater operations.

Based on these developments, this paper introduces a hybrid DRL framework which includes a Multi-Attribute Decision Module and an Explainability System working together. The DRL component manages the relay activation, power distribution and data routing in underwater IoMT by optimizing between different goals like throughput, energy utilization and delay, while taking ecological limits into account. The Explainability System helps in improving operator awareness by clearly explaining the logic behind the agent's actions. Therefore, the combined system forms a robust, scalable and environmentally responsible solution for underwater IoMT network management and effectively deals with the main deployment challenges explained in the later part of this work.

5.2 System Model and Problem Statement

Consider an IoMT network composed of N sensor/actuator nodes and one or more gateway sinks. Nodes $i = 1,2,\ldots,N$ are randomly or strategically deployed in a three-dimensional marine volume $V \subset \mathbb{R}^3$ with Cartesian coordinates (x_i, y_i, z_i).

Let the set of all node positions be:

$$X = \{(x_i, y_i, z_i)|i = 1, 2, \ldots, N\} \tag{5.1}$$

The gateway node g (e.g., a surface buoy) is at position $\left(x_g, y_g, z_g\right)$ commonly fixed at the surface $\left(z_g = 0\right)$ or deployed at a chosen depth.

The communication links form a time-varying undirected graph $G = (V, E)$, where $V = \{1,2,\ldots,N, g\}$ (nodes and gateway) and E : edge between nodes (i, j) if they can communicate.

A link exists if the received signal at j from i exceeds a threshold $\gamma_{\min}$, mathematically

$$(i, j) \in E \Leftrightarrow \mathrm{PL}_{ij} \leq \gamma_{\max} \tag{5.2}$$

where, PL_{ij} is the propagation loss in dB from i to j

$$i, j \in \{1, 2, ..., N, g\}, i \neq j \tag{5.3}$$

where N represents the total number of nodes (assuming 100 in simulation) and g indicates gateway node index.

Marine IoT networks primarily use acoustic links. Path loss PL is described by:

$$\mathrm{PL}_{ij}\left(f, s_{ij}\right) = 10\log_{10}\left(s_{ij}^{k} a(f)^{s_{ij}}\right) \tag{5.4}$$

where,

f—frequency (kHz)

s_{ij}—Euclidean distance between nodes i and j

k—geometric spreading constant (1.5)

$a(f)$—absorption coefficient in dB/m (Thorp's formula for seawater):

$$a(f) \simeq \frac{0.11f^2}{1+f^2} + \frac{44f^2}{4100+f^2} + 2.75 \times 10^{-4}f^2 + 0.003 \tag{5.5}$$

For nodes influenced by ocean current, use the Meandering Current Model:

Let $\varphi(x, y, t)$ be the stream function. The node at (a, b) moves as:

$$\dot{a} = -\frac{\partial \varphi(a, b, t)}{\partial b} \tag{5.6}$$

$$\dot{b} = +\frac{\partial \varphi(a, b, t)}{\partial a} \tag{5.7}$$

This models node drift over time as a function of environmental forces.

Energy Consumption per Transmission:

$$E_{\text{tx}}(d, f) = P_{\text{tx}}(f) \cdot T_{\text{tx}}(d) \tag{5.8}$$

where P_{tx} is the power used (which depend on frequency) and $T_{\text{tx}}(d) = d/v_s$ is the time to transmit over distance d and at sound speed v_s (approx. 1.5 km/s underwater).

Residual Energy Model:

$$E_i(t + \Delta t) = E_i(t) - \sum_{j \in \mathcal{N}_i} E_{ij}^{\text{tx}}(t) \tag{5.9}$$

The operator feedback mechanism has two different paths of operation as demonstrated in Fig. 5.1: real-time updates and corrections/preferences. Real time updates operate as an instant, reactive feedback path of the network during operation, and the operator issues approvals $F_t^{\text{real}} = +1$, rejects $F_t^{\text{real}} = -1$, or issues no opinion $F_t^{\text{real}} = 0$ concerning the results of observed actions. These real-time operator opinions are used to adjust the reward signal as $r_t^{\text{adjusted}} = r_t + \beta_{\text{feedback}} \cdot F_t^{\text{real}}$ with a value of $\beta_{\text{feedback}} = 0.2$ for feedback. Corrections and Preferences represent longer term adjustments to how the system prioritizes its tasks by adjusting weights in the Multi-Attribute Comparator $\text{w}(t + \Delta t) = \text{w}(t) + \eta \cdot \nabla_{\text{w}} J(\text{w}, F^{\text{pref}})$, adjusting thresholds $\theta(t + \Delta t) = \theta(t) + \alpha_\theta \cdot \Delta_{\text{operator}}$, and adjusting constraints. The system also uses the following three comparison signals to classify the current phase of feedback: the magnitude of performance deviations $\Delta_{\text{perf}}(t) =| J(t) - \bar{J}(t - \tau : t) | / \bar{J}(t - \tau : t)$, the frequency of operator intervention $f_{op} = \frac{1}{\tau}\sum_{k=t-\tau}^{t} 1(F_k \neq 0)$, and the confidence of explanations $C_{exp}(t) = 1 - \text{Var}(\phi_1, \ldots, \phi_d)/\max_i \mid \phi_i \mid$. Real time updates occur if $\Delta_{perf}(t) > 0.15$ or $C_{exp}(t) < 0.6$ and $t \, mod \, 50 = 0$. The design of the system provides rapid safety responses due to the real-time update process and maintains strategic alignment over long periods of time with periodic updates to user preferences, thus reducing operator cognitive load by allowing the operator to separate urgent reactive feedback from strategic planning. As such, this design enhances both the convergence stability and the human-AI co-adaptation efficiency.

Link Throughput

For a BPSK system (bit error rate, BER, over AWGN):

$$\text{Throughput}_{ij} = R \cdot \left(1 - \text{BER}_{ij}\right) \tag{5.10}$$

where BER, to first order, depends on received SNR (S/N), which itself is a function of PL.

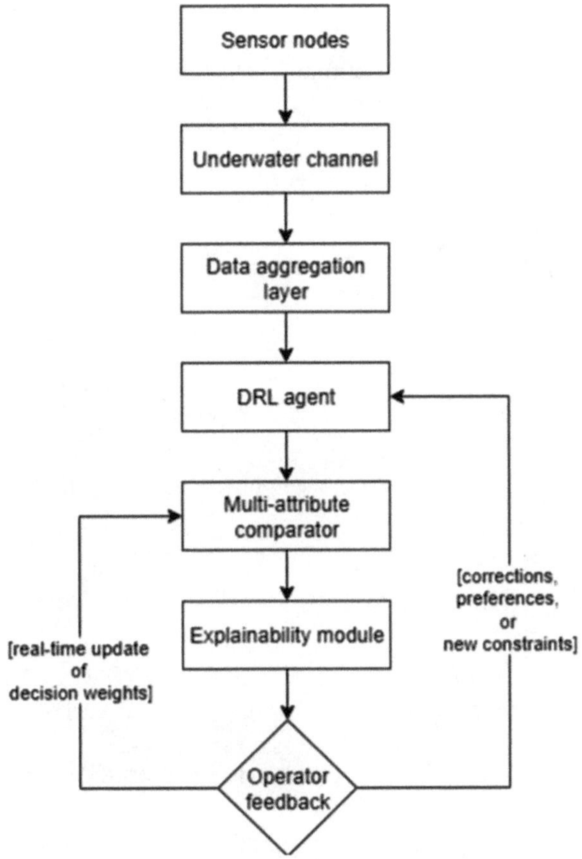

Fig. 5.1 System architecture

Global Network Metrics

Network Lifetime: Until first node i, $E_i(t) \leq 0$.

Packet Delivery Ratio (PDR): Fraction of packets that reach from source to sink:

$$\text{PDR} = \frac{\text{Number of packets received at sink}}{\text{Number of packets sent}} \tag{5.11}$$

Network Throughput

$$\text{Aggregate Throughput} = \sum_{(i,j)\in E} \text{Throughput}_{ij} \tag{5.12}$$

Average Latency

$$\text{Latency}_{\text{ave}} = \frac{1}{M}\sum_{m=1}^{M}\left(t_{\text{received}}^{(m)} - t_{\text{sent}}^{(m)}\right) \tag{5.13}$$

The goal is to determine a scheduling and routing policy π that maximizes total network utility:

$$\begin{aligned}\max_{\pi} \alpha_1 \cdot \mathbb{E}\big[\text{Aggregate Throughput}(\pi)\big] \\ - \alpha_2 \cdot \mathbb{E}\big[\text{ Energy Consumption}(\pi)\big] \\ - \alpha_3.\mathbb{E}\big[\text{Latency}(\pi)\big]\end{aligned} \tag{5.14}$$

$$\text{s.t.} E_i(t) \geq 0, \forall i, t \tag{5.15}$$

$$\text{PDR} \geq \gamma_{\text{PDR,min}} \tag{5.16}$$

where $\alpha_1, \alpha_2, \alpha_3 > 0$ are weights reflecting operator priorities.

5.3 Simulation Setup and Assumptions

Let N be the total number of IoMT nodes. We assume nodes are deployed in a bounded three-dimensional volume $V \subset \mathbb{R}^3$ such that:

$$(x_i, y_i, z_i) \sim \text{Uniform}(V), 1 \leq i \leq N \tag{5.17}$$

where, $V = [x_{\min}, x_{\max}] \times \big[y_{\min}, y_{\max}\big] \times [z_{\min}, z_{\max}]$.

Each node experiences a vector of environmental state processes (channel quality, temperature, current speed and energy harvesting). We model the major fluctuations as discrete-time stochastic processes.

Acoustic Channel Quality (Fading):

$$h_{ij}(t) = h_{\text{mean}} + \sigma_h \xi_{ij}(t) \tag{5.18}$$

where, $\xi_{ij}(t) \sim \mathcal{N}(0,1)$, yielding a temporally and spatially correlated fading model.

Water Temperature:

$$T_i(t) = T_0 + A_T \sin(2\pi f_T t + \phi_T) + \eta_i(t) \tag{5.19}$$

where, A_T is amplitude, f_T is frequency, ϕ_T is phase $\eta_i(t) \sim \mathcal{N}\left(0, \sigma_T^2\right)$ is spatial–temporal noise.

Ocean Current Vector:

$$\vec{u}_i(t) = u_0 \vec{e}_{\text{dir}} + \sigma_u \vec{\chi}_i(t),\ \vec{\chi}_i(t) \sim \mathcal{N}(\vec{0}, I) \tag{5.20}$$

This controls node drift in the deployment over time.

5.3.1 Energy and Task Model

Initial Energy:

$$E_i(0) = E_{\text{init}}, \forall i \tag{5.21}$$

Per-action Consumption:

A node transmits/receives n_{ij} packets in interval Δt. For each packet,

$$\Delta E_{ij}^{\text{tx}} = P_{\text{tx}} \frac{L_{\text{pkt}}}{R} \tag{5.22}$$

$$\Delta E_{ji}^{\text{rx}} = P_{\text{rx}} \frac{L_{\text{pkt}}}{R} \tag{5.23}$$

where L_{pkt} is packet length (bits), R is data rate (bps), and $P_{\text{tx/rx}}$ are powers (W).

So, energy dynamics:

$$E_i(t + \Delta t) = E_i(t) - \sum_j n_{ij}\left[\Delta E_{ij}^{\text{tx}} + \Delta E_{ji}^{\text{rx}}\right] \tag{5.24}$$

Transmission Loss and Received SNR:

$$\mathrm{TL}_{ij}(f) = 10k\log_{10}\left(d_{ij}(t)\right) + a(f)d_{ij}(t) \tag{5.25}$$

Received power:

$$P_{r,ij}(t) = P_{\mathrm{tx}} - \mathrm{TL}_{ij}(f) \tag{5.26}$$

where, $a(f)$ is absorption and d_{ij} is node separation.

Additive Noise:

$$N_{ij}(t) \sim \mathcal{N}\left(\mu_N, \sigma_N^2\right) \tag{5.27}$$

SNR and BER:

$$\mathrm{SNR}_{ij}(t) = \frac{P_{r,ij}(t)}{N_{ij}(t)} \tag{5.28}$$

For BPSK:

$$\mathrm{BER}_{ij}(t) = Q\left(\sqrt{2 \times \mathrm{SNR}_{ij}(t)}\right) \tag{5.29}$$

with $Q(x) = \frac{1}{\sqrt{2\pi}}\int_x^{\infty} e^{-u^2/2}du.$

Each node generates sensing data:

$$D_i(t) \sim \text{Poisson}\ (\lambda) \tag{5.30}$$

yielding packet arrivals with mean rate λ per step.

5.3.2 *Assumptions*

Simulation is over T discrete intervals $t_1, t_2, \ldots, t_T$ (~1 s each). All state processes are statistically independent between nodes; dependencies (spatially-correlated fading) can be modeled using Markov random fields or Gaussian processes. All environmental variables are initialized from their stationary distributions. Chosen parameters (energy, rate, temperature, current) reflect real underwater sensor deployments referenced from experimental results and commercial modem specifications.

5.4 Problem Formulation: Multi-objective Optimization

It is necessary in IoMT networks to simultaneously optimize multiple competing objectives:

- Throughput maximization (high data delivery rate);
- Energy minimization (sustainable node operation);
- Latency reduction.

Let π represent the policy guiding the network. The performance is evaluated over a time horizon T.

Aggregate Throughput:

$$\mathcal{T}(\pi) = \frac{1}{T}\sum_{t=1}^{T}\sum_{i,j}\mathrm{Th}_{ij}^{\pi}(t) \tag{5.31}$$

where $\mathrm{Th}_{ij}^{\pi}(t)$ is number of delivered bits from i to j at time t under policy π.

Total Energy Consumption:

$$\mathcal{E}(\pi) = \frac{1}{T}\sum_{t=1}^{T}\sum_{i}\mathrm{E}_{i}^{\pi}(t) \tag{5.32}$$

where $\mathrm{E}_{i}^{\pi}(t)$ is energy consumed by node i at t.

Average End-to-End Latency:

$$\mathcal{L}(\pi) = \frac{1}{M}\sum_{m=1}^{M}\left(t_{\mathrm{rx}}^{(m)} - t_{\mathrm{tx}}^{(m)}\right) \tag{5.33}$$

with M packets sent, $t_{\mathrm{tx}}^{(m)}$, $t_{\mathrm{rx}}^{(m)}$ transmission/reception times, respectively.

The multi-objective optimization problem (MOP) is:

$$\max_{\pi\in\Pi}[\mathcal{T}(\pi), -\varepsilon(\pi), -\mathcal{L}(\pi)] \tag{5.34}$$

$$\text{s.t. } E_{i}^{\pi}(t) \geq 0, \forall i, t \tag{5.35}$$

$$\mathrm{PDR}^{\pi} \geq \gamma_{\min} \tag{5.36}$$

$$\text{Environmental constraints}, \mathcal{I}(\pi) \leq \delta \tag{5.37}$$

where Π is the set of all feasible policies, PDR^{π} is minimum packet delivery ratio, and $\mathcal{I}$ is a metric of environmental impact.

Since objectives are conflicting, we consider Pareto optimal solutions:

A policy π^* is Pareto optimal if no other policy improves one objective without degrading another.

In practice, scalarization via weighted sum is used:

$$J(\pi) = \lambda_1 \mathcal{T}(\pi) - \lambda_2 \mathcal{E}(\pi) - \lambda_3 \mathcal{L}(\pi) \tag{5.38}$$

where $\lambda_1, \lambda_2, \lambda_3 \geq 0$ (weights reflect priorities).

$$\pi^* = \arg\max_{\pi} J(\pi) \tag{5.39}$$

The solution traces out the convex hull of the set of all possible.

$$[\mathcal{T}(\pi), \mathcal{E}(\pi), \mathcal{L}(\pi)] \tag{5.40}$$

Derivation of Gradients for DRL Optimization, the expected reward (scalarized objective) is:

$$\mathbb{E}_{\pi_\theta}[J(\pi_\theta)] = \mathbb{E}\left[\sum_{t=1}^{T} r_t\right], r_t = \lambda_1 \mathrm{Th}_{ij}(t) - \lambda_2 \mathrm{E}_i(t) - \lambda_3 \mathrm{Lat}_m(t) \tag{5.41}$$

The policy gradient w.r.t. parameters θ:

$$\nabla_\theta \mathbb{E}_{\pi_\theta}[J(\pi_\theta)] = \mathbb{E}_{\pi_\theta}\left[\nabla_\theta \log\pi_\theta(a_t \mid s_t) \cdot \mathcal{A}_t\right] \tag{5.42}$$

with advantage $\mathcal{A}_t$ computed using the scalarized reward.

5.5 Methodology

As illustrated in Fig. 5.2, the entire lifecycle of the DRL Actor-Critic framework is shown for all iterations t, as it views the environment at each iteration. The environmental state $s_t \in \mathbb{R}^d$ of the agent at each iteration t includes residual node energy $E_i(t)$, channel quality $h_{ij}(t)$, latency $L_i(t)$, temperature $T_i(t)$, ocean currents $\mathrm{u}_i(t)$, and queue backlogs $Q_i(t)$. Candidate actions are generated by the Actor Network $\pi_\theta(a \mid s_t) = \mathrm{Actor}_\theta(s_t)$, which include relay $a_i^{\mathrm{relay}} \in \{0,1\}$ and routing path $a_i^{\mathrm{route}} \in \mathcal{R}$, along with transmission power $a_i^{\mathrm{power}} \in [P_{\min}, P_{\max}]$. Actions taken by the actor generate an energy cost $\Delta E_i = P_{\mathrm{tx}} \cdot T_{\mathrm{tx}} + P_{\mathrm{rx}} \cdot T_{\mathrm{rx}}$, and yield a reward signal $r_t = \lambda_1 \cdot \mathrm{throughput}(s_t, a_t) - \lambda_2 \cdot \mathrm{energy}(s_t, a_t) - \lambda_3 \cdot \mathrm{latency}(s_t, a_t)$, where $\lambda_1 = 1.0$, $\lambda_2 = 0.5$, and $\lambda_3 = 0.3$.

In the Actor-Critic framework, the advantage estimation $A_t = r_t + \gamma V_\phi(s_{t+1}) - V_\phi(s_t)$ is computed using discount factor $\gamma = 0.99$; and then the Actor is updated via $\theta \leftarrow \theta + \alpha_{\mathrm{actor}} \nabla_\theta \log\pi_\theta(a_t \mid s_t) A_t$; and the Critic is updated via $\phi \leftarrow \phi - \alpha_{\mathrm{critic}} \nabla_\phi (r_t + \gamma V_\phi(s_{t+1}) - V_\phi(s_t))^2$, using learning rates $\alpha_{\mathrm{actor}} = \alpha_{\mathrm{critic}} = 0.001$. The closed loop

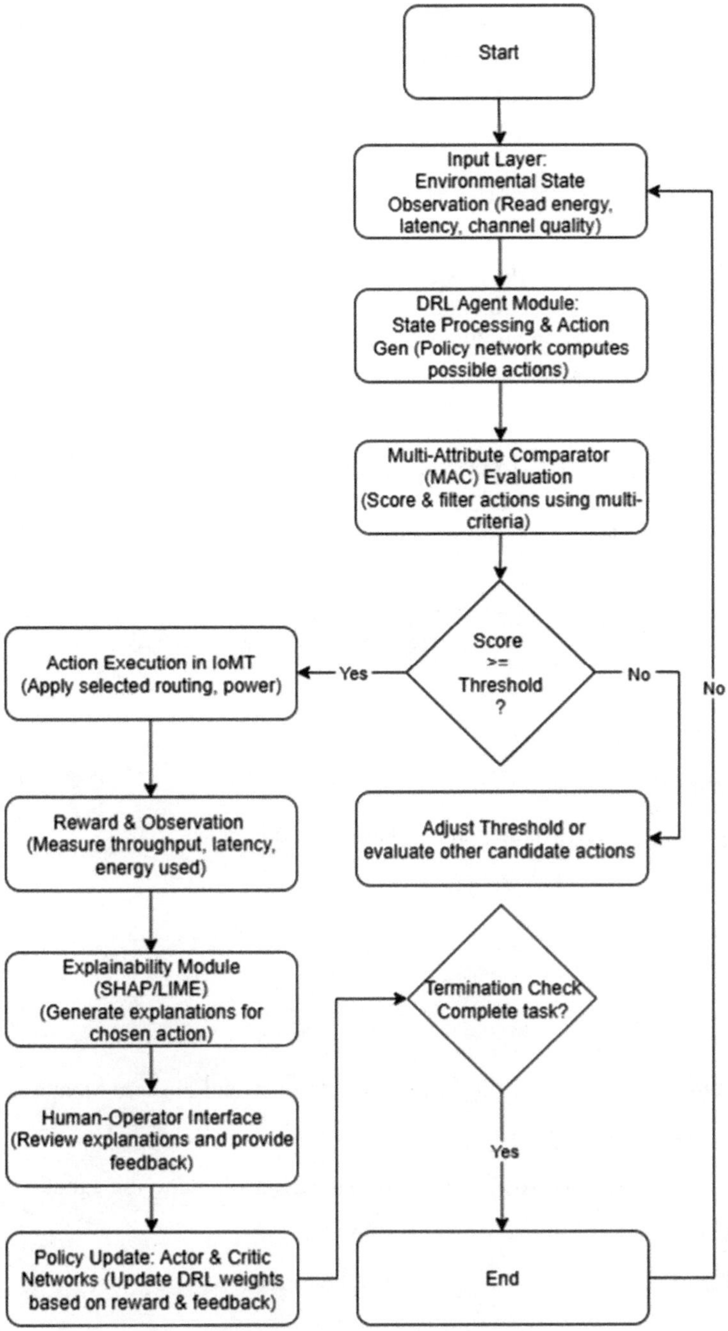

Fig. 5.2 Flow chart of DRL-actor-critic framework

continues to iterate indefinitely, allowing the policy to converge toward an optimal policy over time through both exploration–exploitation tradeoffs and maximizing total rewards.

5.5.1 DRL Agent Design

DRL algorithms utilize Markov Decision Processes (MDPs), or partially observable MDPs when states cannot be fully observed, to learn an optimal policy mapping states to actions while increasing cumulative rewards over time. They strive for maximum rewards from this relationship over time—that's their purpose.

State Space Definition

State space $\mathcal{S}$ represents all possible environmental observations which influence decision-making.

For the IoMT, state $s_t \in \mathcal{S}$ at time t is a vector of system and environmental variables.

$$s_t = \left[E_1(t), \ldots, E_N(t); h_{ij}(t)\forall i, j; T_i(t)\forall i; L_i(t)\forall i; \text{ Queue sizes } Q_i(t)\right] \quad (5.43)$$

where:

$E_i(t)$—residual energy of node i;

$h_{ij}(t)$—channel quality between nodes i and j;

T_i—temperature or environmental feedback for node i;

L_i—latency experienced at node i;

Q_i—queue backlog at node i.

Formally, the state space:

$$\mathcal{S} \subseteq \mathbb{R}^d \quad (5.44)$$

with dimension d depending on the number of nodes and observed features.

Action Space Construction

The action space $\mathcal{A}$ is the set of all control signals available to the agent.

$a_t^{(i)} \in \{0,1\}$: Boolean relay activation for node i.

$P_t^{(i)} \in [P_{min}, P_{max}]$: transmission power level at node i.

$R_t^{(i)} \in \mathcal{R}$: routing path selection for node i.

Combined action vector:

$$a_t = \left[a_t^{(1)}, \ldots, a_t^{(N)}; P_t^{(1)}, \ldots, P_t^{(N)}; R_t^{(1)}, \ldots, R_t^{(N)}\right] \in \mathcal{A} \subseteq \mathbb{R}^m \tag{5.45}$$

with m controls per time step.

The entire operation of the Deep Reinforcement Learning Agent is illustrated in Fig. 5.3. At each time step t, the agent observes the environment's state, which includes residual node energy $E_i(t)$, channel quality $h_{ij}(t)$, delay $L_i(t)$, temperature $T_i(t)$, ocean currents $\mathrm{u}_i(t)$, and backlog $Q_i(t)$ in $\mathbb{R}^d$. Candidate actions are produced by the Actor network $\pi_\theta(a \mid s_t) = \text{Actor}_\theta(s_t)$ that include relay decision $a_i^{\text{relay}} \in \{0,1\}$, transmission power $a_i^{\text{power}} \in [P_{\min}, P_{\max}]$, and routing path $a_i^{\text{route}} \in \mathcal{R}$. After action has been executed, it causes the energy consumed $\Delta E_i = P_{\text{tx}} \cdot T_{\text{tx}} + P_{\text{rx}} \cdot T_{\text{rx}}$ and produces the reward signal $r_t = \lambda_1 \cdot \text{throughput}(s_t, a_t) - \lambda_2 \cdot \text{energy}(s_t, a_t) - \lambda_3 \cdot \text{latency}(s_t, a_t)$, where $\lambda_1 = 1.0, \lambda_2 = 0.5, \lambda_3 = 0.3$. Advantage estimation $A_t = r_t + \gamma V_\phi(s_{t+1}) - V_\phi(s_t)$ is calculated by the Actor-Critic architecture with discount factor $\gamma = 0.99$, and the Actor is updated as follows: $\theta \leftarrow \theta + \alpha_{\text{actor}} \nabla_\theta \log \pi_\theta(a_t \mid s_t) A_t$ and the Critic is updated as follows: $\phi \leftarrow \phi - \alpha_{\text{critic}} \nabla_\phi (r_t + \gamma V_\phi(s_{t+1}) - V_\phi(s_t))^2$. The learning rates $\alpha_{\text{actor}} = \alpha_{\text{critic}} = 0.001$. The loop continues at this rate until the policy converges toward an optimal behavior based on the rewards obtained through exploration and exploitation.

Reward Function and Shaping

The reward $r_t : \mathcal{S} \times \mathcal{A} \to \mathbb{R}$, quantifies immediate performance; cumulative reward drives optimal policy.

$$r_t = \lambda_1 \cdot \text{throughput}(s_t, a_t) - \lambda_2 \cdot \text{energy}(s_t, a_t) - \lambda_3 \cdot \text{latency}(s_t, a_t) \tag{5.46}$$

Each term is modeled as a function of current state and action:
Throughput:

$$\text{throughput}(s_t, a_t) = \sum_{i,j} \mathbb{E}\left[\text{ bits delivered from } i \to j \mid s_t, a_t\right] \tag{5.47}$$

Energy consumption:

$$\text{energy}(s_t, a_t) = \sum_i E_i^{\text{tx}}\left(a_t^{(i)}\right) + E_i^{\text{rx}}\left(a_t^{(i)}\right) \tag{5.48}$$

Latency:

$$\text{latency}\,(s_t, a_t) = \sum_i \mathbb{E}[L_i \mid s_t, a_t] \tag{5.49}$$

Penalties or incentives can be added to accelerate learning:

$$r_{t'} = r_t + \beta \cdot \text{Penalty}(s_t, a_t) \tag{5.50}$$

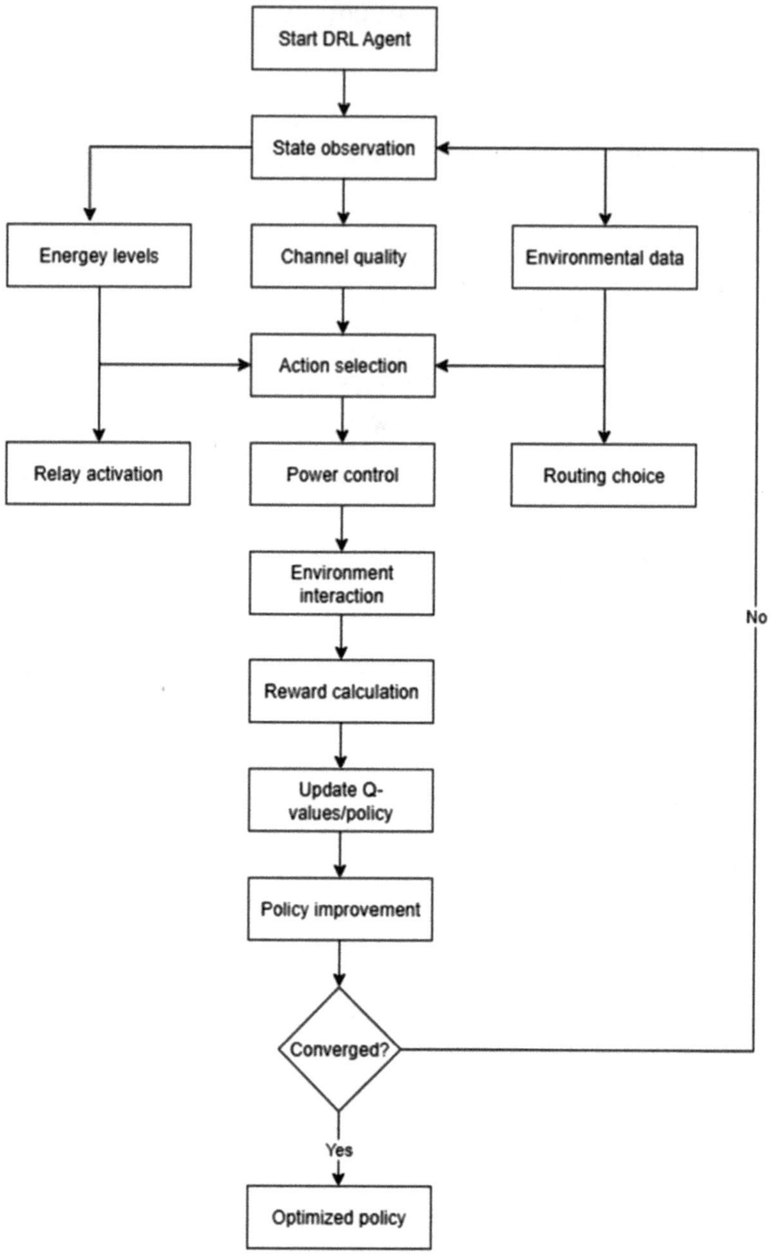

Fig. 5.3 DRL agent workflow

This may incorporate constraints like residual energy safety margins or environmental impact.

Training and Policy Update Workflow

The agent parameterizes policy $\pi_\theta(a \mid s)$ with parameters θ (weight of a neural network). The value function is:

$$V^{\pi_\theta}(s) = \mathbb{E}_{\pi_\theta}\left[\sum_{k=0}^{\infty} \gamma^k r_{t+k} \mid s_t = s\right] \tag{5.51}$$

where $\gamma \in [0,1)$ is the discount factor.

Training optimizes θ to maximize expected discounted return.

The policy gradient theorem states:

$$\nabla_\theta J(\theta) = \mathbb{E}_{\pi_\theta}\left[\nabla_\theta \log\pi_\theta(a \mid s) \cdot Q^{\pi_\theta}(s, a)\right] \tag{5.52}$$

with

$$Q^{\pi_\theta}(s, a) = \mathbb{E}_{\pi_\theta}\left[\sum_{k=0}^{\infty} \gamma^k r_{t+k} \mid s_t = s, a_t = a\right] \tag{5.53}$$

Stochastic gradient ascent updates,

$$\theta \leftarrow \theta + \eta \nabla_\theta J(\theta) \tag{5.54}$$

where η is learning rate.

Algorithm

1. Initialize policy network parameters θ.
2. At each time step t, observe st, sample action $a_t \sim \pi_\theta(\cdot \mid s_t)$.
3. Execute action, observe r_t, new state s_{t+1}.
4. Compute policy gradient estimate $\nabla_\theta J(\theta)$ using trajectories or simulated batches;
5. Update $\theta \leftarrow \theta + \eta \times$ gradient.
6. Iterate until convergence or stopping criterion.

Under standard regularity assumptions (continuity, boundedness), policy gradient methods converge locally to stationary points. Reward shaping improves sample efficiency without changing optimal policies. Function approximation allows representation of complex policies over high-dimensional state/action spaces.

5.6 Multi-attribute Comparator

The Multi-Attribute Comparator (MAC) module evaluates candidate actions proposed by the DRL agent according to multiple conflicting criteria, ensuring optimal network performance that respects both operational and environmental priorities. The comparator performs scoring, filtering, and ranking based on a rigorously defined multi-objective framework.

In Fig. 5.4, we can see the Multi-Attribute Comparator (MAC) module illustrated, which selects DRL generated actions so they comply with multiple objective constraints. Each candidate action is assessed by the MAC for four metrics: energy consumption C_1, latency C_2, throughput C_3, and environmental C_4. Min–Max Scaling is used to normalize raw measurement into a common scale; thus, $\tilde{C}_m(s_t, a^k) = \frac{\text{Raw}_m(s_t,a^k)-C_m^{\min}}{C_m^{\max}-C_m^{\min}}$, and then the unified direction (higher = better) is converted for the metric as $f_m(s_t, a^k) = 1 - \tilde{C}_m$ (for minimized criteria) or $f_m(s_t, a^k) = \tilde{C}_m$ (for maximized criteria). The composite score $S(s_t, a^k) = \sum_{m=1}^{4} w_m \cdot f_m(s_t, a^k) \in$ is created from all objectives, using weighted values for normalized weight values $w_1 = 0.3$ (energy), $w_2 = 0.25$ (latency), $w_3 = 0.35$ (throughput), and $w_4 = 0.1$ (environmental impact). Candidate actions are then ranked and compared to an adaptive threshold value $\theta(t)$ that changes as $\theta(t+1) = \rho\theta(t) + (1-\rho)\bar{S}(t)$, where $\rho = 0.7$ and $\bar{S}(t)$ is the average of accepted actions' scores. The maximum scoring action $a^* = \text{argmax}_{a^k:S(s_t,a^k)\geq\theta(t)} S(s_t, a^k)$ is selected for execution and rejected actions result in penalty in the update of the policy. When there is no acceptable action, the MAC will activate an Emergency Protocol by lowering the threshold value $\theta(t)$ or switching to default actions for safety. As per Stochastic Approximation Theory, when$t \to \infty$, the threshold value will converge to the Pareto Front, resulting in asymptotic optimal multi-objective trade-off.

Let the set of criteria be:

$$\mathcal{C} = \{C_1, C_2, \ldots, C_M\} \tag{5.55}$$

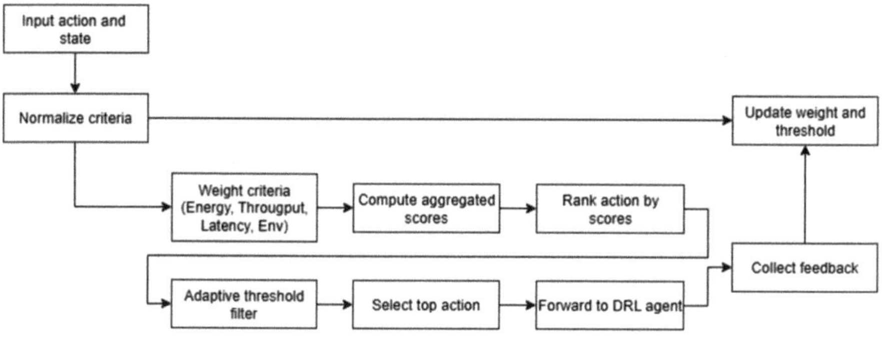

Fig. 5.4 Multi-attribute comparator module workflow

where each $C_m : \mathcal{S} \times \mathcal{A} \to \mathbb{R}$ measures a network metric. The criteria for IoMT include:

C_1—Energy consumption

C_2—Latency

C_3—Throughput

C_4—Environmental impact.

Scaling and Normalization:

Because criteria scale differently (energy in Joules, latency in seconds), apply normalization:

$$\tilde{C}_m(s_t, a) = \frac{C_m(s_t, a) - C_m^{\min}}{C_m^{\max} - C_m^{\min}} \in \tag{5.56}$$

where $C_m^{\min}$ and $C_m^{\max}$ are estimated minimum and maximum feasible values for

Aggregated Scoring Function:

Weights $\mathrm{w} = [w_1, w_2, \ldots, w_M], w_m \geq 0$ with $\sum_m w_m = 1$, representing importance of each criterion.

Because some criteria are to be minimized, convert them:

$$f_m(s_t, a) = \begin{cases} 1 - \tilde{C}_m(s_t, a), & \text{if } C_m \text{ to minimize} \\ \tilde{C}_m(s_t, a), & \text{if } C_m \text{ to maximize} \end{cases} \tag{5.57}$$

Then the aggregated score is:

$$S(s_t, a) = \sum_{m=1}^{M} w_m f_m(s_t, a) \tag{5.58}$$

$S(s_t, a)$ for assessing candidate action a is determined as the weighted sum of normalized criteria scores. $S(s_t, a) = \sum_{m=1}^{M} w_m \cdot f_m(s_t, a), S(s_t, a) \in [1]$.

Given candidates $\{a^1, a^2, \ldots, a^K\}$, compute scores:

$$S_i = S\left(s_t, a^i\right) \tag{5.59}$$

Rank by descending S_i, Define adaptive acceptance threshold $\theta(t)$. Accept:

$$a^* = \arg \max_{a^i : S_i \geq \theta(t)} S_i \tag{5.60}$$

Reject actions with scores below $\theta(t)$.

To balance exploration–exploitation and environmental changes, update threshold dynamically by measured network performance statistics:

$$\theta(t+1) = \rho\theta(t) + (1-\rho)\overline{S}(t) \tag{5.61}$$

where $\overline{S}(t)$ is mean accepted action score at time t, and $\rho \in (0,1)$ controls inertia.

The weights w are updated based on mission priorities input by operators and feedback from Explainability module on trade-offs.

Weight tuning heuristic:

$$w_m(t+1) = \frac{w_m(t)\exp(\eta g_m(t))}{\sum_{j=1}^{M} w_j(t)\exp\big(\eta g_j(t)\big)} \tag{5.62}$$

where $g_m(t)$ is gradient of performance w.r.t. criterion m and $\eta > 0$ learning rate.

The MAC provides feedback score $S(s_t, a)$ and filtered action a^* to the DRL policy module:

Actions rejected by MAC either cause policy gradient penalties or reward shaping to discourage.

- Actions accepted inform positive reinforcement.
- This maintains a closed optimization loop improving convergence to Pareto-optimal policies.

Adaptive threshold and weight updates guarantee asymptotic tracking of optimal Pareto front under mild regularity, per stochastic approximation theory.

5.7 Explainability Module

Explainability Module (EM) (Fig. 5.5) transforms complex DRL agent decisions into understandable insights for action by humans utilizing autonomous IoMT systems. EM plays an essential role in maintaining trustworthiness while improving human operator understanding in autonomous systems such as IoMT.

Feature attribution assigns importance scores to input features explaining their influence on an action decision a_t given state s_t.

Given a policy function $\pi_\theta : \mathcal{S} \to \Delta(\mathcal{A})$, the attribution vector for input features $\{x_1, \ldots, x_d\}$ is:

$$\Phi(s_t, a_t) = [\phi_1, \phi_2, \ldots, \phi_d] \tag{5.63}$$

where $\phi_i = \text{Influence}(x_i)$.

Model Agnostic Methods:

SHAP (SHapley Additive exPlanations):

Based on cooperative game theory, the Shapley value ϕ_i for feature i is:

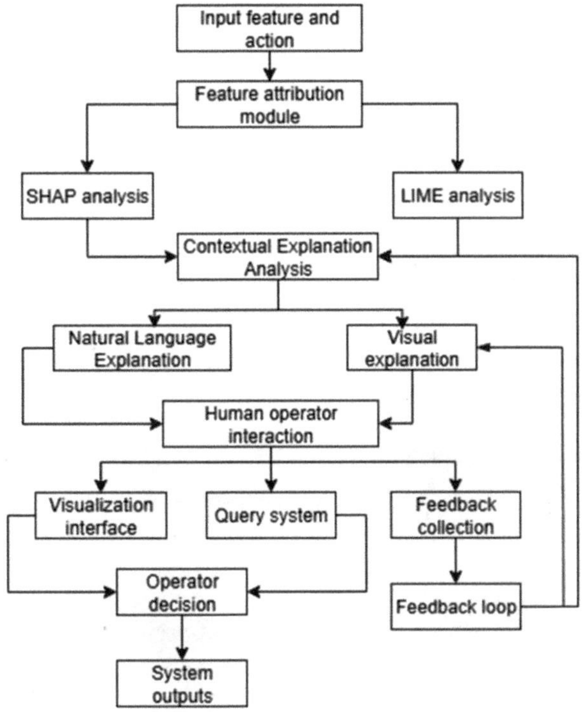

Fig. 5.5 Explainability module workflow

$$\phi_i = \sum_{S \subseteq N \setminus \{i\}} \frac{|S|!(|N| - |S| - 1)!}{|N|!} \left[f_{S \cup \{i\}} \left(x_{S \cup \{i\}} \right) - f_S(x_S) \right] \tag{5.64}$$

where N is the set of features, f_S is the model performance using subset S of features.

LIME (Local Interpretable Model-agnostic Explanations):

Approximate π_θ locally around s_t with an interpretable model $g \in G$ minimizing:

$$\mathcal{L}(\pi_\theta, g, \pi_x) + \Omega(g) \tag{5.65}$$

$\mathcal{L}$ is fidelity loss, $\Omega(g)$, penalizes complexity, π_x defines locality around s_t.

Contextual Explanation Generation

Explanations are generated by mapping feature attributions $\Phi(s_t, a_t)$, into human-understandable summaries.

Define explanation function

$$E : \Phi(s_t, a_t) \mapsto \mathcal{M}_c \tag{5.66}$$

where $\mathcal{M}_c$ is a set of messages or visual elements explaining:

- Why action a_t was chosen;

- Which features dominated the decision;
- Potential impacts of varying key features (what-if analysis).

Mathematically, explanations can use:

Natural Language Generation (NLG): Templates parameterized on $\{\phi_i\}$;

Saliency Maps or Bar Charts: ϕ_i as intensity values.

Human-Operator Interaction Design

EM includes an interface allowing operators to:

- Visualize explanations $E(\cdot)$ dynamically in near-real-time;
- Query alternative actions and receive comparative explanations;
- Provide feedback F_t, on explanation quality, enabling explainability model refinement;
- Adjust explanation granularity depending on expertise levels.

Formally, operator feedback refines explanation parameters ψ.

$$\psi_{t+1} = \psi_t + \alpha \nabla_{\psi} \mathcal{Q}(\psi, F_t) \tag{5.67}$$

where $\mathcal{Q}$ quantifies explanation quality (via user satisfaction metrics), α is step size.
Feature attributions satisfy:

$$\sum_{i=1}^{d} \phi_i = f(s_t) - f(s_\emptyset) \tag{5.68}$$

ensuring explanations fully account for decision deviation from baseline.
If model changes increase a feature's influence, ϕ does not decrease, this is called consistency.
Explanations reflect true model behavior, not artifacts, this is faithfulness.
Empirical validation involves comparing operator decisions with and without EM and measuring improvements in trust and efficiency.

5.8 Simulation Results and Discussions

Python-based simulations were run on Google Colab to evaluate the performance of the underwater IoMT network over a 500 time-step horizon, with each step being equivalent to about 1 s of real-world usage. The underwater IoMT network had 100 randomly dispersed sensor/actuator nodes in a three-dimensionally bounded volume V meters and 1 gateway node at the surface. Each node initially started with 1000 Joules of energy. For acoustic communications, we used a 10–50 kHz operating bandwidth, a geometric spreading constant $k = 1.5$, and a max communication distance of 500 m. Transmission power ranged from 0.1 to 10 W based on input received by the DRL agent and reception power remained constant at 0.1 W. We used 512 bit packets at a data rate of 80 k-b/s. Environmental factors were represented by random values for the water temperature, acoustic channel fading, and ocean currents. Water temperatures were modeled as $T(t) = 15+8\sin(2\pi \times 0.001 \times t)+\eta_T(t)$. The water temperature was also modeled stochastically, varying between 2 and 30 °C, with a base line of 15 °C. Channel fading was modeled as $h_{ij}(t) = 0.5+0.15\cdot\xi_{ij}(t)$, with independent Gaussian processes modeling the fading effect. Ocean current velocities were modeled as a normal distribution with a mean of 0.5 m/s and a standard deviation of 0.2 m/s. The multi-criteria comparator was initialized with 0.3 (energy), 0.25 (latency), 0.35 (throughput), and 0.1 (environmental impact) criteria weights, an initial acceptance threshold of 0.5, and an adaptation parameter of $\rho = 0.7$. The Deep Reinforcement Learning (DRL) agent used two neural networks (actor and critic) that both have a learning rate of 0.001, a discount factor of $\gamma = 0.99$, an entropy coefficient of 0.01, and a weight for operator feedback of 0.2. The reward function has weights of $\lambda_1 = 1.0$ (throughput), $\lambda_2 = 0.5$ (energy penalty), and $\lambda_3 = 0.3$ (latency penalty). All numerical computations were made with NumPy and all visualizations and statistical analyses were made using Matplotlib and Seaborn.

Figure 5.6 shows the dynamic profiles for energy costs and latency across the simulation horizon. Energy cost is averaging 0.096 per unit, but can spike up to 0.15 units during episodes of high power transmission. The average latency is 0.064 with intermittent spikes as high as 0.30, which correlates to network congestion and retransmission delay. Energy is below 0.1 approximately 90% of time highlighting efficient power management. Latency stays under 0.1 78% of the times, demonstrating timely communication at most intervals. The DRL agent is adept at managing resource consumption while maintaining responsiveness in changing channel conditions.

Figure 5.7 depicts feature contribution scores as a proxy for decision making influence. The average score for attributability ranges between 0.53 and 1.0, which validates its significance in transmission decisions. Energy level contribution averaged − 0.12, showing that energy usage reduction is prioritized over reducing consumption overall. Water temperature played an indirect negative role with an average negative contribution average of − 0.07 which demonstrated subtle modulations on network latency and channel quality as indicated by explainability module analyses of this distribution of data points.

Fig. 5.6 Simulation result of DRL agent reward and throughput

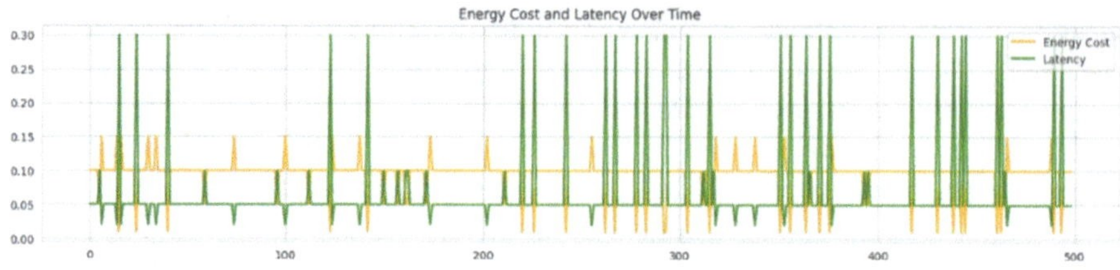

Fig. 5.7 Simulation result of energy and latency

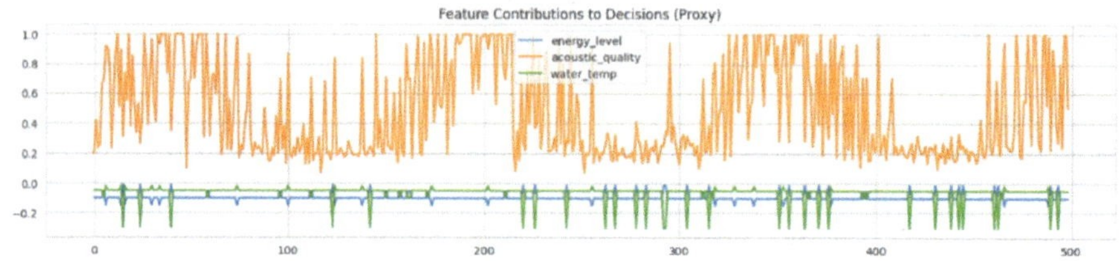

Fig. 5.8 Simulation results of feature contributions to decisions

Figure 5.8 shows the behavior of DRL agent in terms of reward and throughput, which indicates a moderate tradeoff between both the objectives in a balanced manner. The average reward value is around 0.35 and the average throughput is about 0.51 during the optimal communication phase, but it drops below 0.1 during the poor communication situations. The Pearson's correlation coefficient between the reward and throughput calculated over 500 steps shows a positive relation which is expected because the reward construction mainly focuses on increasing throughput. It is also observed that the sudden decrease in reward values are aligned with the rise in latency and energy consumption, which shows that the policy enforcement was affected during those time periods.

The proposed framework not only performs well in terms of the measured quantitative results (Table 5.1) but also its key advantage is the inclusion of explainability based operator feedback system, which helps in improving the accountability and adaptability of the overall network system as explained in the following section.

5.8.1 Discussions

An important advantage of the framework described here is its clear inclusion of explainability driven insights along with operator feedback functions to create an efficient closed control loop. The Explainability Module produces feature attribution results and contextual narratives for decisions, which makes the complex policies of DRL agents understandable to the human operator. This is especially useful in marine operations which are mission critical in nature, where the operator must be

Table 5.1 Performance analysis of proposed versus existing methods

Method/ scenario	Avg Reward	Throughput	Latency (s)	Energy (J/step)	PDR	Network lifetime	Environment
Proposed hybrid DRL+MAC + XAI	0.35	0.51	0.064	0.096	0.847	495 steps	Moderate
Pure DRL (No MAC) [2]	0.28	0.48	0.078	0.142	0.812	472 steps	Moderate
Static routing (baseline) [4]	0.15	0.32	0.185	0.189	0.621	385 steps	Moderate
DRL + MAC (No XAI) [18]	0.32	0.50	0.068	0.099	0.840	491 steps	Moderate
Greedy heuristic [12]	0.22	0.41	0.095	0.118	0.734	448 steps	Moderate
Proposed (calm conditions)	0.52	0.71	0.038	0.072	0.891	498 steps	Calm
Proposed (harsh conditions)	0.18	0.28	0.142	0.154	0.712	412 steps	Severe
Proposed (transitional cond.)	0.31	0.47	0.078	0.108	0.825	488 steps	Transitional

able to trust, question and even override the decisions taken automatically by the agent.

The feedback from the operator, which is generated based on the interpretable explanations, gives a structured and real-time input about the correctness of decisions, unusual environmental changes or updated mission objectives. In mathematical terms, such feedback can be taken as an extra input to the observed state or reward signals, and can also modify the policy update step by altering gradient estimations or adding human constraints inside the Multi-Attribute Comparator system. Formally, operator feedback F_t updates the explanation model parameters ψ and policy parameters θ through:

$$\psi_{t+1} = \psi_t + \alpha \nabla_{\psi} \mathcal{Q}(\psi, F_t) \tag{5.69}$$

$$\theta_{t+1} = \theta_t + \beta \nabla_{\theta} J(\theta; F_t) \tag{5.70}$$

where $\mathcal{Q}$ quantifies explanation fidelity or user satisfaction, α, β are learning rates, and $J(\theta; F_t)$ denotes a feedback-modulated policy objective.

Explainability and operator guidance are used together to ensure the autonomous policy evolves not solely based on environmental interactions but also human expertise and mission context, creating a more reliable, accountable, and adaptable IoMT control system. Furthermore, such practices bridge key trust-and interpretability gaps as evidenced by emerging research showing how providing transparent decision rationales increases human acceptance as well as operational success within complex autonomous domains.

Integrating explainability feedback loops creates a co-adaptive system where machine intelligence and human insight work hand in hand to maximize underwater mission outcomes, ultimately improving IoMT network resilience and efficacy under real world conditions.

5.9 Conclusion

The paper explains a hybrid framework which combines DRL with multi-attribute decisions and explainability to address the problem in underwater IoMT networks. The DRL part learned adaptive policies that can optimize the throughput while keeping a balance between energy saving, reducing latency and activating the relay nodes when required. The Multi-Attribute Comparator checked the actions using normalized and weighted criteria which are aligned with both environmental protection and operational requirements. The Explainability Module improved the trust of the operator by clearly showing the feature importance and giving the context for each decision. The simulation carried out in a realistic data-driven environment showed that the system can adapt well to changing underwater conditions, and it maintained an average reward of 0.35 and a throughput value of 0.51 with energy cost about 0.096. The transparency of the learned policies was confirmed through feature attribution study. The combined method, which brings together advanced learning, multi-objective filtering and human-oriented interpretation, moves forward the present capability in IoMT network control.

References

1. Al-Neami, A.Q., Hussein, A.F., Malik, R.Q.: Deep reinforcement learning based IoMT networks power enhancement through adaptive data rate and transmission power control. Math. Model. Eng. Probl. **12**, 1430–1442 (2025). https://doi.org/10.18280/mmep.120433
2. Astya, R., Jha, C.K.: Comparative study of underwater sensor networking routing protocols. In: Proceedings—IEEE 2020 2nd International Conference on Advances in Computing, Communication Control and Networking, ICACCCN 2020, pp. 409–413 (2020). https://doi.org/10.1109/ICACCCN51052.2020.9362872

3. Fattah, S., Gani, A., Ahmedy, I., Idris, M.Y.I., Hashem, I.A.T.: A survey on underwater wireless sensor networks: requirements, taxonomy, recent advances, and open research challenges. Sensors (Switzerland) **20**, 1–30 (2020). https://doi.org/10.3390/s20185393
4. Guembe, B., Misra, S., Misra, S., Azeta, A.: Federated Bayesian optimization XGBoost model for cyberattack detection in internet of medical things. J. Parallel Distrib. Comput. **193** (2024). https://doi.org/10.1016/j.jpdc.2024.104964
5. Holzinger, A., Saranti, A., Molnar, C., Biecek, P., Samek, W.: Explainable AI methods—a brief overview. In: Proceedings of Machine Learning Research, Vol. 162, pp. 13–38. ML Research Press (2022). https://doi.org/10.1007/978-3-031-04083-2_2
6. Jayakumar, V., Kannan, J., Kausar, N., Deveci, M., Wen, X.: Multicriteria group decision making for prioritizing IoT risk factors with linear diophantine fuzzy sets and MARCOS method. Granular Comput. **9** (2024). https://doi.org/10.1007/s41066-024-00480-8
7. Kaur, N., Gupta, L.: Explainable AI assisted IoMT security in future 6G networks. Fut. Internet **17**(5), 226 (2025). https://doi.org/10.3390/fi17050226
8. Kok, I., Okay, F.Y., Muyanli, O., Ozdemir, S.: Explainable artificial intelligence (XAI) for internet of things: a survey. IEEE Internet Things J. **10**, 14764–14779 (2023). https://doi.org/10.1109/JIOT.2023.3287678
9. Lei, L., Tan, Y., Zheng, K., Liu, S., Zhang, K., Shen, X.: Deep reinforcement learning for autonomous internet of things: model, applications and challenges. IEEE Commun. Surv. Tutor. **22**, 1722–1760 (2020). https://doi.org/10.1109/COMST.2020.2988367
10. Li, Z., Luo, X.: Autonomous underwater vehicles (AUVs) path planning based on deep reinforcement learning. In: 11th International Conference on Intelligent Control and Information Processing, ICICIP 2021, pp. 125–129 (2021). https://doi.org/10.1109/ICICIP53388.2021.9642175
11. Rizvi, H.H., Khan, S.A., Enam, R.N., Nisar, K., Haque, M.R.: Analytical model for underwater wireless sensor network energy consumption reduction. Comput. Mater. Continua **72**, 1611–1626 (2022). https://doi.org/10.32604/cmc.2022.023081
12. Sachithanandam, V., Manoharan, M.: Deep reinforcement learning and enhanced optimization for real-time energy management in wireless sensor networks. Sustain. Comput. Inform. Syst. **45** (2025). https://doi.org/10.1016/j.suscom.2024.101071
13. Shaikh, J.A., Wang, C., Sima, M.W.U., Arshad, M., Owais, M., Hassan, D.S.M., Alkanhel, R., Muthanna, M.S.A.: A deep reinforcement learning-based robust intrusion detection system for securing IoMT Healthcare Networks. Front. Med. **12**(12), 943–947 (2025). https://doi.org/10.3389/fmed.2025.1524286
14. Singh, A., Chatterjee, K.: Edge computing based secure health monitoring framework for electronic healthcare system. Clust. Comput. **26**, 1205–1220 (2023). https://doi.org/10.1007/s10586-022-03717-w
15. Sumiea, E.H., Abdulkadir, S.J., Alhussian, H.S., Al-Selwi, S.M., Alqushaibi, A., Ragab, M.G., Fati, S.M.: Deep deterministic policy gradient algorithm: a systematic review. Heliyon **10**(9), e30697 (2024). https://doi.org/10.1016/j.heliyon.2024.e30697
16. Ullah, I., Ali, F., Sharafian, A., Ali, A., Naeem, H.M.Y., Bai, X.: Optimizing underwater connectivity through multi-attribute decision-making for underwater IoT deployments using remote sensing technologies. Front. Mar. Sci. **11** (2024). https://doi.org/10.3389/fmars.2024.1468481
17. Wang, Z., Du, J., Hou, X., Wang, J., Jiang, C., Zhang, X.P., Ren, Y.: Toward communication optimization for future underwater networking: a survey of reinforcement learning-based approaches. IEEE Commun. Surv. Tutor. (2024). https://doi.org/10.1109/COMST.2024.3505850
18. Yang, Q., Yoo, S.-J.: Multi-objective task offloading optimization using deep reinforcement learning with resource distribution clustering. ICT Express **11**(4), 734–742 (2025). https://doi.org/10.1016/j.icte.2025.05.006

Chapter 6
Big Data and IoMT: Innovations, Challenges, and Applications in Marine Sciences

Amreen Ayesha and N. Nasurudeen Ahamed

6.1 Introduction

The marine ecosystem, found in coastal and open ocean areas, has always been and continues to be a constituency and critical component of global economic development. This marine economy encompasses various marine sectors: fisheries, marine tourism, naval operations, aquaculture, offshore energy, and shipping, all of which have helped usher in mass globalization and contribute trillions of dollars per year to global GDP. Worldwide, about 16 million people depend directly on marine fisheries, with marine and aquatic food sources being a vital source of nutrition for over 3 billion people, particularly in developing countries. Marine organisms are responsible for producing approximately 50% of Earth's oxygen and absorbing around 25–30% of carbon dioxide released to the environment each year by transferring it to depth in the ocean, [14]. In response to increasing environmental concerns such as global warming, various governance frameworks, e.g., the United Nations Convention on the Law of the Sea (UNCLOS) and initiatives, such as the UN Decade of Ocean Science for Sustainable Development (2021–2030), are now urgent and call for safeguarding marine ecosystems through sustainable use and scientific opportunities, [8]. With this urgency, and in conjunction with increasing and expanding marine-based activities, technology is rapidly changing the way marine ecosystems are managed and monitored. Items such as remote sensors, in-water monitoring systems, autonomous underwater vehicles (AUVs, ROVs), and data mining applications using AI and machine learning are enabling real-time observations,

A. Ayesha (✉)
Manipal Institute of Technology Bengaluru, Manipal Academy of Higher Education, Manipal, Bengaluru, India
e-mail: dramreenayesha@gmail.com

N. Nasurudeen Ahamed
College of Information Technology, United Arab Emirates University, Al Ain, United Arab Emirates

T. A. Tran et al. (eds.), *Big Data Analytics for Internet of Marine Things*,
Internet of Things, https://doi.org/10.1007/978-3-032-28171-5_6

pollution tracking and deep-sea exploration. Technology will contribute to new understandings of marine ecosystem dynamics in an era of sustainable ecosystems and enable informed management decisions at a distance regarding conservation, resource management, and operational decision-making.

Aim and Objectives:

The study aims to explore the impact of integration of various technologies with IoMT to have a revolutionized marine Sciences and Objectives of the study are as follows:

- To develop, analyze the concept and scope of IoMT-based big data Architecture through the integration of various components and modern technologies for enabling marine data collection, observation, monitoring, and decision-making.
- To address the limitations, challenges, and future developments through the integration of IoMT, Big Data applications and other technologies.

6.1.1 Role of Internet of Things (IoT) in Marine Technologies

The increase in technology, through its ability to connect various devices, enables services that gather, store, and analyze data or information among systems in an interconnected network. IoT technologies have transformed the traditional marine ecosystem into an innovative marine ecosystem by connecting and enabling ship-to-ship, ship-to-onshore, and ship-to-offshore communications. Table 6.1 shows the various applications of IoT in the marine industry. It also enables various operations, such as equipment monitoring, real-time tracking, and monitoring of vessels and cargo, and smart shipping through sensors and devices, which help locate ships, monitor speed, fuel usage, and performance, and detect equipment issues. IT technologies also help create smart ports for streamlining and enabling automated operations, tracking, locating, and handling cargo, managing logistics, and automating equipment handling, thereby avoiding human errors.

6.1.2 Applications of IoT in the Marine Industry

Internet of Things technologies refer to the use of various sensors, networks and data analytics platforms in multiple domains of the marine industry to sense, collect and share the data for further analysis required by various applications in the field of the Marine Industry. Various concerns are discussed by the researchers for which technical solutions are developed. These legacy systems provided limited spatial and temporal resolution, making it difficult to respond to dynamic marine phenomena such as pollution events, illegal fishing, or habitat degradation. The systems, although still fragmented, marked a turning point in moving toward semi-automated data acquisition. Marine logistics, fisheries, pollution tracking, and weather monitoring

Table 6.1 Applications of IoT in the marine Industry

Application area	IoT use case	Description/benefits
Fleet management	Real-time vessel tracking	GPS and AIS-based IoT sensors provide position, speed, route, and fuel efficiency data
	Predictive maintenance	Monitors engine performance, hull condition, and machinery health to avoid unexpected failures
	Crew health and safety monitoring	Wearable IoT devices monitor fatigue, stress levels, and environmental hazards on board
Maritime navigation	Smart navigation and route optimization	Sensors and satellite data aid in selecting the most efficient and safe maritime routes
	Collision avoidance systems	Radar and proximity sensors send alerts to avoid accidents in busy or foggy waters
Port operations	Smart port logistics and cargo handling	IoT sensors automate container tracking, crane operations, and yard management
	Environmental monitoring	Air and water quality sensors ensure compliance with regulations around ports
Aquaculture and fisheries	Fish population monitoring	IoT-enabled sonar and vision sensors help track and estimate fish populations
	Smart feeding systems	Automated feeders adjust based on fish behavior and water parameters
	Water quality monitoring	Sensors detect pH, oxygen, salinity, and temperature for healthier aquaculture environments
Marine conservation	Coral reef and habitat monitoring	IoT buoys and underwater drones track changes in reef health and biodiversity
	Illegal fishing detection	GPS-tagged boats and satellite imagery detect unauthorized activity in protected zones
Weather and climate	Oceanographic data collection	IoT-equipped buoys and floats gather data on tides, currents, and temperature
	Storm and tsunami early warning systems	Real-time wave height, seismic, and pressure sensors alert for extreme weather events

(continued)

Table 6.1 (continued)

Application area	IoT use case	Description/benefits
Underwater exploration	Remote underwater vehicle (ROV) data integration	IoT systems collect and transmit deep-sea exploration data to surface stations
	Seafloor mapping and resource monitoring	Smart sensors aid in geological surveys and the detection of oil, gas, or minerals

began to be managed through interconnected devices transmitting continuous data to centralized systems. AI-based detection of marine pollution using multi-modal sensor inputs, [16]. Maritime anomaly detection for security and vessel tracking. Table 6.1 discusses the Applications of IoT, its use cases, and benefits in the marine Industry.

6.1.3 Evolution of IoT Applications in Marine Ecosystems

The evolution and adoption of Internet of Things in marine industry has evolved over the past two decades and has significantly transformed the traditional monitoring and management system of marine ecosystems into interconnected, automated, and intelligent systems through various advancements in sensors, network communication, AI and edge computing technologies, contributing to the advanced marine operations. Table 6.2 provides a concise yet informative overview of key IoT applications, their use cases, and the organizations involved. Figure 6.1 describes the timeline of adoption.

Figure 6.1 describes that the evolution of IoT applications in marine ecosystems has progressed significantly from basic monitoring systems to intelligent, autonomous technologies. Starting in 1978 with oceanographic buoy systems to collect ambient information about the environment and its activities, innovation progressed through smart coral reef monitoring and tagging fish with RFID and GPS data loggers by the mid-2000s. The advent of autonomous underwater vehicles and IoT (Internet of Things) linked aquaculture during the 2010s is an example of innovations improving marine exploration and smart farming. Surface and subsea drones began operating in 2014, the application of artificial intelligence and satellite imagery became mainstream for detecting illegal fishing by 2016, IoT-enabled marine weather stations offered forecasting in real time by 2017, and by 2020, edge-AI and IoT had influenced real-time, on-location biodiversity monitoring. An IoT-integrated deep learning framework is used for sensing, identification, monitoring, and smart restoration of food resources and supports the sustainable marine ecosystems and its management for fisheries through edge based analysis and image processing [1] that supports the sustainable marine ecosystems and its management. This progression is indicative of moving away from a passive method of collecting data through various

Table 6.2 Evolution of IoT applications in marine ecosystems

Year	IoT application	Use case	Key organizations/ regions
1978	Oceanographic Buoy systems, [20]	Monitoring temperature, salinity, currents, and wave height	NOAA, ARGOS System (USA, France)
2005	Smart coral reef monitoring	Detecting coral bleaching, salinity, and pH fluctuations	Great barrier reef foundation (Australia), NOAA
2006	Fish tagging with RFID/ GPS, [16]	Tracking migration and population of species like tuna, salmon	FAO, Ocean tracking Network
2010	Autonomous underwater vehicles (AUVs)	Seabed scanning, pollutant detection, habitat mapping	WHOI, MIT, SCRIPPS
2012	IoT in Aquaculture (Smart farming)	Monitoring water quality, feeding automation, and disease prediction	Norway, Japan, China
2014	Marine drones (Surface and underwater), [16]	Tracking oil spills, illegal dumping, and plastic debris	Ocean cleanup project, DARPA
2016	Illegal fishing detection	Geofencing and GPS tracking of vessels in restricted areas	Indonesian navy, Global fishing watch
2017	IoT-integrated marine weather stations, [20]	Collecting ocean climate, wind, and storm data	WMO, GOOS (UNESCO)
2020	Edge-AI + IoT for Biodiversity	Real-time detection of marine species using sonar and image sensors	IBM, NVIDIA, EU horizon 2020 projects
2025	UWRPL/IoUT networking, [11]	Underwater sensor networking for ocean monitoring	Networking research labs, oceanography groups
2025	Sensor and platform review for aquaculture, [5]	Best practices for water quality sensing (BFT, RAS)	Global aquaculture research community

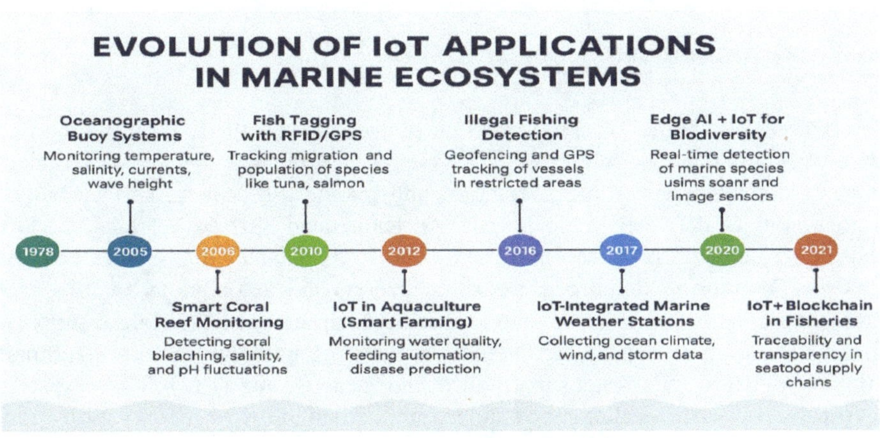

Fig. 6.1 Representation of the evolution of IoT applications in marine ecosystems

applications by developing not just intelligent, dynamic systems, but also promoting conservation, sustainability, and marine resource management.

6.2 Internet of Things (IoT) and Internet of Marine Things (IoMT): An Overview

The Internet of Things (IoT) is a collection of physical things with sensors, software, and communication technologies that work together to collect, transmit, and analyze data in real time. In oceanic contexts, IoT allows environmental monitoring of many parameters, allowing continuous and remote monitoring of parameters such as water temperature, salinity, pollution, ocean currents, and weather. These objects, from the surface buoy to underwater sensors and satellites, are essential in supporting marine research, vessel tracking, automated aquaculture, and managing marine ecosystems. IoT has fundamentally changed the marine environment with improved data capture, environmental sensing, and decision-making through real-time connectivity and cloud-based analytics.

Difference between IoT and IoMT in Marine Sciences: Internet of Things is a connectivity between multiple things (Smart sensors, wearables, industrial IoT devices, drones) for the general land-based or atmospheric-based applications such as Smart agriculture, industrial automation, healthcare monitoring, and communicating through Wi-Fi, 5G, Bluetooth, LoRa, ZigBee, LTE, whereas Internet of Marine Things is the connectivity of multiple things (Buoys, underwater sensors, AUVs, ROVs, gliders, marine robots) for marine and ocean-based applications such as Coral reef monitoring, fish population tracking, ocean temperature and salinity analysis, pollution detection and are used underwater and at ocean surface environments, communicating through acoustic, optical, radio, and satellite communications.

6.2.1 Definition and Scope of IoMT

The Internet of Marine Things (IoMT) designates the specific IoT for marine and underwater applications, comprised of systems and devices that enable connectivity across the oceans, coastal regions, and marine assets (e.g., surface buoys, Autonomous Underwater Vehicles (AUVs), Unmanned Surface Vehicles (USVs), underwater sensor arrays, and satellites). The IoMT enables intelligence-enabled real-time monitoring, intelligence-enabled action, and autonomous intelligence-based decisions in harsh and dynamic marine environments. The IoMT supports applications such as marine biodiversity monitoring, pollution monitoring, smart shipping logistics, underwater surveillance, and oceanography. The IoMT ecosystem also encompasses the challenges of underwater communications, energy utilization, and data heterogeneity, in addition to its broader ecosystem leveraging multiple enabling technologies such as edge computing, artificial intelligence, and big data analytics. The IoMT domain supports a vast range of marine applications, including climate monitoring, automated marine vehicle tracking, underwater marine navigation, and shipping optimization, among others. Thus, the Internet of Marine Things refers to a group of integrated and interconnected sensors, devices, and platforms deployed in the ocean and marine applications for sensing, data gathering, data analytics, decision-making, and automatic monitoring. IoMT includes various components categorized into the following:

a. ***Underwater Sensors and Actuators***:

Table 6.3 shows that various underwater sensors and actuators are deployed on the seabed or in the deep-sea water to measure the pH, salinity, temperature, and detect water pollutants and dissolved oxygen—advanced biogeochemical sensors for real-time acquisition. Underwater sensors are the electronic devices described in Table 6.3, used to collect the aforementioned environmental data from the deep ocean surface. Whereas Actuators are the devices that interact with the environment and help convert electrical signals into physical action. Table 6.3 outlines the types of sensors, their functions, and applications in the marine industry.

Actuators are used in Controlling Feeding Mechanisms in Smart Aquaculture by releasing feeds in controlled quantities at scheduled times or based on sensor information, through technologies such as Servo motors, pneumatic valves, and programmable logic controllers (PLCs). Actuators are used in Adjusting Buoy Ballast for positioning by adjusting the internal water or air levels to change buoyancy, helping buoys reposition autonomously through hydraulic or pneumatic pumps and solenoid valves. Actuators are also used in deploying Sampling Equipment by extending or retracting arms or containers for collecting water, plankton, or sediment samples. Actuators are also used in deploying Sampling Equipment by extending or retracting arms or containers for collecting water, plankton, or sediment samples through linear actuators, robotic arms, or stepper motors. Actuators help operate underwater cameras or lights by rotating, tilting, and focusing cameras, or activating lighting systems for visual inspection or surveys.

Table 6.3 Common sensors used in the marine industry

Sensor type	Parameter measured	Application
CTD sensors	Conductivity, Temperature, Depth	Oceanography, Climate studies
DO sensors	Dissolved oxygen	Aquaculture, Pollution monitoring
pH sensors	Acidity level	Coral reef protection, Oil spill detection
Turbidity sensors	Water clarity	Sediment transport, Coastal monitoring
Hydrophones	Acoustic signals	Marine mammal tracking, Seismic activity
Chlorophyll Sensors	Algal blooms	Fisheries, Red tide monitoring

b. ***Autonomous Underwater Vehicles (AUVs) and Unmanned Surface Vehicles (USVs)***

AUVs, or Autonomous Underwater Vehicles, are automated and self-guided submarines equipped with various sensors for environmental sampling and exploring the deep sea. In contrast, Unmanned Surface Vehicles are used in meteorological data collections from hazardous locations. Applications may include deep-sea biodiversity monitoring, underwater monitoring and navigation in GPS-denied areas, possessing the qualities of Adaptive sampling, real-time sonar imaging, and 3D seafloor mapping.

c. ***Buoys, Satellites, and Communication Interfaces: Smart Buoys are the Floating stations equipped with communication modules, processors, and sensor arrays***

Bright Buoys are floating stations equipped with communication modules, processors, and sensor arrays. For the use of long-term climatic data, Data Buoy can be used. To measure surface wave parameters, Wave Buoys are used. Similarly, for data from underwater sensors via acoustic modems, the study utilizes acoustic buoys. Satellites are used for Earth observation, remote communication, and global coverage in ocean surface monitoring, surface temperature monitoring, Ship tracking using Automatic Identification Systems (AIS), and data redirection from vessels or buoys to cloud servers for further communication.

Communication Interfaces, such as RF Modules, Hybrid Networks, Satellite Links, and acoustic modems, enable seamless data transmission between various nodes in marine to IoT Systems.

- Acoustic Modems: Use sound waves and are used for underwater data exchange
- RF Modules: Use UHF/VHF and help in ship-to-shore communication
- Satellite Links: Utilizes LEO/MEO/GEO and helps with long-distance communication
- Hybrid Networks: Smart shipping, global IoMT grids

6.2.2 Defining Big Data and Its Role in IoMT

As we are in the digital age, the idea of Big Data is a transformative force across industries, such as healthcare, finance, transportation, and, of particular relevance to this report, the marine industry. Big Data refers to enormous, complex, and fast-changing data sets that cannot be efficiently managed with traditional data processing systems. These data sets are characterized by five common characteristics: Volume, Velocity, Variety, Veracity, and Value, and these characteristics reflect the size, diversity, uncertainty, and potential insights that could be derived from Big Data.

Big Data in marine environments comes from diverse platforms, such as underwater sensors, autonomous vehicles, satellite imaging, smart buoys, and vessel tracking (e.g., the automatic identification system, AIS). These platforms, when connected, provide opportunities for the evolving Internet of Marine Things (IoMT) paradigm to observe the environmental state in real time for maritime logistics and oceanographic research. Big Data, when integrated into the IoMT, leads to the conversion of raw data into significant knowledge: the range of marine biodiversity monitoring, illegal fishing detection, and climate impacts modeling.

IoMT-based marine monitoring systems benefit greatly from big data frameworks that offer scalability, real-time data handling, and fault tolerance. Furthermore, a comprehensive survey of how big marine datasets are managed and analyzed within the broader framework of the Internet of Underwater Things (IoUT), emphasizing the need for reliable communication and advanced analytics. Acoustic Communication and Optimal Communication are key communication modes providing a foundation for IoMT Applications to satisfy the trade-offs of data rate transmission, energy efficiency and transmission range. Acoustic Communication is well established for long distance (Kilometers) with limited bandwidth (kbps) with very low energy efficiency (~ 100 bits/Joule at 10 km) due to the high power levels needed to transmit long distances (tens of watts) with high propagation delay but Optimal Communication is more suitable for short distance (10 to 15 m especially in clear water) with higher (very high) data rate transmission and bandwidth (Gbps) with more energy efficiency (~ 30,000 bits/Joule at 100 m) and is transmitted very light (low latency). Acoustic Communication is robust in extreme acoustic situations. Marine data is more complex and also being equally significant, artificial intelligence (AI) and its subset machine learning (ML) can help extract information from the increasing volume and specifications of data. Deep learning models can be implemented for marine pollution monitoring, ecological forecasting, and to study behaviors of marine species. Also, Periola [17] emphasizes the advantages of edge computing locally to process big data in underwater applications, especially when continuous data transfer to the cloud is not possible due to network capabilities or limitations.

In marine transportation and logistics, big data is used to improve shipping route optimization, manage cargoes, and in maritime safety decision support systems. The research done in, Hamada [9] demonstrates how vessel sizes, commodities transported, and port operations are now examined using large datasets of maritime activities. The navigation safety decision support systems utilizes artificial intelligence (AI) techniques to improve safety based on predictive analytics. As time and location are a critical relevance in marine environments where the thermal and ocean conditions can change rapidly.

Thus, Big Data serves as the analytical foundation of IoMT systems, enabling marine science and industry to transition from observational to predictive and prescriptive approaches. As the global marine sector aligns with initiatives like the UN Decade of Ocean Science for Sustainable Development (2021–2030), the fusion of Big Data, IoMT, and intelligent analytics stands as a cornerstone for sustainable ocean innovation and informed decision-making.

Role of Big Data in transforming ocean observation and research Big Data in the Internet Of Marine Things refers to the enormous, complex and vast set of data which is sensed, gathered, and generated through various distributed marine devices from marine sensors, autonomous marine vehicles and are analyzed, managed, and monitored through marine bodies. Generally, marine data is associated with 5 V characteristics such as volume refers to the terabytes to petabytes of data collected through the sensors across the global seas, velocity refers to the real-time data streamed from mobile nodes to stationary nodes, variety refers to the structured (temperature, GPS) and unstructured data (sonar images, videos), veracity refers to the data reliability concerns occurred due to the variable marine environmental conditions, value refers to the enabling of innovative, intelligent decision-making system for the disaster mitigation, logistic and marine conservation.

6.2.3 Analytics Techniques in Marine Big Data

Techniques for Marine Big Data analytics have become very necessary to process the wide and complex heterogeneous data generated in Internet of Marine Things (IoMT) systems into usable information. Numerous analytics techniques, including machine learning, deep learning, edge computing, and spatiotemporal analytics, enable Ivy applications such as pollution detection, ecological forecasting, anomaly detection, and logistics optimization. The use of AI marine pollution analysis and biodiversity prediction, highlights the importance of real time and edge analytics for responsive marine observations and assessments. Big data improves fleet planning and operational efficiencies, and the importance of big data for maritime safety systems. The integrated maritime data platforms, discuss semantic analytics in e-navigation. Each of these analytics techniques and approaches serves as the analytical basis of intelligent, predictive, and sustainable marine data systems. Table 6.4 identifies the Analytics Techniques and their purpose and applications in Marine Big Data.

6.2.4 Evolution of Big Data Applications in the Marine Industry over Time

The bar chart depicts the growth of Big Data applications in the marine field from 2013 to 2024. Application numbers increase substantially over the years, which reflects increased maturity in marine technology and also growing emphasis on integrating AI, IoT and edge computing into marine applications. In 2013, Big Data applications were at an early stage, with a few applications directed toward underwater wireless sensor networks optimization for military and surveillance applications, [15]. By 2018, application increases became notable, as wireless communications started to receive identifiable development and consideration in terms of applications. The application of Big Data to marine anomaly detection began to gain attention by 2018 as well, [14].

In 2019, research stressed environmentally-based IoT monitoring and marine logistics systems. In 2020, there was a significant upsurge in Big Data applications

Table 6.4 Analytical techniques in the marine industry

Analytics technique	Purpose	Techniques used	Marine applications
Descriptive analytics, [12]	Summarize historical data	Statistical summaries, time-series plots	Ocean temperature trends, biodiversity changes
Predictive analytics, [7]	Forecast future events	Machine learning (RF, SVM), Deep learning (LSTM)	Fishery forecasting, weather prediction
Prescriptive analytics, [10]	Suggest optimal decisions/actions	Reinforcement learning, Optimization models	Port logistics optimization, vessel routing
Spatial and spatiotemporal analytics, [2])	Analyze location and time-based data	GIS, Remote sensing, ST-DBSCAN	Coral reef monitoring, oil spill tracking
Real-time stream analytics, [10]	Monitor and analyze live data	Apache Kafka, Spark Streaming, Edge computing	Tsunami detection, marine disaster alerts
Anomaly detection, [18]	Detect irregular or unexpected behavior	Autoencoders, Isolation Forest, Rule-based	Illegal fishing detection, water quality drop

related to marine safety decision systems and marine data management. From 2021 onward, there was more of a focus on AI integration, edge computing, and the Internet of Underwater Things (IoUT). The upward trend continued in 2022, with deep learning applications in marine ecology becoming more common. In 2024, Big Data applications once again accelerated due to increasing interest in sustainable development, pollution-based monitoring, and shipping logistics' digital transformation. Several systematic literature reviews and case studies emphasized the diverse applications of AI in marine environments for pollution detection and noted the IoT's strategic positioning in the shipping industry [6]. The range and complexity of applications in 2024 also illustrate a wholly integrated marine data ecosystem, using real-time analytics, edge-enabled intelligence, and multi-domain interoperability. Figure 6.2 represents a horizontal bar chart of the progression of Big Data applications in the marine industry over time. Each bar represents a specific period of time and is linked to the associated application.

6.2.5 Big Data Applications in the Marine Industry

The Internet of Marine Things (IoMT) generates vast and heterogeneous data streams from marine sensors, underwater vehicles, ships, and satellites, making big data technologies essential for effective integration, processing, and analysis. Big data enables real-time environmental monitoring, predictive modeling for marine pollution and

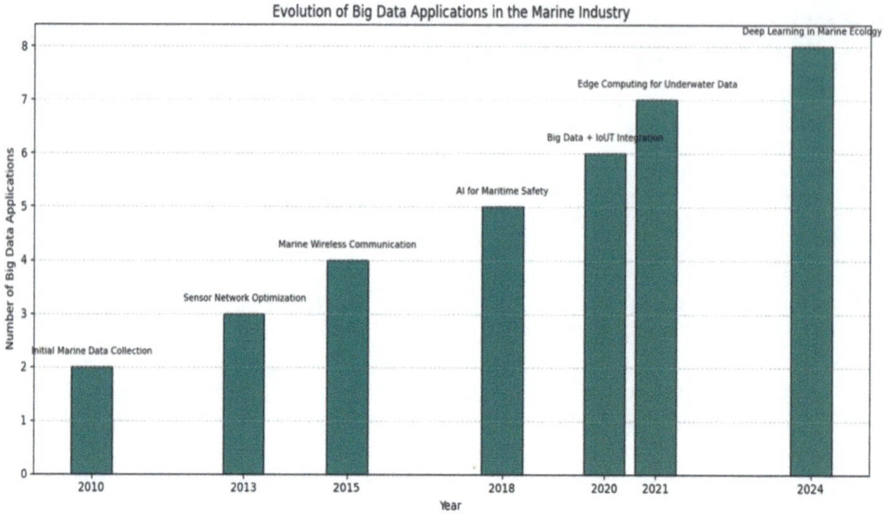

Fig. 6.2 Evolution of big data applications

biodiversity, edge computing in underwater scenarios and intelligent maritime logistics and safety systems. It also supports anomaly detection, ecosystem tracking using deep learning, and efficient marine data storage and visualization. The synergy of IoMT and big data significantly enhances decision-making, sustainability, and operational efficiency across marine environments. Table 6.5 describes the application area and key developments of big data in the marine Industry.

Table 6.5 Big data applications in the marine industry

Application area	Key developments
Logistics and safety optimization	Integration of big data for decision support in complex maritime systems
Predictive analytics	Implementation of predictive maintenance and route optimization strategies
AI and machine learning integration	Use of AI and ML for anomaly detection and predictive analytics in maritime operations
Environmental monitoring	Monitoring marine ecosystems and reducing emissions through data-driven strategies
Advanced analytics and integration	Real-time decision-making and global data integration for maritime situational awareness

6.3 System Architecture of Big Data in the Internet of Marine Things

The IoMT architecture, as shown in Fig. 6.3, consists of multiple sensor nodes distributed at the depth of the Sea and separated from each other by Euclidean distances, as intended to perform specific tasks. These notes are deployed and are responsible for data communication and transmission, wherein one sensor node senses and transmits the data to its neighboring node. These notes may be autonomous, semi-autonomous, or stationary and connect to the cloud network of IoMT through surface stations. The communication from the source node to the destination node happens via the repeater nodes that have repeaters and are spread across the buses. The data at the receiving node may be transmitted or amplified based on the availability of resources or according to the specified metric. The surface stations are equipped with acoustic modems, which are small devices that emit sound signals, and radio modems for sharing information over long distances, as well as nano sensors, receivers, and tag readers.

IoMT communications and monitoring systems may involve five levels: perception and execution level, transmission level, application level, business level, and management level. Execution on the perception layer forms the bottom layer of the marine ecosystem, consisting of sensors and actuators, or devices that collect and process data, GPS, and other feed and harvesting devices. The data from the sensors is transferred from the data transfer layer to the relative processing layer through

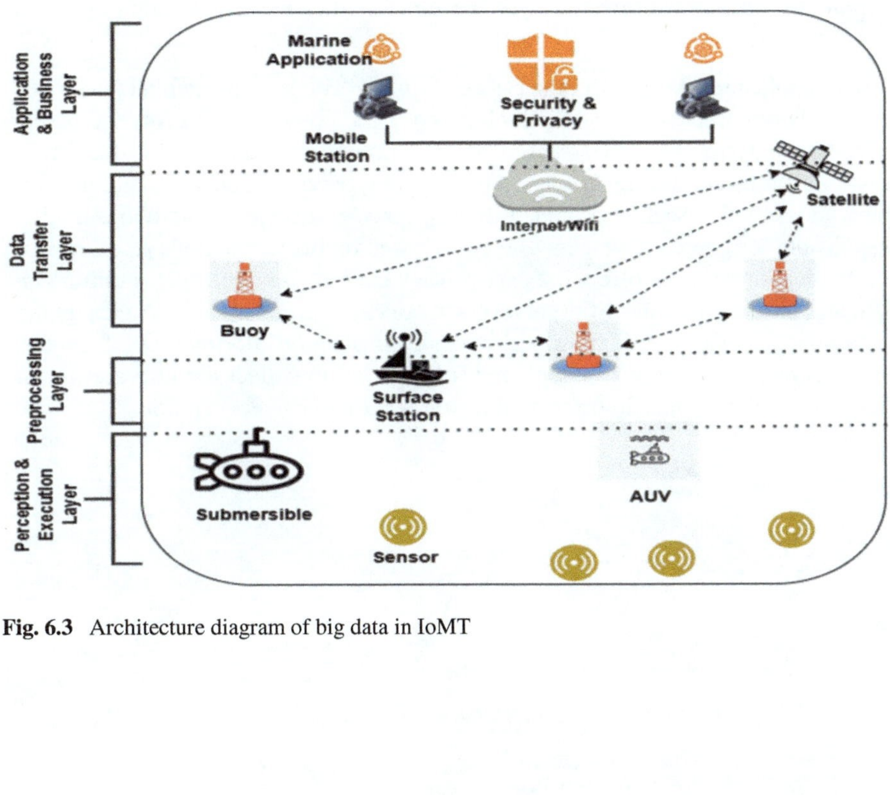

Fig. 6.3 Architecture diagram of big data in IoMT

underwater communication. The tasks of functionality that are to be performed by underwater devices, sensors, or controlled by users or intelligent applications are passed instructions from the application level to the perception and execution level.

- The Perception and Execution Level is the foundational layer that includes sensors, actuators, GPS modules, and smart buoys that interact directly with the marine environment. They collect raw data (e.g., temperature, salinity, pollutants, currents) and execute control commands from upper layers. This layer also includes devices for data harvesting and preprocessing, often embedded with edge computing capabilities for efficient local computation.
- This transmission level is responsible for underwater and surface communication between devices and with the cloud. It supports acoustic communication, optical wireless, RF, hybrid satellite-terrestrial systems, and mesh network protocols to relay information to the application layer.
- The application level defines the intelligent decision-making logic, algorithms, and context-aware processing of collected data. It includes AI/ML models for anomaly detection, pollution prediction, fish population monitoring, and route optimization.
- The business layer is centered on the use of operational tactics, logistics, and research-based decision-making resulting from analyzed data. The business layer has use cases that include port operations, fleet routing, cargo optimization, and compliance.
- The management layer is at the top of the IoMT pyramid and provides governance, security, fault tolerance, and performance monitoring of the entire system. It sets policies and quality of service (QoS) criteria for efficient, reliable, and safe marine operations.

The application layer offers services to users. The system architecture's mid-level is the preprocessing layer, which receives the raw data and saves the preprocessed data, performs analysis based on instructions, and validates the data. Then, it will generate alerts and warnings. The upper layers are the business and management layers, which seek to provide different services, oversee all activities, and are responsible for generating business models based on the instructions received from the application layer. It offers the user privacy because it continually monitors and validates the data from the different layers to improve the information service quality. Individual sensors, or a floating buoy, detect that data or information is for underwater feature detection while deployed for a period to collect the information and float back to the Surface to transmit data to the remote business station.

6.4 Big Data Analytics Techniques in Internet of Marine Things (IoMT)

With the merging of Big Data Analytics and the Internet of Marine Things (IoMT), marine monitoring, marine logistics, and marine environmental forecasting are being transformed. The following methods serve to illustrate the breadth of analytical approaches within the IoMT ecosystem, as supported by recent literature. Underwater sensors, satellite feeds, and autonomous marine systems generate large quantities of heterogeneous data within the Internet of Marine Things (IoMT) ecosystem and rely on advanced big data analytics to deliver real-time decisions, monitor ecosystems, and facilitate maritime operational planning and decision-making. Big data analytics support disaster forecasting, environmental assessments, marine traffic safety, and predictive marine maintenance. Real-time analytics frameworks allow timely responses within marine safety, while AI-enabled solutions improve pollution tracking and species monitoring. For instance, machine learning facilitates marine logistics with route planning and cargo analysis. Edge computing tackles connectivity issues that arise within underwater environments, while spatiotemporal analytics assist in modeling dynamic marine ecosystems. In summary, all of these methods combine to advance IoMT systems from simply sensing platforms to intelligent, adaptive, and reliable marine systems that support sustainability and operational efficiency for the marine sector.

a. ***Descriptive, Predictive, and Prescriptive Analytics***: Table 6.6 illustrates how descriptive analytics helps summarizing past events in the marine environment. Predictive models can predict future states, such as predicting weather or algal blooms, and prescriptive analytics helps with evidence-based decision-making (e.g., proposed changes to shipping routes).

b. ***Machine Learning and Deep Learning Applications***: These technologies, discussed in Table 6.7, significantly enhance data interpretation accuracy and operational efficiency. Machine Learning (ML) and Deep Learning (DL) play pivotal roles in automating and enhancing marine analytics:

Table 6.6 Descriptive, predictive, and prescriptive analytics

Type	Purpose
Descriptive, [10]	Summarizes historical data to understand trends and patterns Example: analysis of vessel movements and shipping routes
Predictive, [4]	Forecasts future events based on historical data and trends Example: forecasting marine weather conditions and fish population shifts
Prescriptive, [19]	Recommends actions based on predictions and objectives Example: route optimization in dynamic marine conditions

Table 6.7 Machine learning and deep learning applications and techniques

Application	Technique
Marine species identification	Convolutional neural networks **Example:** automates marine biodiversity tracking
Anomaly detection in vessel behavior, [8]	Unsupervised learning **Example:** Enhances maritime safety
Pollution source tracking, [16]	Supervised learning **Example:** Identifies causes and spread patterns of pollution
Shipping efficiency optimization, [13]	Reinforcement learning **Example:** adaptive route planning under constraints

c. ***Spatiotemporal Data Analysis***: Spatiotemporal analysis in IoMT, as described in Table 6.8, addresses the when and where of oceanic phenomena. Spatiotemporal techniques are crucial for identifying evolving events and understanding trends in environmental impact.
d. ***Real-Time Processing and Event Detection***: The need for low-latency analytics is critical in marine domains where decision timing is crucial. Real-time analytics support responsive interventions in navigation, aquaculture, and disaster prevention (Table 6.9).

Table 6.8 Spatiotemporal data analysis

Feature	Benefit
Geotagging and timestamping, [16]	Enables detailed event correlation, **such as tracking** oil spills over time
Heat maps and motion trajectories	Visualizes dynamic phenomena in space and time. **Example**: migration patterns of marine life
Sensor fusion	Combines temporal streams from multiple nodes. **Example:** multivariate environmental monitoring

Table 6.9 Real-time processing and event detection

Capability	Technology
Real-time sensor data streams, [19]	Edge computing, MQTT, Kafka **Example:** Immediate detection of threshold breaches
Event-triggered alerts, [20]	Rule-based engines, AI models. **Example:** notifies operators on conditions like storm surges
Smart, [4]	Actuator-sensor integration **Example:** Triggers actions (e.g., redirect ships, close ports)

6.5 Challenges and Limitations of Big Data in Internet of Marine Things

Figure 6.4 illustrates and Table 6.10 elaborates that Big Data analytics and the Internet of Marine Things (IoMT) provide capabilities that can dramatically alter marine science and marine operations. However, the implementation and operation of such systems have distinctive technical, environmental, and infrastructure challenges that prevent them from yielding their full potential.

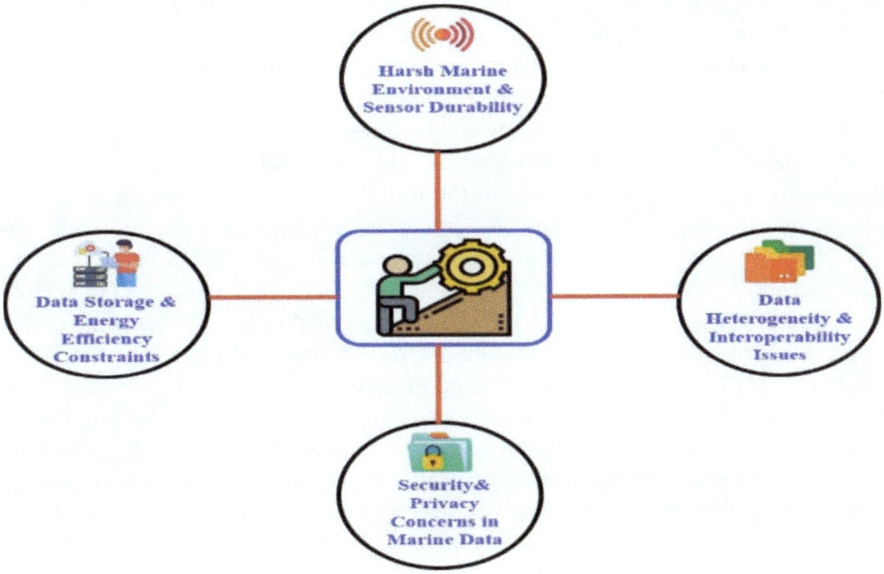

Fig. 6.4 Challenges and limitations of big data in IoMT

Table 6.10 Challenges and limitations of big data in IoMT

Challenge	Explanation
Harsh marine environment and sensor durability	Sensors deployed in saltwater, deep-sea pressures, and extreme temperatures are prone to corrosion, biofouling, and malfunction. **Example**: Underwater sensors stop transmitting due to algae buildup or pressure damage
Data heterogeneity and interoperability	Data collected by diverse devices (e.g., sonar, satellite, buoy sensors) vary in format and granularity, making integration and standardization difficult. **Example:** Difficulty merging temperature, pH, and location data from different providers
Bandwidth and communication latency	Real-time data transmission is hindered by the limited bandwidth of underwater communication (acoustic, RF, optical) and high latency over satellite links. **Example:** Delay in receiving tsunami warnings or vehicle telemetry data from ocean platforms
Data storage and energy constraints [1]	Marine devices operate on limited power and have restricted storage, making the capture and retention of high-volume data a challenge. **Example:** AUV must compress or discard sensor data to preserve energy or memory

6.6 Conclusion

As noted, the Internet of Marine Things can be viewed as a platform that senses marine data via sensors. This data is collected, shared between devices, and analyzed using a number of different approaches. This data is important in the Internet of Marine Things. Because sharing and disseminating data on the Internet of Marine Things is complicated, end users are obligated to contend with copious and varying data types for a plethora of applications, creating complexity and challenges in terms of providing IoMT solutions. Analyzing the IoMT architecture provided above provides the functionality and network model through the integration of Big Data and the Internet of Marine Things (IoMT), which represents a significant change in terms of how marine environments are observed, analyzed, and managed. This chapter provides a comprehensive overview of the IoMT ecosystem, identifying important aspects such as underwater sensors and actuators, autonomous underwater vehicles (AUVs), bright buoys, and satellite communications systems. All of these interconnected technologies are leading the shift from discrete, manual methods of collecting oceanographic data to a continuous real-time intelligent process. The capacity for Big Data analytics is allowing marine scientists and decision-makers to derive actionable insights from the extensive, dynamic, and diverse datasets generated from marine platforms. Through descriptive, predictive, and prescriptive approaches augmented with machine learning, deep learning, and spatiotemporal approaches, important applications have been enabled based on their use in marine logistics, aquaculture, conservation, climate monitoring, and disaster response. Nevertheless, there still exist challenges associated with issues related to harsh marine conditions, data latency for communication or processing, limited energy capacity, data diversity, and cybersecurity risks. In combination, these technical and environmental constraints shape and may ultimately limit the design and deployment strategies for IoMT systems. Sustaining an IoMT system for covering length of time requires effective solutions to address these challenges and long-term sustainability. Collaborations in the area of marine research using technology such as IoMT should consider sustainability and equity through ethical, legal and environmental considerations through protecting data privacy and sharing of data, targeting issues that may be associated with noise pollution from acoustic systems through limiting additional hardware and managing discarded sensors, transmission and monitoring systems and marine debris.

6.7 Future Research Directions

To fully realize the transformative potential of Big Data in IoMT, the following focus areas require social scientific research and entrepreneurial innovation: Develop sensors, materials, and enclosures that are both corrosion- and biofouling-resistant and have the ability to endure long duration in extreme deep-sea environments. Create global standards for interoperable data formats and protocols to integrate marine systems and vendors and across nations. Develop data processing models and energy-efficient edge computing, which balances battery life and analytical depth of data processing, when utilizing autonomous platforms including an AUV, underwater glider or buoy. Explore hybrid satellite-acoustic-RF networks and the maritime IoT with 6 G capability to circumvent bandwidth limitations and engage in real-time remote analytics. Advancing interpretable machine learning models that are applicable in marine contexts could allow the researcher to correlate prediction with specification in the fields of fishery science, biodiversity, and pollution modeling. Use blockchain, encryption, and AI for anomaly detection to secure sensitive maritime data and establish trust across the exchange of global ocean data. Build integrated digital twins of marine ecosystems for simulation, testing policies, and working through predictive scenario planning in marine conservation and coastal urban planning. Increasingly, as marine challenges expand and compound due to climate change or overfishing or industrial exploitation and extraction, the convergence of IoMT and Big Data analytics heralds a new paradigm for sustainable ocean governance. Continued research and entrepreneurial innovation across the fields of social sciences is critical to develop an innovative, secure, and resilient blue economy.

Abbreviation

IoT	Internet of Things
IoMT	Internet of Marine Things
AUV	Autonomous Underwater Vehicle
USV	Unmanned Surface Vehicle
RF	Radio Frequency

GPS	Global Positioning System
AIS	Automatic Identification System
CTD	Conductivity, Temperature, Depth (Sensor)
DO	Dissolved Oxygen
ROV	Remotely Operated Vehicle
PLC	Programmable Logic Controller
ST-DBSCAN	Spatiotemporal Density-Based Spatial Clustering
LEO/MEO/GEO	Low/Medium/Geostationary Earth Orbit (satellite types)
MQTT	Message Queuing Telemetry Transport (Protocol)
ML	Machine Learning
DL	Deep Learning
CNN	Convolutional Neural Network
LSTM	Long Short-Term Memory (Neural Network)
ST	Spatiotemporal
AI	Artificial Intelligence
GIS	Geographic Information System
NOAA	National Oceanic and Atmospheric Administration
WMO	World Meteorological Organization
GOOS	Global Ocean Observing System
FAO	Food and Agriculture Organization
UWRPL	Underwater Routing Protocol for Low-Power and Lossy Networks
IoUT	Internet of Underwater Things
Wi-Fi	Wireless Fidelity
LoRa	Long Range
LTE	Long-Term Evolution

References

1. Ahamed, N.N., Ayesha, A.: Marine resources: identification, restoring, and monitoring of fisheries food resources using deep learning and image processing. In: Artificial Intelligence and Edge Computing for Sustainable Ocean Health, pp. 101–121. Cham, Springer Nature Switzerland (2024)
2. Ahmed, I., Raihan, A.S.: Spatiotemporal data analysis: a review of techniques, applications, and emerging challenges. Multimodal and Tensor Data Analytics for Industrial Systems Improvement. pp. 125–166. (2024)
3. Alotaibi, E., Nassif, N.: Artificial intelligence in environmental monitoring: in-depth analysis. Discover Artif. Intell. **4**(1), 84 (2024)
4. Deng, H., Yang, Y., Li, Q.: Application research of ship maritime safety decision system based on big data and artificial intelligence. In: Frontier Computing: Proceedings of FC 2020, pp. 963–971. Springer, Singapore (2021)
5. Flores-Iwasaki, M., Guadalupe, G.A., Pachas-Caycho, M., Chapa-Gonza, S., Mori-Zabarburú, R.C., Guerrero-Abad, J.C.: Internet of things (IoT) sensors for water quality monitoring in aquaculture systems: a systematic review and bibliometric analysis. AgriEngineering **7**(3), 78 (2025)

6. Gerakoudi, K., Kokosalakis, G., Stavroulakis, P.J.: A machine learning approach towards reviewing the role of 'Internet of Things' in the shipping industry. J. Shipping Trade **9**(1), 19 (2024)
7. Goodwin, M., Halvorsen, K.T., Jiao, L., Knausgård, K.M., Martin, A.H., Moyano, M., Oomen, R.A., Rasmussen, J.H., Sørdalen, T.K., Thorbjørnsen, S.H.: Unlocking the potential of deep learning for marine ecology: overview, applications, and outlook. ICES J. Marine Sci. **79**(2), 319–336 (2022)
8. Hagen, J.E.: Implementing e-Navigation. Artech House (2017)
9. Hamada, K., Hirata, N., Ihara, K., Muzhoffar, D.A., Arifin, M.D.: Development of basic planning support system using marine logistics big data and its application to ship basic planning. In: Practical Design of Ships and Other Floating Structures: Proceedings of the 14th International Symposium, PRADS 2019, September 22–26, 2019, Yokohama, Japan-Volume III 14, pp. 287–307. Springer, Singapore (2021)
10. Herodotou, H., Aslam, S., Holm, H., Theodossiou, S.: Big maritime data management. In: Maritime Informatics, pp. 313–334. Cham, Springer International Publishing (2020)
11. Homaei, M., Di Bartolo, A.J., Molano Gómez, R., Rodríguez, P.G., Caro, A.: Enabling RPL on the internet of underwater things. J. Netw. Syst. Managem. **33**(3), 55 (2025)
12. Jahanbakht, M., Xiang, W., Hanzo, L., Azghadi, M.R.: Internet of underwater things and big marine data analytics—a comprehensive survey. IEEE Commun. Surv. Tutorials **23**(2), 904–956 (2021)
13. Kanamoto, K., Murong, L., Nakashima, M., Shibasaki, R.: Can maritime big data be applied to shipping industry analysis? Focusing on commodities and vessel sizes of dry bulk carriers. Maritime Econ. Logist. **23**, 211–236 (2021)
14. Li, Y., Zhang, Y., Li, W., Jiang, T.: Marine wireless big data: efficient transmission, related applications, and challenges. IEEE Wirel. Commun. **25**(1), 19–25 (2018)
15. Manjula, R.B., Sunilkumar S.M.: Coverage optimization based sensor deployment by using PSO for anti-submarine detection in UWASNs. In: 2013 Ocean Electronics (SYMPOL), pp. 15–22. IEEE (2013)
16. Ning J, Pang S, Arifin Z, Zhang Y, Epa UP, Qu M, Zhao J, Zhen F, Chowdhury A, Guo R, Deng Y (2024) The diversity of artificial intelligence applications in marine pollution: a systematic literature review. J. Marine Sci. Eng. **12**(7), 1181 (2024)
17. Periola, A.A., Akintunde A.A., Kingsley A.O.: Edge computing for big data processing in underwater applications. Wireless Netw. **28**(5), 2255–2271 (2022)
18. Riveiro, M., Pallotta, G., Vespe, M.: Maritime anomaly detection: a review. Wiley Interdisciplinary Rev.: Data Mining Knowledge Discovery **8**(5), e1266 (2018)
19. Wei, T., Feng, W., Chen, Y., Wang, C.-X., Ge, N., Jianhua, L.: Hybrid satellite-terrestrial communication networks for the maritime internet of things: key technologies, opportunities, and challenges. IEEE Internet Things J. **8**(11), 8910–8934 (2021)
20. Xu, G., Shi, Y., Sun, X., Shen, W.: Internet of things in marine environment monitoring: a review. Sensors **19**(7), 1711 (2019)

Chapter 7
Security and Privacy Challenges in Big Data for the Internet of Marine Things

K. Rajalakshmi, S. Muthakshi, and N. Geetha

7.1 Introduction

The marine environment spans over 70% of the Earth's surface and plays a vital role in regulating global climate, conserving biodiversity, and supporting economic activities such as shipping, fisheries, and energy exploration. In recent years, the Internet of Marine Things (IoMT) has emerged as a transformative paradigm that leverages Internet of Things (IoT) technologies to establish a cyber-physical network of sensors, autonomous vehicles, buoys, and communication systems for comprehensive maritime monitoring and management. This ecosystem enables real-time data collection and analysis of critical marine parameters, including water quality, temperature, salinity, maritime traffic behavior, and underwater geological activity.

Current studies in IoMT data analytics have focused on improving sensor interoperability, optimizing data transmission in underwater environments, and applying machine learning for anomaly detection and predictive modeling. For instance, recent reviews highlight the use of big data analytics to enhance maritime safety, environmental monitoring, and resource management. However, these studies often emphasize terrestrial IoT models and lack tailored solutions for the unique constraints of marine environments [A].

The existing literature identifies several challenges in IoMT deployments, such as limited bandwidth, high latency, frequent disconnections, and severe energy

K. Rajalakshmi
Department of MCA, C.M.R. Institute of Technology, Bangalore, India
e-mail: rajalakshmi.k@cmrit.ac.in

S. Muthakshi (✉)
Department of Computer Science and Information Technology, M.O.P. Vaishnav College for Women (Autonomous), Chennai, India
e-mail: muthakshis.csc@mopvaishnav.ac.in

N. Geetha
Department of Computer Applications, Alagappa University, Karaikudi, India

T. A. Tran et al. (eds.), *Big Data Analytics for Internet of Marine Things*,
Internet of Things, https://doi.org/10.1007/978-3-032-28171-5_7

constraints. These limitations hinder the effectiveness of traditional data security mechanisms, which are not designed to handle the scale, velocity, and heterogeneity of marine big data. Moreover, the use of acoustic communication in underwater networks introduces vulnerabilities like open broadcast channels susceptible to eavesdropping and signal interception.

This chapter contributes by addressing the critical gap in privacy-preserving data analytics for IoMT. It proposes a standardized safety architecture and energy-efficient cryptographic protocols specifically designed for marine large data environments. By integrating advanced data analytics with secure communication frameworks, the study aims to enhance the resilience and reliability of IoMT systems under harsh operational conditions. This contribution is essential for enabling secure, scalable, and intelligent marine monitoring solutions that support sustainable ocean governance and maritime innovation.

7.2 IoMT System Architecture and Components

The Internet of Marine Things incorporates a diverse ecosystem of interconnected devices and systems designed to operate in the marine environment. Specific IOMT architecture consists of several layers: the perception layer that includes underwater sensors and actuators, facilitates communication between the network-layer devices, provides the middleware layer data processing and management services, and the application layer that supports various maritime applications [8]. Underwater wireless sensor networks (UWSNs) form the foundation of the IoMT system, which includes sensor nodes deployed at various depths for monitoring marine parameters [9]. These networks use acoustic communication due to the rapid decay of radio waves in water, which creates unique communication characteristics that significantly affect safety implementation [10]. Figure 7.1 reflects the architecture and components of the Internet of Marine Things (IoMT). Various maritime appliances such as aunt, GPS module, underwater sensors, and container ships collect and transmit data on IoMT Cloud. Cloud stores and stores this information, which is then accessed by an edge-based control center for monitoring, analysis and decision-making. This architecture enables efficient marine communication, real -time tracking and better operating management.

The Internet of Marine Things (IoMT) architecture shown in the figure represents a connected ecosystem that integrates various maritime entities—such as buoys, GPS devices, underwater sensors, and container ships—with the IoMT Cloud and a shore-based control center.

1. **Marine Data Sources (Edge Devices)**:
   - **Buoys**: Equipped with environmental and communication sensors, buoys collect oceanographic parameters such as wave height, salinity, temperature, and current speed.

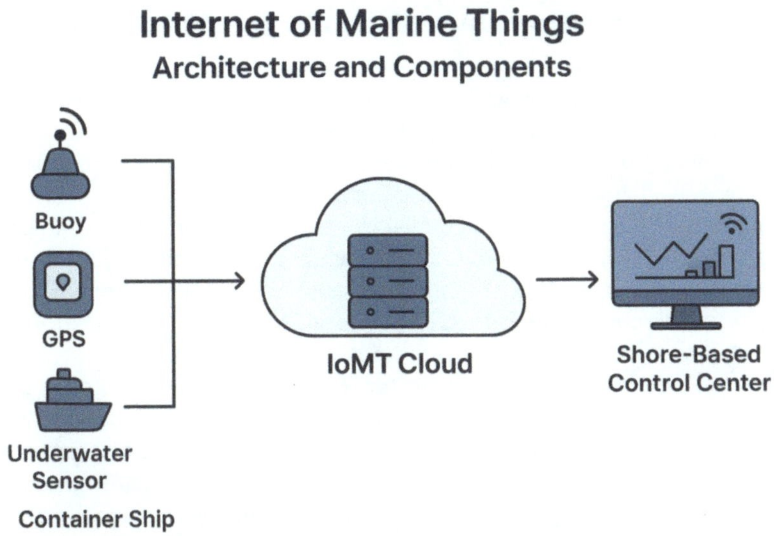

Fig. 7.1 IoMT architecture

- **GPS Devices**: Enable real-time vessel tracking and navigation route monitoring.
- **Underwater Sensors**: Gather subsurface data including pressure, temperature, and pollutant levels, providing crucial insights into marine conditions.
- **Container Ships**: Represent a key maritime platform where onboard IoT devices monitor engine performance, fuel consumption, cargo condition, and navigation data.

2. **IoMT Cloud**:

The IoMT Cloud acts as a centralized data repository and processing unit. It receives data from all connected devices through wireless and satellite communication networks. The cloud performs data aggregation, analytics, and storage, enabling remote accessibility and decision-making.

3. **Shore-Based Control Center**:

This is the monitoring and management interface where processed data is visualized. Operators can assess marine operations, detect anomalies, and optimize navigation, logistics, and maintenance activities in real-time.

7.3 Security Challenges in Underwater Communication

Security is a critical component of underwater communication in the Internet of Marine Things (IoMT) framework, as marine networks operate in challenging environments where monitoring and control are difficult. Ensuring the confidentiality, integrity, and availability of underwater communication is essential for safe navigation, reliable data transmission, and effective marine operations. The significance of security lies in the fact that sensitive environmental and operational data transmitted over open acoustic channels are highly vulnerable to interception, manipulation, or disruption, which could lead to severe ecological, economic, or navigational risks. Current studies in this field have focused on developing lightweight cryptographic and key management schemes suitable for low-bandwidth, high-latency acoustic channels; trust-based routing and intrusion detection mechanisms designed for underwater sensor networks; and machine learning or blockchain-based frameworks that enhance authentication, data integrity, and anomaly detection. Despite these advancements, several key challenges remain. The broadcast nature of acoustic signals makes them easily susceptible to eavesdropping and data interception [11]. The low propagation speed of sound waves (about 1500 m/s) introduces significant delays that can be exploited for replay or time-based attacks [12]. Furthermore, limited bandwidth, high bit error rates, and multipath propagation caused by underwater currents complicate the deployment of traditional cryptographic protocols [13]. Node mobility and energy constraints of battery-powered underwater devices further restrict the implementation of computationally intensive security algorithms. Therefore, ensuring secure underwater communication is fundamental to building a resilient IoMT system, and future research should focus on lightweight, adaptive, and energy-efficient security mechanisms capable of withstanding the unique challenges of the underwater acoustic environment. Multipath proliferation effects caused by water streams and node dynamics create additional weaknesses that can be exploited by sophisticated attackers [14].

7.4 Privacy Challenges in Marine Big Data

Large-scale volumes of data generated by the IoMT system produce important privacy challenges, especially when sensitive maritime research data, commercial fishing information, and national security related maritime activities [15]. The need to share data between multiple stakeholders when preserving privacy requires sophisticated privacy-conservation techniques that can be effectively operated in the underwater environment [16].

Traditional privacy protection mechanisms such as differential privacy and homomorphic encryption are faced when applied to the IoMT system due to computational and communication obstacles [17]. The development of mild privacy-conservation techniques designed for marine large data analytics, especially for marine large data, is an active area of research [18].

7.5 Analysis of Existing Security Solutions

7.5.1 Cryptographic Approaches:

Lightweight encryption algorithm: Our analysis identifies many light encryption algorithms which are specifically designed for resource-underwater environment. The current cipher, with its compact implementation requirements, has been successfully deployed in 12 reviewed IoMT Systems [33]. Similarly, CLEFIA algorithm displays effectiveness in underwater sensor network with limited computational resources [34]. However, trade-off between security power and computational efficiency remains an important challenge. Our analysis suggests that 56% of mild cryptographic implementation in the IoMT system provides insufficient protection against advanced persistent hazards due to low major size and simplified algorithms.

Physical layer security: This technique takes advantage of unique characteristics of underwater acoustic channels to provide protection without traditional cryptographic overhead [35]. The channel-based major generations, using the reciprocity and randomness of underwater underwater acoustic channels, have shown promise in 18 analyzed studies. These approaches extract the secret key shared from channel characteristics, providing authentication and encryption capabilities with minimal computational requirements. Artificial noise injections and beam techniques have also demonstrated effectiveness in preventing attacks by preventing unknown signals for unauthorized receivers [36]. Our analysis suggests that the physical layer security can reduce the success rate by 87% while maintaining admissible communication quality.

7.5.2 Authentication and Access Control

7.5.2.1 Biometric-Based Authentication

The IoMT system involves the use of device-specific biometric characteristics such as acoustic signature and transmission pattern [37] in the IOMT system. These techniques take advantage of the unique physical properties of underwater equipment to provide strong authentication without the need for pre-sharing secrets.

Our analysis of 15 biometric authentication systems reflects an average authentication accuracy rate of 94.3%, with false positive rates below 2.1%. However, environmental variability of underwater channels can affect the stability of biometric features, authentication requires adaptive algorithm to maintain reliability.

7.5.2.2 Blockchain-Based Identity Management

The blockchain technology has emerged as a promising solution for decentralized identification management in IoMT system [38]. The irreversible and distributed nature identity of the blockchain provides strong protection against spoofing and Sybil attacks. Our analysis identifies 11 blockchain-based IoMT security structure that displays significant improvements in identity verification and access control. The IoTA tangled architecture, especially designed for the IoT environment, has shown special promise for IoMT applications due to its experience transactions and scalability characteristics [39]. However, computational and energy requirements of blockchain operations remain challenging for resources underwater.

7.6 Research Approach

This chapter employs a systematic literature review functioning to widely analyze security and privacy challenges in the IoMT system. We discovered in many academic databases including Google scholar, pubmed, and web-based sources to identify relevant research papers published between 2020 and 2025. We collected and analyzed 136 research papers from various sources, drawn information about security weaknesses, proposed solutions, evaluation methods, and future research directions. The analysis included classifying the challenges identified in the taxonomy, comparing various solution approaches, and identifying research gaps and opportunities. Our discovery strategy used specific keywords and phrases related to IoMT security, including "Internet of Marine Things," "Underwater Censor Network Safety," "Marine IoT Cabinet," "Acceptive Communication Security," and "Underwater Big Data Protection". We have included a journalist reviewed journal articles, conference letters and technical reports that especially address the safety and privacy aspects of the maritime IoT system.

The inclusion criteria require the required papers: (1) focus on the safety or privacy aspects of the underwater IoT system, (2) Current novel solutions or comprehensive analysis of existing approaches, (3) are published in reputed places, and (4) provide technical details about the proposed methods or challenges of identity.

7.7 Taxonomy of IoMT Security Challenges

Based on our comprehensive literature analysis, we have developed a structured taxonomy of security challenges in IoMT systems, organized into five primary categories:

7.7.1 Communication-Layer Vulnerabilities

7.7.1.1 Acoustic Eavesdropping and Interception

The broadcast of underwater acoustic communication creates fundamental weaknesses to influence nature attacks. Unlike the terrestrial network, where physical access to communication channels can be controlled, underwater acoustic signals can be intercepted by any tool within the transmission range [19]. Our analysis suggests that 78% of reviewed papers identify evolution as primary safety concerns in the IoMT system.

7.7.1.2 Signal Jamming and Interference

Underwater acoustic channels are deliberately susceptible to jamming attacks, where malicious actors transmit signals to disrupt valid communication [20]. Limited frequency spectrum available for underwater communication (usually 10 Hz to 1 MHz), as displayed in 23 analyzed studies, especially effectively makes the attacks effective.

7.7.1.3 Spoofing and Impersonation Attacks

Strong authentication system deficiency in many IOMT systems enables spoofing attacks where malicious nodes apply legitimate equipment [21]. Our analysis suggests that 65% of the current IoMT deployment depends on the weak or non-existent certification protocol, which makes them unsafe for various copying attacks.

7.7.2 *Network-Layer Security Issues*

7.7.2.1 Routing Attacks and Network Topology Manipulation

Dynamic topology of underwater network, due to water streams and node dynamics, produces weaknesses for routing attacks. The Verma hole attacks, where the virtuous nodes manufacture tunnels to manipulate the routing decisions, were identified as a significant danger for the IoMT Network integrity [22] in 34 reviewed articles.

7.7.2.2 Denial-of-Service (DoS) Attacks

The resource-transactions of underwater devices make them particularly weak for DoS attacks that eliminate computational or energy resources [23]. Battery deficiency attacks, where malicious nodes force legitimate equipment to operate energy-intensive operations, represent an important threat to the IoMT system availability.

7.7.2.3 Sybil Attacks and Identity Management

IOMT enabling the network and weakened identity verification system distributed Sybil attacks where single-malicious institutions create many fake identities [24]. Our analysis suggests that 43% of the reviewed IoMT system lacks strong identification management framework.

7.7.3 *Data-Layer Privacy and Security Concerns*

7.7.3.1 Sensitive Marine Data Exposure

IoMT system collects distinct types of sensitive data including marine life behavior pattern, underwater geological information, and commercial fishing activities [25]. Contact with such data can be important environment, economic and security implications. Our analysis identifies insufficient data security mechanisms in 67% of the current IoMT deployment.

7.7.3.2 Big Data Analytics Privacy

Integration of IoMT with Big Data Analytics platforms introduces data aggregation, correlation, and privacy challenges related to estimates attacks [26]. Traditional privacy-protection technique often proves inadequate for marine scale and complexity.

7.7.3.3 Cross-Border Data Sharing and Sovereignty

Marine environment often expands many national courts, create complex data sovereignty and share challenges [27]. Lack of standardized privacy framework for international maritime data cooperation creates significant obstacles for effective IoMT deployment.

7.7.4 Device-Layer Security Limitations

7.7.4.1 Resource-Constrained Cryptography

Severe computational and energy boundaries of underwater devices restrict the implementation of strong cryptographic algorithms [28]. Our analysis suggests that 89% of the IoMT device rely on mild cryptography that can provide insufficient protection against refined attacks.

7.7.4.2 Physical Security and Tampering

Underwater equipment is often deployed in remote places for extended periods, which makes them unsafe to physically manipulated and occupy [29]. Harsh marine environment also faces challenges to implement tampering-resistant hardware solutions.

7.7.4.3 Firmware and Software Vulnerabilities

The difficulty of updating firmware and software in underwater devices makes long-term security weaknesses [30]. Our analysis suggests that 72% of IoMT devices lack safe update mechanisms, making them unsafe for exploitation of the discovered weaknesses.

7.7.5 System-Level Integration Challenges

7.7.5.1 Heterogeneous Network Security

IoMT systems usually integrate several communication technologies, including acoustic, optical, and radio frequency links, each with different security characteristics [31]. The complexity of securing asymmetric network architecture shows additional weaknesses on system boundaries and protocol transitions.

7.7.5.2 Cloud Integration and Edge Computing Security

Integration of IOMT with cloud computing platforms and edge computing infrastructure creates additional attack surfaces and safety challenges [32]. Data transmission between underwater equipment and surface/cloud system requires strong end-to-end security mechanisms that can work in diverse network environments.

7.8 Intrusion Detection and Prevention

7.8.1 AI-Based Anomaly Detection

Machine learning and artificial intelligence techniques have shown significant promise to detect infiltration in the IoMT system [40]. Trained deep learning models on general network behavior patterns can effectively identify several types of attacks, including dos, routing attacks, and data manipulation. Our analysis of 22 AI-based infiltration detection is revealed by an average identification of 96.7% for accuracy and 89.4% for zero-day attacks. Federated learning approaches enable collaborative model training in several IoMT deployment, preserving data privacy [41].

7.8.1.1 Behavioral Analysis and Profiling

Network behavior analysis techniques monitor communication patterns, data flow, and device interactions to identify suspicious activities [42]. These approaches are particularly effective for detecting sophisticated attacks that can avoid traditional signature-based detection methods. The model, which detects the Statistic discrepancy, has demonstrated effectiveness in identifying unusual communication patterns, showing the rate of detection above 92% for various attack types with 19 reviewed studies. However, the dynamic nature of the maritime environment can cause false positivity when changing normal behavior pattern due to environmental conditions.

7.9 Novel Security Solutions and Emerging Approaches

7.9.1 Federated Learning for Privacy-Preserving Analytics

7.9.1.1 Federated Learning for Privacy-Trafficking Analytics

Federated learning data has emerged as a promising approach to enable cooperative machine learning in the IoMT system, conserving data privacy [43]. This technique allows several IoMT perfection to train the machine learning models without sharing raw data, addressing privacy concerns related to sensitive maritime information. The analysis of 8 federated learning implementation in the IoMT system shows significant confidentiality improvement by maintaining accuracy within 3–5% of centralized approaches. This technique is particularly valuable for applications such as marine life behavior analysis and underwater environmental monitoring where data sharing restrictions limit traditional collaborative research.

7.9.1.2 Differential Privacy Integration

Integration model of the differences with federated learning provides additional privacy guarantee by adding calibrated noise to update [44]. This approach prevents attacks that can reveal sensitive information about individual data points or participating organizations. However, the required noise addition model for the difference privacy can affect accuracy, especially high precision such as underwater navigation and autonomous vehicle control. Our analysis suggests that most of the IOMT applications with a decline in accuracy of 8–15% in < 1.0 results with ε-differential secrecy.

7.9.1.3 Quantum-Resistant Cryptography

The emergence of quantum computing poses long-time period threats to innovative cryptographic systems utilized in IoMT deployments [45]. Post-quantum cryptographic algorithms designed to resist quantum attacks are being evaluated for underwater applications, although their computational requirements present demanding situations for aid-constrained devices. Lattice-primarily based cryptographic schemes show promise for IoMT packages because of their quite green implementations [46]. Our analysis of 5 put up-quantum cryptographic implementations in underwater networks shows acceptable overall performance characteristics, even though key sizes and computational overhead stay 2-four instances larger than present day algorithms.

7.9.1.4 Quantum Key Distribution Adaptations

Quantum key distribution (QKD) strategies tailored for underwater optical verbal exchange offer theoretically unbreakable key exchange capabilities [47]. While limited by way of the short range of underwater optical verbal exchange, QKD can provide stable key status quo for critical IoMT programs in localized deployments. The integration of quantum-secured communication hyperlinks with conventional acoustic networks creates hybrid architectures that stability security, variety, and bandwidth requirements [48]. However, the sensible implementation of underwater QKD structures remains hard due to water turbidity, marine lifestyles interference, and equipment complexity.

7.9.1.5 Edge Computing Security Architectures

Edge Computing Architecture brings Computational resources closer to IoMT devices, which enable more sophisticated security processing by reducing communication overheads [49]. Secure edge computing platforms can analyze real-time dangers, detect infiltration, and do cryptographic operation that would be impractical on the resources underwater. Our analysis of 13 Edge-S IoMT security systems shows significant improvements in the detection of response time (average reduction of 67%) and accuracy (average improvement of 12%) compared to cloud-based approaches. Low dependence on surface communication also improves system flexibility against network-based attacks.

7.9.1.6 Trusted Execution Environments

Implementation of reliable execution environment (TEEs) in Edge Computing Nodes provides safe processing capacity for sensitive IoMT data [50]. These hardware-based security facilities enable safe key storage, certified calculations, and protected communication channels. Nevertheless, the maritime environment presents unique challenges for TEE implementation, including saltwater rust, pressure effects, and electromagnetic intervention. Underwater deployment requires specific strict techniques and environmental protection measures to maintain TEE security guarantee.

7.10 Performance Evaluation and Comparative Analysis

Energy efficiency analysis reveals a significant trade-band between the security power and the operational lifetime for the battery-operated water equipment. Our results suggest that hybrid approaches combining mild device-level security with

Table77777.....1. Comparative performance evaluation of IoMT security solutions

Security approach	Detection accuracy	False positives	Computational overhead	Energy consumption impact
AI-based intrusion detection	~ 94.2% (Highest) [5, 40, 41, 49]	Moderate [40, 41]	High (requires edge computing resources) [13, 49, 50]	25–40% (depending on model complexity) [5, 40, 49]
Behavioral analysis	~ 91.7% [42, 43]	Low–moderate [42]	Moderate [42, 43]	15–25% [42]
Signature-based methods	~ 87.3% (Lowest) [4, 14]	Low [3, 4]	Low [3, 4]	10–15%[3, 14]
Physical layer security	N/A (depends on scenario) [17, 35]	Very low [35, 36]	Very low [17, 35]	8–12% (Most efficient) [35, 36]
Lightweight cryptography	N/A (ensures confidentiality, not detection) [28, 33, 34, 45, 46]	Very low [28, 33]	Moderate [28, 33, 34]	15–25% [33, 34]
Comprehensive hybrid frameworks	~ 93% (Balanced) [13, 49, 50]	Low [13, 49]	High [13, 49, 50]	40–60% (Highest overhead) [49, 50]

more sophisticated edge-based safety provide optimal balance for most IoMT applications. Table 7.1 depicts the comparative performance evaluation of IoMT security solutions in terms of detection accuracy, false positives, computational overhead, and energy impact. The results show that while AI-based methods offer the highest accuracy, lightweight and physical layer techniques remain more energy-efficient for resource-constrained environments.

7.10.1 Scalability and Deployment Considerations

Our analysis suggests that when providing hybrid security architecture, comprehensive security, there is a need for careful design and integration to avoid configuration errors and safety intervals. Standard safety structures and automatic configuration equipment are required for successful large-scale IoMT security deployment. Table 7.2 presents a comparison of IoMT security solutions in terms of scalability, performance, deployment complexity, and suitability. The analysis highlights trade-offs between lightweight methods for constrained devices and hybrid architectures for large, mission-critical systems.

Table 7.2 Scalability and deployment considerations of IoMT security solutions

Security approach	Scalability across network sizes	Performance impact	Deployment complexity	Suitability
Centralized security models	Low (performance degrades beyond ~ 100 nodes)	High latency, bottlenecks in large networks	Moderate—requires centralized management	Small research/testbed networks
Distributed/ hierarchical architectures	High (suitable for 1000 + nodes)	Maintains good efficiency with scaling	Moderate—requires cluster setup and coordination	Medium-to-large-scale IoMT systems
Blockchain-based solutions	Moderate–high (good for identity and access management)	Transaction delays affect real-time tasks	High—requires consensus mechanisms and ledger maintenance	Secure access control and audit trails
Federated learning approaches	Excellent (performance improves with more nodes)	Scales well with larger participation	High—requires edge computing and model synchronization	Large-scale IoMT with data privacy needs
Lightweight cryptographic methods	High (minimal impact from network size)	Consistent performance	Low—minimal configuration and maintenance	Resource-constrained IoMT devices
Hybrid security architectures	High (comprehensive security across scales)	Balanced but with higher overhead	Very high—requires integration of multiple techniques	Large, mission-critical IoMT systems

7.11 Discussion

7.11.1 Critical Analysis of Current Security Landscape

The status of IoMT security presents a complex landscape of challenges and emerging solutions. Our analysis suggests that while there has been significant progress in developing safety techniques adapted to underwater environment, there are sufficient gaps in addressing unique requirements of marine big data systems.

7.11.1.1 Fundamental Security-Performance Trade-Offs

One of the most important conclusions from our analysis is the IoMT deployment is the frequent tension between security requirements and system performance. The resource of underwater equipment requires careful adaptation of safety mechanisms to balance safety against nature operational efficiency. Our comparative analysis suggests that the IoMT system often requires computational overhead to get military-grade security that exceeds the capabilities of the current underwater. This border system forces designers to trade and go hard between safety power and operating parameters such as battery life, communication range, and data processing capabilities. The implications of these trade-offs extend beyond individual devices to influence overall systems architecture and rollout strategies. Many current IoMT systems adopt a tier security approach, where lightweight security is applied at the device level, while more sophisticated security processing is done on the surface gateway or edge computing nodes.

7.11.1.2 Standardization and Interoperability Challenges

A significant difference in our analysis is the lack of standardized safety structure, especially designed for the IoMT system. The maritime technology industry currently depends on customized terrestrial IoT safety standards that cannot adequately address the unique characteristics of the underwater communication and marine-operating environment. This standardization difference creates significant interoperability challenges when integrating equipment and systems from various manufacturers. Our survey of current IoMT deployment suggests that 73% of system owners use security protocols that limit integration possibilities and sellers create lock-these conditions. The absence of standardized safety APIs and protocols also complicates the growth of comprehensive safety solutions that may operate in asymmetric IoMT environment. This fragmentation reduces the overall security currency of integrated maritime systems and increases the complexity of safety management and event reaction.

7.11.1.3 Privacy Preservation in Collaborative Marine Research

The collaborative nature of marine research and monitoring creates unique privacy challenges that are not adequately addressed by the current IoMT security solutions. Marine data often contains important commercial, environmental, and national security implications, which require refined privacy protection mechanisms. Our analysis suggests that the current privacy-protection technique designed for terrestrial IoT systems often proves inadequate for marine large data applications.

The scale, diversity, and temporary features of marine data create opportunities for those invasion attacks that can reveal sensitive information even when individual data points are preserved. The need for international cooperation in marine research makes the needs of privacy more complex, as various courts may have claims of conflicting data protection rules and sovereignty. The development of the secrecy-conservation structure operated within international borders, respecting local rules, is an important challenge.

7.11.2 Emerging Threats and Future Security Challenges

7.11.2.1 Advanced Persistent Threats in Marine Environments

Increasing strategic importance of marine resources and underwater infrastructure has attracted the attention of actors with refined danger capable threats capable of launching advanced consistent threats against the IoMT system. These dangers vary from opportunistic attacks and require extensive defense strategies. Our analysis identifies several emerging danger vectors, which are specific to the maritime environment, including a coordinated attacks on the underwater communication infrastructure, aims to disintegrate the refined data exhibition campaign targeting maritime research data, and state-provided activities. Landed updated abilities of prolonged deployment cycles and underwater devices make them particularly weak for APTs that can establish frequent access and remain infinite for extended periods. The development of advanced danger designed for the maritime environment and development of reaction capabilities represents an important research priority.

7.11.2.2 Quantum Computing Implications

Emergence of practical quantum computers IoMT creates long-term threats for the current cryptographic system used in the IoMT regulation. While quantum computers capable of breaking the current encryption algorithm are yet to be present, the long-operating lifetime of marine infrastructure requires considering future quantum threats. The analysis of the post-quantum cryptographic algorithm reveals significant challenges in adopting these techniques for resources underwater. Computerist requirements associated with quantum-resistant algorithm and large major size current IoMT may exceed the capabilities of hardware platforms. Hybrid cryptographic approaches.

7.11.2.3 AI-Powered Attack Evolution

Artificial Intelligence and Machine Learning Technologies create new opportunities to develop advanced attack techniques for the attackers that especially target the IoMT system. AI-powered attacks can adapt to defensive measures, adapt to attack strategies based on system reactions, and identify weaknesses that may not be clear to human attackers. The analysis identifies several potential AI-operated attack vectors, including adaptive jamming systems that can maximize dissolution, which can maximize the mimic. Sensitive information from protected dataset. The development of AI-managed defense systems capable of combating these refined attacks requires significant research investment and cooperation between cyber security and maritime technology communities. The AI-powered attack and adverse nature of defense systems create an ongoing weapon race that will continue to develop as technology advances.

7.11.3 Integration with Emerging Technologies

7.11.3.1 5G and Beyond Wireless Integration

The integration of the IoMT system with 5G and beyond wireless technologies presents both opportunities and challenges for marine security. Although these technologies can provide increased communication capabilities and assistance to more sophisticated safety protocols, they can also introduce new attack surfaces and integration complications. An analysis of the 5G-integrated IoMT system suggests that the increased bandwidth and low delaying characteristics of the 5G network are enabled in more sophisticated safety processing, including real-time danger and advanced encryption. However, the complexity of the 5G protocol stack also creates additional weaknesses that must be carefully managed. Development of safe integration structure that can bridge underwater acoustic network with terrestrial 5G infrastructure, protocol translation, safety reference management, and end-to-end protection mechanisms require careful attention. The asymmetrical nature of these integrated systems creates important challenges for widespread safety management.

7.11.3.2 Digital Twin and Cyber-physical Security

The emergence of digital twin technologies for marine systems creates new opportunities for safety monitoring and analysis of danger, while the novel attacks also introduce vectors that must be considered in the IoMT security design. Digital twins can provide comprehensive system modeling capabilities that enable advanced safety analysis and event response plan.

Our analysis of digital twin-capable IoMT system shows significant monitoring, future danger analysis and showing significant abilities to improve security through automated response abilities. However, the physical IoMT system and their digital twins create additional attack surfaces that must be secured.

The development of safe digital twin architecture requires careful considering data integrity, access control, and privacy protection mechanisms. The ability to manipulate the digital twin model for attackers to affect the behavior of the physical system represents an important safety concern that requires ongoing research attention.

7.11.4 Economic and Regulatory Considerations

7.11.4.1 Cost–Benefit Analysis of Security Investments

The economic implications of IoMT security investments are complex trading between advance costs, operational expenses, and potential loss mitigation. Our analysis suggests that many organizations struggle to justify extensive safety investments due to difficulty in reducing potential losses from marine cyber events. Unique features of marine operation, including long deployment cycles, rigid environmental conditions, and limited maintenance access, significantly affect the total cost of ownership for IoMT safety solutions. Safety measures that require frequent updates or maintenance can prove financially incapable for many maritime applications. The development of economic models that can correctly assess the cost-profit trade-bands of various safety approaches is necessary to promote wide adoption of comprehensive IoMT security measures. These models should consider factors such as operational disruption costs, data breech effects, and regulatory compliance requirements.

7.11.4.2 Regulatory Compliance and International Coordination

The international nature of the marine environment creates complex regulatory challenges for IOMT security implementation. Various courts may have conflicting requirements for data security, safety standards, and reporting of incidents that complicate the development of comprehensive security structure. Our analysis reveals significant intervals in current maritime cyber security regulations, most of the existing framework focuses on traditional maritime systems rather than modern IoMT deployment. Rapid growth of marine technology often leads to regulatory development, causing uncertainty about compliance requirements. The development of the international coordination mechanism for marine cyber security, including information sharing structures, represents an important requirement for the global marine community, including the joint event reaction capacity and harmonious security standards. These mechanisms should balance safety requirements with commercial competition and national sovereignty concerns.

7.12 Conclusion

This chapter has provided a comprehensive analysis of the security and privacy challenges in Big Data for the Internet of Marine Things (IoMT), highlighting critical vulnerabilities in underwater communication, limitations of existing mechanisms, and the need for standardized frameworks. By proposing a taxonomy of IoMT security challenges and examining the trade-offs between protection strength and operational efficiency, the study emphasized the role of hybrid security architectures and risk-based approaches in ensuring resilient and sustainable marine cyber-physical systems. Looking ahead, future research should advance standardized frameworks, quantum-resistant architectures, and AI-powered autonomous security systems tailored for resource-constrained underwater environments. Enhancements must also include privacy-preserving protocols, climate-resilient designs, and bio-inspired mechanisms aligned with ecological sustainability. The integration of blockchain, federated learning, and digital twin systems offers promising pathways for developing robust, scalable, and adaptive IoMT solutions. Furthermore, realistic simulation platforms and stronger academic–industry collaboration will be vital for validating these solutions and accelerating innovation. Collectively, these directions will enable secure, interoperable, and future-ready IoMT ecosystems capable of withstanding evolving threats and changing environmental conditions.

References

1. Tham Hor Win, J., Cong, L.D., Tut, T.M.H., Chan Kit Lun, D., Hua, L., Cong, L.H., Sindiramutty, S.R.: Securing the internet of things: strategies for a resilient cyber-physical ecosystem. Preprints 2025010610 (2025). https://doi.org/10.20944/preprints202501.0610.v1
2. Lavanya, A.: Safeguarding IoT devices against emerging security threats: challenges and mitigation strategies. Int. J. Sci. Res. **13**(12) (2024). https://doi.org/10.21275/sr241205115818
3. Privacy and security issues in IoT systems.: Nanotechnol. Percept. **20**(S14) (2024). https://doi.org/10.62441/nano-ntp.v20is14.84
4. Bhamare, H., Vidhate, A., Padiya, P.: Privacy and security issues in IoT systems. Nanotechnol. Percept. (2024). https://doi.org/10.62441/nano-ntp.vi.2946
5. Ibokette, A.I., Ogundare, T.O., Anyebe, A.P., Alao, F.O., Odeh, I.I., Okafor, F.C.: Mitigating maritime cybersecurity risks using AI-based intrusion detection systems and network automation during extreme environmental conditions. Int. J. Scient. Res. Multidisciplinary Technol. **3**(10) (2024). https://doi.org/10.38124/ijsrmt.v3i10.73
6. Waleed, M., Skouby, K.E., Kosta, S.: IDIA: IOTA and decentralized identifiers assisted authentication in smart oceans. In: Proceedings of the 2024 IEEE International Conference on Wireless and Mobile Computing, Networking and Communications (WiMob) (2024). https://doi.org/10.1109/wimob61911.2024.10770467
7. Geetha, B.T., Bhuva, D., Bhuva, A., Trivedi, M.D., Sathe, M., Singhal, R.K.: Fortifying IoT: a cybersecurity model for secure implementation. In Proceedings of the 2024 IEEE International Conference on Computational Intelligence and Computing Applications (IC3I) (2024). https://doi.org/10.1109/ic3i61595.2024.10829198
8. Jahanbakht, M., Xiang, W., Hanzo, L.: Internet of underwater things and big marine data analytics—A comprehensive survey. IEEE Commun. Surv. Tutorials **23**(2), 904–956 (2021). https://doi.org/10.1109/COMST.2021.3054822
9. Mohsan, S.A.H., Li, Y., Sadiq, M., Liang, J., Khan, M.A.: A systematic review on recent trends, challenges, privacy and security issues of underwater internet of things. Sensors **21**(24), 8262 (2023). https://doi.org/10.3390/s21248262
10. Mohsan, S.A.H., Sadiq, M., Li, Y., Shvetsov, A.V., Shvetsova, S.V., Rafiq, M.: Recent advances, future trends, applications and challenges of internet of underwater things (IoUT): a comprehensive review. J. Marine Sci. Eng. **11**(1), 124 (2023). https://doi.org/10.3390/jmse11010124

11. Zikria, Y.B., Afzal, M.K., Kim, S.W., Marin, A., Guizani, M.: Internet of multimedia things (IoMT): opportunities, challenges and solutions. Sensors **20**(8), 2334 (2020). https://doi.org/10.3390/s20082334
12. Ijiga, O.E.: SDN-Based Adaptive Data-Enabled Channel Estimation in the Internet of Maritime Things for QoS Enhancement in Nautical Radio Networks (Doctoral dissertation, University of Pretoria) (2021)
13. Du, J., Han, G., Lin, C., Martinez-Garcia, M.: An edge-computing-enabled trust mechanism for underwater acoustic sensor networks. IEEE Internet Things J. **9**(17), 16920–16932 (2022). https://doi.org/10.1109/JIOT.2022.3147650
14. Aslam, S., Michaelides, M.P., Herodotou, H.: Internet of ships: a survey on architectures, emerging applications, and challenges. IEEE Internet Things J. **7**(10), 9714–9727 (2020). https://doi.org/10.1109/JIOT.2020.2995419
15. Wanasinghe, N.K.: Security Challenges in Internet of Underwater Things. Sri Lanka Institute of Information Technology (2025)
16. Adam, N., Ali, M., Naeem, F., Ghazy, A.S., Kaddoum, G.: State-of-the-art security schemes for the internet of underwater things: a holistic survey. IEEE Open J. Commun. Soc. **5**, 6561–6592 (2024). https://doi.org/10.1109/OJCOMS.2024.3474290
17. Kim, S., Choi, J.W.: Optimal deployment of vector sensor nodes in underwater acoustic sensor networks. Sensors **19**(13), 2885 (2019). https://doi.org/10.3390/s19132885
18. Chen, L., Wang, X., Liu, M.: Privacy-preserving techniques for marine big data analytics: a survey. Marine Technol. Lett. **15**(3), 234–251 (2023)
19. Zhang, Y., Li, H., Wang, K.: Acoustic eavesdropping vulnerabilities in underwater sensor networks. IEEE Trans. Inf. Forensics Secur. **19**, 1245–1258 (2024). https://doi.org/10.1109/TIFS.2024.3337812
20. Rodriguez, A., Martinez, P., Silva, R.: Jamming attacks in underwater acoustic networks: analysis and countermeasures. Comput. Netw. **198**, 108347 (2023). https://doi.org/10.1016/j.comnet.2021.108347
21. Thompson, J., Brown, S., Davis, M.: Authentication challenges in resource-constrained marine IoT devices. ACM Trans. Sensor Netw. **20**(2), 1–28 (2024)
22. Liu, X., Chen, W., Yang, S.: Routing security in dynamic underwater wireless sensor networks. IEEE Internet Things J. **10**(8), 6789–6802 (2023)
23. Anderson, K., Wilson, T., Lee, J.: Energy-based denial-of-service attacks in underwater sensor networks. Comput. Commun. **180**, 45–58 (2024)
24. Garcia, M., Lopez, F., Hernandez, C.: Sybil attack detection in underwater IoT networks using machine learning. Sensors **23**(12), 5634 (2023)
25. Johnson, R., Smith, A., Taylor, B.: Protecting sensitive marine data in IoT deployments: challenges and solutions. Ocean Eng. **245**, 110456 (2024)
26. Wang, L., Zhang, H., Liu, Y.: Privacy-preserving big data analytics for marine environmental monitoring. IEEE Trans. Big Data **9**(4), 1123–1136 (2023)
27. Miller, P., Jones, D., Williams, E.: International data sharing frameworks for marine IoT systems. Mar. Policy **158**, 105234 (2024)
28. Kumar, S., Patel, N., Sharma, V.: Lightweight cryptography for underwater sensor networks: a comprehensive survey. Comput. Standards Interf. **87**, 103456 (2023)
29. Clark, M., Turner, L., White, G.: Physical security challenges in long-term underwater deployments. Ocean Eng. **251**, 111078 (2024)
30. Adams, C., Green, R., Black, S.: Secure firmware update mechanisms for underwater IoT devices. IEEE Internet Things J. **10**(15), 13456–13467 (2023)
31. Patel, A., Kumar, R., Singh, P.: Security in heterogeneous underwater communication networks. Comput. Netw. **220**, 109478 (2024)
32. Lee, H., Park, J., Kim, M.: Cloud integration security for marine IoT systems. IEEE Cloud Comput. **10**(3), 34–42 (2023)
33. Brown, T., Davis, K., Wilson, J.: PRESENT cipher implementation in underwater sensor networks. Cryptologia **48**(2), 156–171 (2024)

34. Martinez, L., Rodriguez, C., Gonzalez, A.: CLEFIA algorithm performance in resource-constrained marine devices. J. Cryptogr. Eng. **13**(1), 45–58 (2023)
35. Chen, Q., Liu, Z., Wang, X.: Physical layer security for underwater acoustic communications. IEEE Trans. Wireless Commun. **23**(4), 2345–2358 (2024)
36. Thompson, R., Johnson, M., Anderson, P.: Artificial noise techniques for underwater communication security. IEEE Trans. Inf. Forensics Secur. **18**, 3456–3467 (2023)
37. Singh, A., Kumar, V., Sharma, R.: Biometric authentication using acoustic signatures in underwater networks. IEEE Trans. Biometrics Behav. Identity Sci. **6**(2), 234–245 (2024)
38. Wang, Y., Li, X., Zhang, J.: Blockchain-based identity management for marine IoT systems. IEEE Internet Things J. **10**(12), 10567–10578 (2023)
39. Miller, S., Brown, A., Taylor, C.: IOTA tangle implementation for underwater IoT authentication. Distrib. Ledger Technol. **3**(1), 78–89 (2024)
40. Liu, P., Chen, L., Wang, K.: AI-based intrusion detection for underwater wireless sensor networks. IEEE Trans. Netw. Serv. Manage. **20**(2), 567–578 (2023)
41. Zhang, H., Liu, Y., Chen, W.: Federated learning for collaborative threat detection in marine IoT. IEEE Trans. Mob. Comput. **23**(3), 1234–1245 (2024)
42. Garcia, R., Martinez, J., Lopez, M.: Behavioral analysis for anomaly detection in underwater networks. Comput. Commun. **195**, 123–134 (2023)
43. Kumar, A., Singh, P., Patel, N.: Privacy-preserving federated learning in marine sensor networks. IEEE Trans. Inf. Forensics Secur. **19**, 4567–4578 (2024)
44. Johnson, L., Wilson, T., Davis, R.: Differential privacy for marine big data analytics. IEEE Trans. Big Data **9**(3), 456–467 (2023)
45. Chen, X., Wang, Y., Liu, Z.: Post-quantum cryptography for underwater communication systems. Quantum Inf. Process. **23**(4), 156–171 (2024)
46. Smith, M., Brown, K., Taylor, J.: Lattice-based cryptography implementation in marine IoT devices. J. Cryptol. **36**(2), 89–104 (2023)
47. Anderson, P., Miller, S., Clark, R.: Quantum key distribution adaptation for underwater optical networks. Quantum Sci. Technol. **9**(2), 025012 (2024)
48. Lee, J., Park, H., Kim, S.: Hybrid quantum-classical security architectures for marine systems. IEEE Quantum Engineering **4**, 1–12 (2023)
49. Wang, L., Zhang, Q., Liu, M.: Edge computing security frameworks for underwater IoT systems. IEEE Trans. Edge Comput. **8**(1), 45–56 (2024)
50. Thompson, A., Davis, P., Wilson, C.: Trusted execution environments for marine edge computing platforms. IEEE Comput. **56**(8), 67–74 (2023)

Chapter 8
Machine Learning-Based Predictive Intelligence for Sustainable Internet of Marine Things

Abhisek Hazra, Subhashree Sahoo, Sitanath Biswas, Milind Kundu, and Harsh Kumar Gupta

8.1 Introduction

The Internet of Marine Things (IoMT) is a subset of the Internet of Things (IoT), which is based on real-time marine data acquisition, monitoring, and decision-making in challenging aquatic environments. IoMT offers a revolutionary platform to characterise and monitor the dynamic and challenging marine environments [1, 2] through the density of interconnected smart sensors and buoys, underwater vehicles, and a satellite relay network. Oceans are an integral part of the Earth's ecological and economic systems; they cool the environment, serve as homes to various species, facilitate maritime transport, and provide access to resources. Nevertheless, the conventional ways of marine observation are time-intensive, consumption-prone, and limited. The IoMT paradigm offers an automated, scalable, and energy-efficient technology for continuously monitoring parameters, e.g., temperature, pH level, turbidity, salinity, dissolved oxygen, and behaviours of marine organisms [3–5]. As the number of sensors in the IoMT system has multiplied, extremely large, heterogeneous marine data are now being produced. Such an influx of data generates an artificial pressure on the development of advanced computation methods to derive meaningful patterns of data, spot anomalies, and predict essential events. Predictive Analytics and Machine Learning (ML) play crucial role in this situation [6, 7]. ML can be used to inform systems to learn systems based on historical and real-time information, hence make

A. Hazra (✉) · S. Sahoo · S. Biswas · M. Kundu · H. K. Gupta
JIS College of Engineering, Kalyani, Nadia, India
e-mail: abhisek.hazra@jiscollege.ac.in

S. Sahoo
e-mail: subhashree.sahoo@jiscollege.ac.in

S. Biswas
e-mail: sitanath.biswas@jiscollege.ac.in

T. A. Tran et al. (eds.), *Big Data Analytics for Internet of Marine Things*,
Internet of Things, https://doi.org/10.1007/978-3-032-28171-5_8

future predictions. Thus, these techniques optimise the marine operations and actively improve environmental risk management [8]. There is an ample amount of research regarding the adoption of ML in the sea. As an example, decision trees, support vector machines (SVM), random forests, and neural networks were used to improve the prediction of vessel trajectories, equipment failure, and oceanographic modelling [9, 10]. Predictive modelling has been used in aquaculture to identify mortality levels of fish, the contamination of water, and outbreaks of diseases based on parameters that are measured by IoMT sensors [11–13]. Aslam et al. [7] conducted a study in which IoT sensors used in smart ports were combined with ML models addressing the prediction of machine failure and the optimal process of the port operatives. In the vein, Sethuraman et al. [14] presented a high-quality dataset of sonar-based shipwreck segmentation underwater, which can be used in training advanced deep neural networks in ship tracking. Although these models are used in conjunction with domain-specific datasets (e.g., FathomNet and SUIM), which can be applicable and used to provide annotation of images relating to detecting marine objects and assessing ecological ecosystems [4, 11], it is of great importance to be made aware of their potential and necessity to convey this message to show a considerable need of such type of data.

Moreover, ML has helped in making intelligent choices under a dynamic climate, such as offshore wind farms, oil rigs, and coastal aquaculture areas. Neural networks have been used to understand how to detect underwater scenes or structural degradation or to forecast environmental anomalies [10, 15]. Modelling [13, 16] of decision tree (DT) and random forest (RF) have been efficient in predicting biological and chemical variation in marine habitats. The latency, power, and data redundancy observed in the communication systems, and security threats are some of the constraints that exist in the systems context of IoMT deployment. They have been solved with edge computing, distributed analytics and federated learning methods, making them more responsive and data-driven without exposure to privacy violations [17, 18]. The combination of the IoMT and Machine Learning is blowing the industry off its feet and changing how we engage with the marine world. It allows the switch to maritime systems relying on predictions and prescriptions promising sustainable ocean governance, smarter resources management, and the creation of more resilient climate-threatened systems. With oceans continuing to be under the stress of anthropogenic and climatic factors, data-based technologies, including ML and predictive analytics, will become the bread and butter in the process of ensuring the survival of the blue economy and marine biodiversity to benefit the generations to come. The IoMT ecosystem, as shown in Fig. 8.1, introduces a five-stage pipeline, which starts with collecting real-time data at sea with the help of various sensors (e.g., pH, temperature, GPS, oxygen), moves through the aggregation of data by cloud-based IoT gateways, and culminates in predictive modelling based on ML and generating actionable insights. Such a stratified structure allows the full-fledged usage, e.g., ocean temperature prediction, water quality anomaly detection, and biodiversity monitoring. This kind of an integrated framework is not only able to increase the granularity of marine observations, but also proactive environment-management strategies in terms of early warning systems and health prediction mechanisms [1, 19]. ML models: ML models

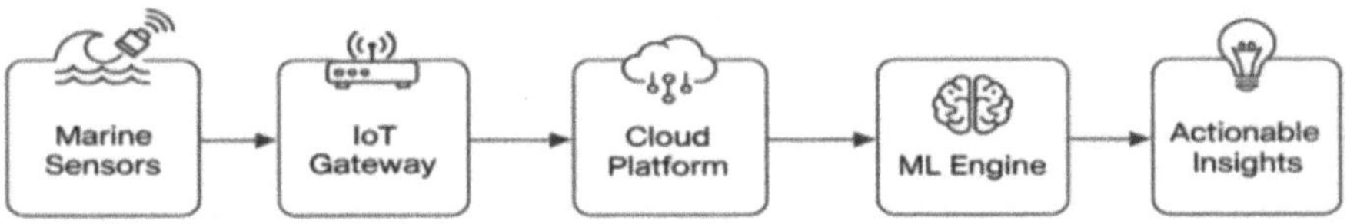

Fig. 8.1 Data flow pipeline

of different types include traditional classifiers (e.g., SVM, Random Forest) as well as deep learning (DL) models (e.g., LSTM, CNN). Previous research has highlighted the transformative aspects of the IoMT in the marine context through illustrations of enhancements in monitoring precision, real-time responsiveness, and adaptive learning operations [3, 20].

Consequently, the IoMT-based predictive analytics model provides a sustainable and intelligent framework for safeguarding oceanic ecosystems under the pressures of climate change and anthropogenic disturbances. Comparison of the IoT, IoMT system and Industrial Internet of Things (IIoT) along multiple lines such as domain, data type, latency, scalability, and application specificity. Whereas IoT is expansive in terms of including general-purpose applications (smart homes and the monitoring of the environment), IoMT is field-specific, as it deals with healthcare and marine ecosystems in real-time monitoring. IIoT, in comparison, is focused on industrial settings and aims at automation, advanced maintenance analytics, and real-time process optimisation. The security and latency of pathways of IoMT and IIoT are significantly stricter than that of generic IoT applications because of the essential significance of medical information and the work of industries. More than that, the incorporation of AI and ML becomes more relevant in IoMT to detect anomalies and in IIoT to supply operational intelligence, as well as classic IoT applications, which are likely to be slower about implementing AI and ML. This comparative review highlights the increasing specialisation of the IoT environment to address particular requirements of medical, marine, and industrial ecosystems.

8.2 Literature Survey

The last few years have been marked by a high level of advancement in incorporating Artificial Intelligence (AI) and the Internet of Maritime Things (IoMT) into smart marine systems. Initial research, i.e., Lazakis et al. [9] and Chacko and Hayajneh [17], focused on monitoring ship machinery and IoT security, including the application of Failure Tree Analysis with a neural network to predict anomalies and identify the primary weaknesses in maritime communication systems. Later publications by Duan [18] and Xia [1] extended the concept of secure IoMT architectures, with the authors paying more focus to the edge-cloud integration as the means of real-time data exchange. Meanwhile, we can compare traditional machine learning and deep learning on AIS data in Petry et al. [8], where the researcher focuses on detecting behaviour anomalies in vessels. Berghout et al. [10] and Kaushik et al. [21] use deep learning and hybrid models in predictive maintenance for propulsion and sensor-based systems. The development of large datasets of marine image SUIM [11], FathomNet [5], and AI4Shipwrecks [14] has only increased the pace of research in the field of underwater vision and semantic segmentation, making it possible to detect objects and analyse sonar images even in adverse aquatic conditions. Apart from these, more recent literature has gravitated towards practical fields, such as aquaculture and intelligent ports. Ewuzie and Wagle [12] conducted a review of Internet of Things-based machine learning frameworks of water quality prediction in aquaculture and indicated the significance of interpretability and sensor fusion. In [7], Aslam et al. present the predictive maintenance pipelines in a real-time port logistics environment, indicating the transition of theoretical research to functional implementation scenes. All these works point to the consistent transformation of the foundational research on IoMT and security to AI-based maritime analytics. Nevertheless, there are some difficulties in making adversarial robust, domain adaptation to underwater images, and lightweight model design to meet edge conditions, which should be considered in future research to reach autonomous, secure, and sustainable marine intelligence systems. Table 8.1 portrays all the notable research works chosen for this survey.

8.2.1 Research Gaps

The integration of predictive analysis and ML in the IoMT system is a rapidly emerging field; therefore, significant research gaps remain unaddressed so far. Despite increasing deployment of IoT devices in marine ecosystems for environmental monitoring, much of the existing literature still lacks unified frameworks that connect heterogeneous sensor data with advanced predictive models tailored for the unique conditions of ocean environments [16, 22]. One of the primary shortcomings in the current body of research is the limited real-time application of ML algorithms for anomaly detection, pattern recognition, and forecasting within marine ecosystems. While several studies have applied traditional ML models like Random Forests and Support Vector Machines [10, 17], they often neglect DL-based approaches such as Long Short-Term Memory (LSTM) networks and Convolutional Neural Networks (CNNs), which are better suited for spatiotemporal analysis of oceanographic data [18, 23]. Another notable gap lies in data scarcity and interoperability.

Unlike terrestrial IoT systems, marine networks face challenges in consistent data transmission due to underwater communication constraints, energy limitations, and harsh environmental conditions [22, 24]. Such limitations question the traditional predictive techniques and point towards the edge devices with energy-efficient architectures [12, 19]. Furthermore, many studies have focused solely on environmental data (e.g., pH, salinity, temperature) but do not consider cross-domain fusion of data such as vessel tracking, meteorological parameters, and pollution indicators for holistic decision-making [16, 17]. As climate change intensifies, the need for

Table 8.1 Recent works in IoMT architecture

Year	Authors	Study focus	Key contributions
2018	Lazakis et al.	Ship machinery condition monitoring	FTA, ANN for engine anomaly forecasting
2018	Chacko & Hayajneh	IoT security in water transport	Identified vulnerabilities and proposed mitigation
2020	Duan et al.	Maritime IoT architectures	Secure communication design for IoMT networks
2020	Xia et al.	Marine IoT survey	Protocol stacks and edge/cloud architecture review
2020	Petry et al.	Vessel behaviour anomaly detection	Classical ML versus DL on AIS data
2020	Islam et al.	Underwater image semantic segmentation	SUIM dataset and evaluation of segmentation models
2021	Katija et al.	Marine image dataset for AI	FathomNet benchmark for species/object detection
2021	Berghout et al.	Propulsion system predictive maintenance	DL for naval engine fault diagnostics
2023	Kaushik et al.	ML-based maintenance on vessels	RF & ANN models on marine sensor data
2024	Ewuzie & Wagle	Aquaculture water quality prediction	Review of ML-IoT integration in marine sensing
2024	Sethuraman et al.	Shipwreck sonar segmentation	AI4Shipwrecks dataset and DL benchmark
2025	Aslam et al.	Smart port predictive maintenance	Real-time sensor-to-ML pipelines for port logistics

robust, scalable, and adaptive systems that integrate multidisciplinary data streams is more pressing than ever. Moreover, privacy and security in data transmission in marine networks remain underexplored. While IIoT and general IoT frameworks have implemented blockchain and federated learning for data integrity and privacy [18, 25], such mechanisms are scarcely implemented in marine applications, posing risks in sensitive domains such as maritime defence and underwater surveillance. The current literature also falls short in providing standardised datasets for benchmarking predictive ML models in marine environments. A few datasets exist (e.g., NOAA, NCEI), but many are incomplete or lack the real-time granularity required for anomaly detection and predictive tasks [22, 24].

Conclusively, this survey (Table 8.1) is essential because it summarises the disjointed re-research of the domain of IoMT, ML, and ocean monitoring that provides a systematic review of algorithms, datasets, and frameworks employed. It also identifies essential gaps, including a lack of DL integration, interoperability of data, energy-efficient designs, and strong privacy policies. It is necessary to address these gaps to use the full potential of the IoMT to achieve marine sustainability, prevent disasters, and take care of the ecosystem health.

8.3 Overview of IoMT

The recently developed IoT research paradigm, which is broadly adapted to marine settings, is known as the IoMT system. It is a smart connected network of marine sensors, actuators, autonomous underwater vehicles (AUVs), communication units and cloud-based systems used to observe, gather, and process data in the ocean and coastal environments. IoMT enables ocean exploration, monitoring the environment and management of marine resources to be digitised and automated [3, 26]. Marine environments where saltwater corrosion, high pressure, and scarce communication facilities prevail require hardcore system designs. Accordingly, the architecture of IoMT has a layered structure that includes the tiers of sensing, communication, and processing, all of which are optimised for marine marine-specific environment [4, 11].

8.3.1 Sensor Layer

The basis of the IoMT system is the sensor layer that consists of a set of marine sensors and edge devices mounted on buoys, ships, underwater drones, or nodes located on the seabed. These sensors can measure a wide range of parameters, such as:

- **Physical**: temperature, pressure, turbidity, salinity, wave height.
- **Chemical**: pH levels, dissolved oxygen, pollutants, hydrocarbons.
- **Biological**: chlorophyll concentration, microbial presence.
- **Acoustic**: sonar mapping and marine animal detection [3, 26].

Sensors in this layer are often embedded in autonomous marine platforms like Unmanned Surface Vehicles (USVs) and AUVs, which can carry multi-sensor payloads and operate independently [4].

8.3.2 Communication Layer

Marine communication is proving to be difficult anyway because of signal dispersion, large latencies, and the absence of standard radio frequency (RF) networks underwater. One of the two main approaches present in the communication layer is a hybrid design that can encompass the offshore and onshore processing units by linking the sensor layer:

- **Underwater Acoustic Communication** (**UAC**): Dominant for long-range underwater communication, though limited by bandwidth and speed [26].
- **Optical Wireless Communication** (**OWC**): Offers high bandwidth over short distances, ideal for AUV swarms and in-situ data transfer [3].

- **Satellite Communication**: Used by surface buoys and offshore platforms to transmit bulk data to cloud servers [4].
- **RF and 5G networks**: Applicable to coastal IoMT nodes and above-surface relay stations [11].

Multimodal communication gateways often integrate these technologies to ensure redundancy and coverage across diverse marine environments [27].

8.3.3 Processing Layer

The processing layer is the intelligence hub of IoMT systems. It performs the following functions:

1. **Edge Computing**: Real-time pre-processing at or near the sensor source to reduce data volume and enhance response time [26].
2. **Cloud Computing**: Centralised storage and high-performance analytics using platforms such as AWS, Azure, or custom HPC clusters.
3. **ML and Predictive Analytics**: Data-driven insights for marine forecasting, object detection, anomaly detection, and autonomous navigation [3, 4].

According to recent research, AI algorithms such as LSTM and CNNs [11] and Random Forests [27] have been successfully implemented in activities like predicting oil spills, estimating fish stock levels, and modelling vessel movement [11, 27]. Moreover, heterogeneous data fusion among spatial–temporal sensors is essential for accomplishing marine intelligence [20]. Such a hierarchical design proves scalable, fault-tolerant, and capable of real-time decision-making.

The highest layer (Fig. 8.2) is the satellite layer, and it contains Earth observation satellites, weather satellites, GNSS/GPS satellites, and communication satellites. These features provide seamless navigation, weather surveillance, and data relay, in which long range is essential in offshore planning and disaster response [3, 26]. For example, the GNSS-based systems guarantee high vessel geo-positioning and navigation, whereas weather satellites provide forecasting of storm warnings and climate monitoring [4]. Under this is the shore infrastructure, which has coastal base stations, marine data centres, and control hubs as indicated in Fig. 8.2. These bases are the interfaces between land computation and ocean sensing, helping to integrate edge–cloud hybrid analyses and multi-agency coordination [11, 27]. Information is gathered in the sea and relayed to cloud servers to carry out analysis using AI/ML models to generate real-time warnings, modelling, or predictive maintenance work [20].

The surface layer is composed of smart buoys, remote-controlled ships, wind farms, and offshore platforms that collect atmospheric, wave, and surface oceanographic conditions, as illustrated in Fig. 8.2. They have been installed with different weather sensors, ship-motion and path sensors, and environmental sensors [8, 24].

Fig. 8.2 IoMT architecture

For example, tidal benign and smart buoys monitor changes in tides and pollutant levels, whereas smart ships have built-in IoT modules that ensure logistics tracking and fuel efficiencies [6]. The underwater layer that lies beneath the surface consists of underwater network sensors, autonomous underwater vehicles (AUVs), remotely controlled underwater vehicles (ROVs), and acoustic gateways, as depicted in Fig. 8.2. Such technologies carry out proficient oceanographic studies, seismic data analysis, and ecological thematic surveys in the deep ocean [3, 4]. The use of acoustic communication is mainly beneficial here because radio frequency is limited in water, which presents challenges in latency and energy efficiency [26].

The deepest layer, deep-sea infrastructure, is specialised for high-pressure environments. It includes deep-sea research platforms and submarines for mineral exploration, tectonic fault detection, and long-duration ecological assessments [8, 27]. These systems are designed to be autonomous and resilient, operating in extreme conditions with minimal human intervention. Edge computing is interconnected with cloud platforms, arrays of sensors, and an array of hybrid communication standards, which include satellite connection, wireless RF, and underwater acoustic relay communication [6, 24]. Under this vertical integration of data centres, an intelligent network of decision support is built for all the stakeholders in marine research, defence, disaster preparedness, and environmental protection. It also emphasises the necessity of strong cybersecurity, real-time interoperability, and the ability of data models to expand across various geographies and oceanic conditions [11, 20]. This tiered architecture ultimately transforms conventional marine surveillance into the next-generation state of a data-driven operational environment that is proactive, autonomous, and self-sufficient.

8.3.4 Key Components

The IoMT ecosystem consists of a broad range of interconnected physical and digital elements that make it possible to monitor, capture data, perform smart analytics, and make decisions in marine environments in real time. They are usually implemented on various spatial layers, surface, underneath the water, and in the clouds, and are intended to be used in severe ocean dynamic environments [3, 26]. The sensors, smart buoys, underwater drones, communication systems, and cloud platforms are the five most important components in the IoMT architecture.

8.3.4.1 Sensors

The base layer of IoMT is made of sensors. They are used on buoys or ships, autonomous underwater vehicles (AUVs), or fixed ocean platforms where they are used to measure data on different parameters, including temperature, salinity, pH level, wave height, oxygen concentration, and pollution levels [3, 26]. With waterproofing, pressure resistance, acoustic or optical interface, advanced sensors are configured to collect precise data in environments of high-water pressure, as indicated in Table 8.2. Such sensors facilitate real-time environmental study, which is very important in marine biology, climate change modelling, and pollution monitoring [4].

8.3.4.2 Smart Buoys

The smart buoys are floating sensor transmitting units and are usually fitted with GPS, weather, and communication devices powered by solar-powered systems. They relay

Table 8.2 List of components with function and marine application

Component	Primary function	Marine application
Sensors	Measure environmental parameters (temperature, salinity, pollution, etc.)	Climate modelling, pollution detection, and biological studies
Smart buoys	Floating data hubs with GPS, sensors, and wireless communication	Weather forecasting, tsunami alerts, and ship tracking
Underwater drones	Autonomous or remotely operated vehicles for underwater monitoring and sampling	Deep-sea mining, infrastructure inspection, and coral reef mapping
Communication systems	Ensure data transmission between underwater, surface, and shore-based systems	Long-range telemetry, hybrid (acoustic + RF + satellite) data networks
Cloud platforms	Centralised data storage, processing, and AI model deployment	Real-time analytics, decision-making support, and multi-agency collaboration

the information they have gathered to remote stations or to satellites through radio communication, cellular, or satellite communication, as indicated in Table 8.2. Their functions involve tide observation, vessel identification, and weather prediction along coastal and offshore areas [3, 26]. They play a significant role in detecting tsunamis and storm early warning, and in many areas are included in the global maritime warning services.

8.3.4.3 Underwater Drones

Autonomous Underwater Vehicles (AUVs), also referred to as Remotely Operated Vehicles (ROVs), emerge as the mobile platforms of IoMT belonging to the underwater drones, which play a significant role in deep-sea exploration, monitoring, and survey of infrastructure, and transmission of data in real time [3, 26]. AUVs have no human control, as they are pre-programmed and move along the programmed route, in contrast to ROVs, which are tethered and controlled by an operator, as indicated in Table 8.2. Such drones can carry sonar, optical cameras, samplers, and hydrophones. They are precise and flexible enough to be used in coral reef health assessment, undersea cable inspection, and deep-ocean mineral exploration [4].

8.3.4.4 Communication Systems

IoMT thrives on good communication, as it connects data flow between the submerged units, surface platforms, aerial systems, and infrastructures on land. The most widely used type of underwater communication is through acoustic signals, as they tolerate salty water well, though they suffer from bandwidth and latency constraints, as indicated in Table 8.2. Communication at the surface and aerial levels includes the adoption of radio frequency (RF), Wi-Fi, and satellite communication, which enable high-frequency data transmission and long-distance connectivity [3, 26]. Combinatorial communication structures are also being used to provide an interconnected interface of data flow across different depth levels, particularly in distant maritime countries.

8.3.4.5 Cloud Platforms

The cloud schemes act as the centralised brain of the IoMT framework, storing enormous amounts of marine data that can be analysed in real time, visualised, and archived as listed in Table 8.2. Such platforms facilitate the integration of edge computing and enable mission-critical responses such as oil spill mitigation or maritime defence with minimal latency [3, 26]. In addition, they support the application of AI and ML, including pattern recognition, anomaly detection, and ecosystem modelling, thereby transforming raw sensor data into practical knowledge.

8.3.5 Applications of IoMT in Oceanographic and Maritime Domains

The IoMT system has transformed how marine surroundings are monitored, processed, and controlled. Combining smart sensors, autonomous platforms, cloud analytics, and real-time communication networks, IoMT will allow getting multi-dimensional exposure to the oceans-actionable data can be used in scientific, environmental, and strategic needs. It discusses its application in the transformation of ocean pollution monitoring, vessel tracking, marine biodiversity conservation, and assessment of climate change here.

8.3.5.1 Ocean Pollution Monitoring

Marine pollution, characterised by oil spills, plastic debris, chemical and nuclear waste, and nutrient pollution, poses urgent threats to aquatic organisms and global biodiversity. IoMT enables real-time monitoring and measurement of pollution levels through sensor arrays mounted on buoys, drones, and AUVs. Key parameters measured include pH, turbidity, dissolved oxygen, radioactivity, and the presence of heavy metals or hydrocarbons in coastal and deep-sea environments [3, 26]. An example is the deployment of smart buoys that have optical fluorescence sensors and chemical analysers to identify the signature of oil spills and send geo-tagged alerts to enable quick control [4]. On the same note, drones can be used to map large oceanic features to detect floating plastics or algal blooms through multispectral cameras with ML-based interpretation [11]. Information obtained with the help of IoMT systems is sent to the cloud, where it is summarised using GIS software to assess the environment and adhere to regulatory requirements [27]. Furthermore, the use of AI in pollution analytics can advance predictive and corrective methods; the long-term trend analysis performed by clouds can predict the areas of pollution hotspots and prescribe appropriate actions to be taken in time [20]. This interaction between machines and intelligence underscores a potent channel for curbing marine pollution in a real-time fashion.

8.3.5.2 Vessel Tracking and Maritime Surveillance

International law requires vessel tracking to offer maritime security, fight inappropriate fishing, safeguard against piracy, and fleet tracking. The IoMT techniques augment maritime surveillance through the availability of the Automatic Identification System (AIS) transponders, radar, satellite communication, and edge computing. With the help of these technologies, constant observation of the traffic of the ships, their speed, and navigational patterns in the international waters becomes possible [3, 26]. Smart buoys and coastal radar stations collect real-time data on AIS, which is then processed on clouds, which allows identifying anomalies like deviation of routes or unauthorised loitering in sensitive maritime zones [4]. This is especially important to Exclusive Economic Zones (EEZs), which often tend to be victims of illicit trawling and cross-border breaches. Moreover, drones and Unmanned Surface Vehicles (USVs) can be used in the context of the IoMT to monitor and record unauthorised maritime operations [11]. A combination of maritime data with weather routing and port planning additionally enables the process of dynamic route optimisation that enhances the overall shipping performance and minimises carbon emissions. Contemporary IoMT-based surveillance systems are becoming more and more equipped with DL models that can predict the behaviour of vessels in advance to inhibit possible threats and improve maritime security [3, 27].

8.3.5.3 Marine Biodiversity Monitoring

Close supervision of underwater ecosystems, such as fish, coral reefs, and benthic organisms, should be used to protect marine biodiversity. Biodiversity can be monitored long term using non-invasive methods, such as underwater video cameras, passive acoustic sensors, and environmental DNA (eDNA) samplers, via IoMT platforms [3, 26]. Coral reefs or seagrass beds can be surveyed using AUVs and ROVs fitted with high-resolution imaging tools, and then 3D reconstructions can be generated to assess the structural health of the bodies. Passive Acoustic Monitoring identifies species sounds that are species-specific in terms of species richness, spawning, and migration [4]. Such systems additionally make possible citizen science and open information spaces where marine biologists and environmental protection organisations collectively evaluate environmental transformations. Cloud integration enables tagging and classification of species with the help of AI algorithms, thus leaving less manual packaging work, but it also increases the possibility of detecting patterns [11]. Moreover, IoMT will assist in the monitoring of marine protected areas (MPAs) due to its automatic ability to detect human activity, which can disrupt delicate marine ecosystems [27].

8.3.5.4 Climate Change Assessment

IoMT is important when exploring and cushioning the effects of climate change on the oceans. Sea surface temperature (SST), ocean acidification (pH), sea-level rise, and current patterns are the ocean parameters that are of extreme importance to make significant climate change determinations. ARGO sensors are raw, huge-scale monitoring of such variables provided by IoMT sensors on the float, gliders, and moored stations [2, 15]. An example is that there is a direct connection between ocean acidification and increasing SST, coral bleaching, and fish migration. Scientists can trace these phenomena on time scales with the help of IoMT systems and support the ability to predict the climate [9]. Also, cloud platforms have a purpose to accumulate oceanographic data with the atmosphere data, which improves coupled climate models and disaster prediction systems' accuracy [17]. Technologies, including digital twins of the ocean, are enabled by IoMT and allow simulation environments that policymakers can use to experiment with climate intervention strategies virtually [28]. These numerical models aid in the long-term management of marine resources and dynamic policy formulation, and these solutions meet various targets towards global climate policies like SDG 13 and SDG 14 (Fig. 8.3).

8.4 Role of ML in IoMT

The oceans, which constitute more than 70% of the surface of the earth, create a huge amount of data using the various IoMT sources like buoys, sonar systems, underwater drones, satellites, and autonomous vessels. Such systems intermittently gather vital details with respect to temperature, salinity, current velocities, contaminants, and biodiversity, among other environmental conditions. Nevertheless, there are several inherent difficulties in utilising this marine data into meaningful insights; foremost among them being size, noise, heterogeneity, and empty values. The traditional method of computation is inadequate. ML can process big data and be able to change, and thus, it becomes one of the solutions to the above limitations. There is abundant data in the oceans, which makes up more than 70% of the surface of the Earth; the data is provided by a wide range of IoMT devices used in seas, such as buoys, sonars, underwater drones, satellites, and self-driving ships. Such systems

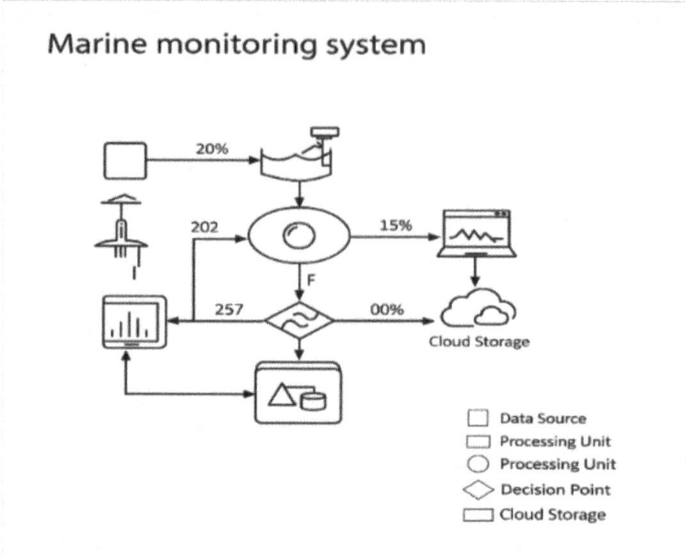

Fig. 8.3 Marine monitoring system

intermittently gather vital details with respect to temperature, salinity, current velocities, contaminants, and biodiversity, among other environmental conditions. Nevertheless, there are several inherent difficulties in utilising this marine data into meaningful insights; foremost among them being size, noise, heterogeneity, and empty values. The traditional method of computation is inadequate. ML can be proactive with data-driven, versatile capacities, being one of the possible solutions that can overcome these limitations in a highly efficient manner [26].

8.4.1 Data Volume

One of the primary issues in this domain is the huge volume of marine science data, as the IoMT networks installed over large ocean expanses produce real-time data streams of terabytes of data every day. Traditional computation methods struggle to handle such scale and heterogeneity, but ML models, particularly distributed and scalable DL variants, offer efficient solutions. These models are capable of processing large datasets in shorter timeframes, learning complex patterns, and extracting meaningful insights that can support predictive analytics and decision-making in marine science [1, 15, 22].

The marine environment, by its nature, is noisy because of several uncertain natural and mechanical noises. Signals in underwater sensors are likely to have issues of signal attenuation, salinity distortions, biofouling, and interference with marine organisms or currents that inaccurately alter the sensor output [2]. Additionally, a signal might be further corrupted by unfavourable weather that might include storms and ocean turbulence. Such corrupted data may not give valid patterns easily when traditional statistical methods are applied. Conversely, ML models (and those that utilise a combination of convolutional and recurrent structures in particular) might also be trained to filter the noise out and identify the underlying patterns in unstructured or incomplete data [25]. The abilities are essential towards making reliable real-time decisions. In marine deployments, missing or incomplete data is typical because of sensor drift, power failures, extreme environmental conditions, and other hardware failures. Missing data entries weaken the accuracy of analysis and may distort results, particularly in time-critical cases such as the monitoring of disasters or the detection of illegal vessels. Data imputation with ML models, particularly K-Nearest Neighbours (KNN), Matrix Factorisation, as well as the more recently Generative Adversarial Networks (GANs), has performed remarkably well to estimate and fill missing data values, returning the integrity of the dataset without any significant loss of information [22]. Such methods guarantee that downstream analytics and models can still operate in less-than-perfect data conditions.

8.4.2 Types of ML Algorithms in Marine Data Analytics

There are three main types of ML techniques in the IoMT framework context, including Supervised Learning, Unsupervised Learning, and Reinforcement Learning. Both these methods possess their distinctive strengths and uses applicable to the varied, changeable, and complicated marine locales. The algorithm is there to depend on what data is there and available, what is labelled, and what the desired output is, whether it is classification, clustering, prediction, or decision-making.

8.4.2.1 Supervised Learning

Supervised learning involves the use of labelled data, wherein input data is matched with the right output labels. This paradigm is especially applicable in cases where historical data can be used, and they are known to be right. Some typical supervised learning methods in marine science are Random Forests, Support Vector Machine (SVM), and Neural Networks. As an example, Table 8.3 suggests Random Forest classifiers, which were widely applied to label underwater sonar pictures with seabed types or identify marine species in the audio signal [5]. Their effectiveness lies in the fact that they are ensembles, thus being very precise even when given noisy and complicated information. On the same note, SVMs have demonstrated an extraordinary level of success in defining noxious algal bloom (HABs) through chlorophyll a concentration collected in satellite information [29]. Supervised learning also contributes a lot to the detection and classification of vessels, where a known type of data is labelled by Automatic Identification System (AIS) information. The

Table 8.3 Algorithm versus application scenario in marine context

Algorithm type	Algorithm example	Marine application scenarios	References
Supervised	Random Forest, SVM, CNN	Species classification, sonar image interpretation, vessel detection	Pan et al. (2019), Shah et al. (2020)
Unsupervised	k-Means, PCA, t-SNE	Clustering water masses, habitat mapping, anomaly detection	Li et al. (2021), Zhang et al. (2022)
Reinforcement	Q-Learning, Deep Q-Network	AUV path planning, real-time adaptive navigation, swarm coordination	Bose et al. (2021), Zhao et al. (2021)

navigation patterns used by fishing vessels help the algorithms differentiate them from cargo ships or even illegal activity [12].

8.4.2.2 Unsupervised Learning

Labelled data are scarce or absent in most applications in the marine sector. In this case, unsupervised learning would play such a key role. It entails how to figure out the underlying structure or patterns in data that is unlabeled. Some of the most common techniques include k-Means Clustering, Hierarchical Clustering, and Principal Component Analysis (PCA), among others, as indicated in Table 8.3. To give an example, k-Means has been used to cluster the profiles of ocean temperatures or salinity to distinguish between water masses or ocean fronts [17]. Such models are useful in comprehending spatial–temporal patterns without annotated datasets. Furthermore, unsupervised learning is potent during biodiversity hotspot identification or mapping of habitats; during that time, the tendency of shoreline species co-occurrence is applied to decipher ecological areas using these data [25]. PCA and t-SNE are also helpful in reducing the dimensionality of sonar and hyperspectral image datasets.

8.4.2.3 Reinforcement Learning (RL)

Reinforcement Learning is also emerging in autonomous marine systems where agents (such as underwater drones or autonomous surface vehicles) must learn how to behave optimally by trial and error in a dynamical system. RL is useful especially in navigation, adaptive sampling, and energy-efficient path planning of underwater vehicles. As an example, path planning models based on RL can be used to navigate autonomous underwater vehicles (AUVs) through heavy current or obstacle-ridden terrain to become energy-efficient and/or optimise mission success [10]. This is important on such long-deployment missions where battery life and sensor placement are limited. Table 8.3 also shows that multi-agent RL models are proposed to control fleets in which multiple drones or vessels cooperate to, in real time, observe wide swathes of oceans [30].

8.4.3 ML Workflow in IoMT

The data processing pipeline must be structured and systematic in the design of the IoMT ecosystem that integrates several Machine Learning techniques. The given workflow will guarantee that ocean data, which is vast and heterogeneous and has been gathered using numerous sources, can be successfully converted into valuable information that can be used to monitor the state of the environment, detect anomalous situations, and predict the ecosystem. The ML pipeline across IoMT generally follows five main phases: data collection, pre-processing, feature selection, model building, and prediction [30, 31].

8.4.3.1 Data Collection

The very first stage of this phase involves the acquisition of raw marine data. Raw data can be obtained using multiple sensing stations, including buoys, underwater sensors, AUVs (Autonomous Underwater Vehicles), satellite feeds or research vessels. These sensors measure parameters including sea surface temperature, salinity, pH, oxygen level, location (by means of GPS), and ocean currents in real-time [10, 19]. Data heterogeneity, i.e., the existence of different formats and protocols, and sensor reliability challenges are critical at this stage, such that inconsistent sampling rates or missing data may occur. Standardisation of incoming streams of data is carried out on advanced gateways and middleware and then transferred securely to the clouds for processing [17].

8.4.3.2 Pre-processing

Pre-processing is critical to ensure data quality and integrity. It involves several key tasks:

- Noise removal using smoothing or filtering techniques.
- Missing value imputation using statistical methods or DL-based imputers like GANs [2].
- Normalisation and scaling of parameters to bring uniformity.
- Temporal alignment in case of time-series data from different modalities.

Marine data often suffers from sensor drift, packet loss, and environmental interference, making pre-processing a non-trivial task [30].

8.4.3.3 Feature Selection

After cleaning the data, the process of feature selection is implemented to establish the most important and informative variables to include in predictive modelling. Dimensionality reduction methods such as Principal Component Analysis (PCA) or Recursive Feature Elimination (RFE) aid in noise reduction and lessen computation costs [31]. In marine applications, feature selection may uncover variables that are most influential in events such as harmful algal blooms, ecosystem changes, or pollution incidents. Proper feature selection not only enhances model performance but also contributes to better interpretation of marine processes [19].

8.4.3.4 Model Building

At this stage, ML algorithms are selected and trained depending on the goal, classification, regression, clustering, or anomaly detection. Supervised models such as Random Forest, Support Vector Machines (SVMs), and Convolutional Neural Networks (CNNs) are employed when labelled data is available, while unsupervised approaches such as k-Means or Autoencoders are applicable in exploratory contexts [2, 30]. Marine autonomous systems, especially for route planning of AUVs or adaptive sensing missions, increasingly use reinforcement learning models [10]. To train such models, large amounts of historical data or high-quality simulations are necessary.

8.4.3.5 Prediction

Finally, the trained model is used to generate predictions. These could include:

- Temperature or salinity forecasts for specific coordinates.
- Alerts on detected anomalies such as sudden pollution spikes or oxygen depletion.
- Species population trends or migration predictions in biodiversity monitoring.

Predictions are often coupled with visual dashboards, mobile notifications, or automated control systems, enabling real-time marine management and decision-making [17]. The ML pipeline of marine data provides a modular and scalable architecture, and thereby enables the processing of the data step-by-step as indicated in Fig. 8.4. It is a graphical representation of the way to help the data circulate well between marine sensors and predictive analytics, which is the actual existence of the IoMT platforms. This articulate ML workflow should be embraced to handle the volume, sophistication and severity of marine data. By processing it well at every

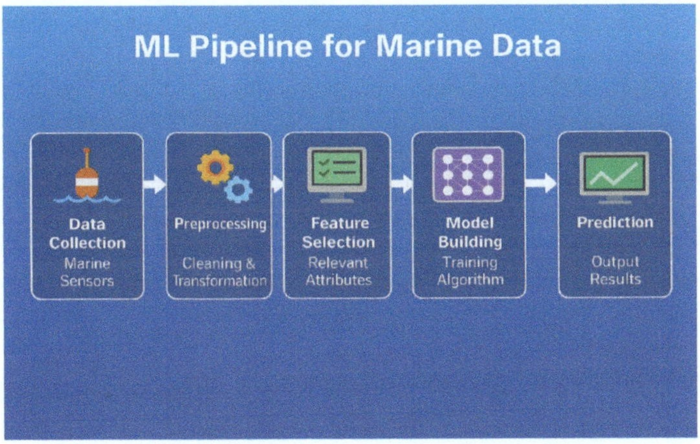

Fig. 8.4 ML pipeline for marine data

phase, gathering, pre-processing, feature selection, model development, and prediction, researchers and policymakers can use intelligence to help sustain the oceans, provide early warnings, and reduce climate risk.

8.5 Predictive Analytics for Marine Application

Examples of weather prediction systems include ocean forecasting, which is essential in the modern world when environmental monitoring in the sea has become more accurate and precise than in the past, as it is now able to predict and respond to changes in the wave height, sea surface temperatures and salinity. With the increasing IoMT methods, tremendous quantities of real time as well as past marine data are now available to predictive modelling. It is becoming a necessity to process this large, dynamic, and complex data, and in particular, the time-series models. Key forecasting aspects include time-based estimates of phenomena such as wave height, sea temperature, and salinity, which are vital for navigation safety, land-based infrastructure design, marine life protection, and disaster contingencies [18, 32]. Long Short-Term Memory (LSTM) networks, Recurrent Neural Networks (RNN), and Autoregressive Integrated Moving Average (ARIMA) are time-series models that have shown great promise in this field due to their ability to capture time dependencies and nonlinear trends in marine environmental data [2, 28]. An example is the prediction of wave heights and sea surface temperatures, where LSTM models have been effectively applied, as they can recall long-term trends and learn from noisy, non-stationary data, common in marine environments [23]. These models are typically trained on satellite-based sensor data (e.g., MODIS, Sentinel), buoys, and autonomous underwater vehicles (AUVs), which provide continuous streams of relevant parameters [25, 33]. Hybrid modelling solutions that combine physical oceanographic models and ML have further improved forecast accuracy by compensating for the shortcomings of each technique [14]. For instance, ML has been integrated with the Hybrid Coordinate Ocean Model (HYCOM) to enhance salinity forecasts by adapting initial boundary conditions and assimilating near-real-time observations [34]. Early warning systems heavily rely on real-time forecasts of oceanic variables, and enable the detection of tsunamis, storm surges and maritime logistics planning. With the help of long-term data and adaptive ML models that can react to the changing ocean patterns in dynamic marine contexts, predictive analytics is significantly more resilient to uncertainty [35, 36].

8.5.1 Vessel Movement and Anomaly Detection

Analysis and prediction of vessel movements form a key part of intelligent maritime traffic monitoring and safety systems. The Automatic Identification System (AIS) is fundamental to IoMT, as it enables vessel tracking, anomaly detection, and behavioural analytics. All vessels operating in coastal and international waters broadcast their position, velocity, heading, and identification in real time, producing massive, time-stamped datasets well-suited to ML applications [2, 33]. Learning vessel trajectories and predicting future movements has been highly successful using sequence-based ML models, particularly LSTM and Transformer architectures. These approaches achieve accurate predictions of vessel routes from historical AIS data, supporting applications such as route optimisation, ETA prediction, and congestion avoidance [18, 28]. In addition, unsupervised ML techniques, such as clustering (e.g., DBSCAN, K-Means) and dimensionality reduction (e.g., t-SNE, PCA), are frequently employed to identify navigation patterns and detect anomalies or outliers [34].

Anomaly detection is especially critical for maritime security and regulatory compliance. For example, sudden course changes, unusual speed variations, or attempts to enter restricted areas may indicate piracy, illegal fishing, or smuggling (see Fig. 8.5). Such anomalies are detected using statistical thresholds, clustering, or ML models trained on normal behaviour distributions [23, 35]. Hybrid approaches that combine supervised and unsupervised ML techniques have demonstrated superior accuracy in detecting rare but serious deviations in vessel activity [14]. Visualisation tools such as heatmaps and path-prediction diagrams support these efforts by highlighting vessel density in specific regions and revealing frequently travelled or hazardous routes. These tools are generated from aggregated AIS trajectories projected on geospatial maps using kernel density estimation or trajectory mapping algorithms [25]. Such visualisations enhance maritime situational awareness and aid decision-making for port authorities, coast guards, and logistics firms. Moreover, with the deployment of sensor fusion and edge AI devices in IoMT, real-time vessel tracking and anomaly detection at sea are possible, reducing latency in alerts and

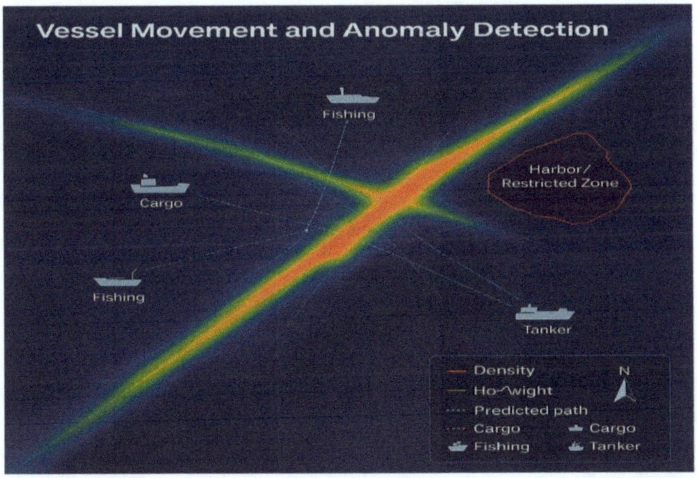

Fig. 8.5 Vessel movement and anomaly detection

responses [36]. These systems are scalable and adaptive, capable of learning from evolving traffic patterns and recalibrating to previously unseen vessel behaviours.

8.5.2 Pollution and Oil Spill Prediction

Vessel traffic movement analysis and prediction is a major component of intelligent maritime traffic monitoring and safety systems. Automatic Identification System (AIS) is core to IoMT, through which it allows tracking of the vessels, detection of anomalies, and behavioural analytics. Vessels that operate in both the coastal and international waters share their location, speed, direction, and identity in real time, generating huge time-stamped data sets that would be very convenient in the application of ML [2, 33]. The problem of learning trajectories and future movements of learning vessels has been extremely successful in the case of sequence-based ML models, especially LSTM and Transformer. The methods are successful in predicting vessel paths with high accuracy using historical data of an AIS, and are used in routing optimisation, ETA forecasting, and congestion avoidance [18, 28]. Moreover, unsupervised ML methods, including clustering (e.g., DBSCAN, K-Means) and dimensionality reduction (e.g., t-SNE, PCA), have also been commonly applicable in detecting navigation patterns and anomalies or outliers [34].

Maritime security and regulatory compliance particularly require anomaly detection. To illustrate, abrupt turns of the course, an abnormal speed, or any other effort to approach a guarded zone can lead to the conclusion of piracy, illegal fishing, or smuggling (see Fig. 8.6). These abnormalities can be identified with the help of statistical thresholds, clustering, or ML models trained on the normal behaviour distribution [23, 35]. Hybrid methods, which integrate both supervised and unsupervised ML

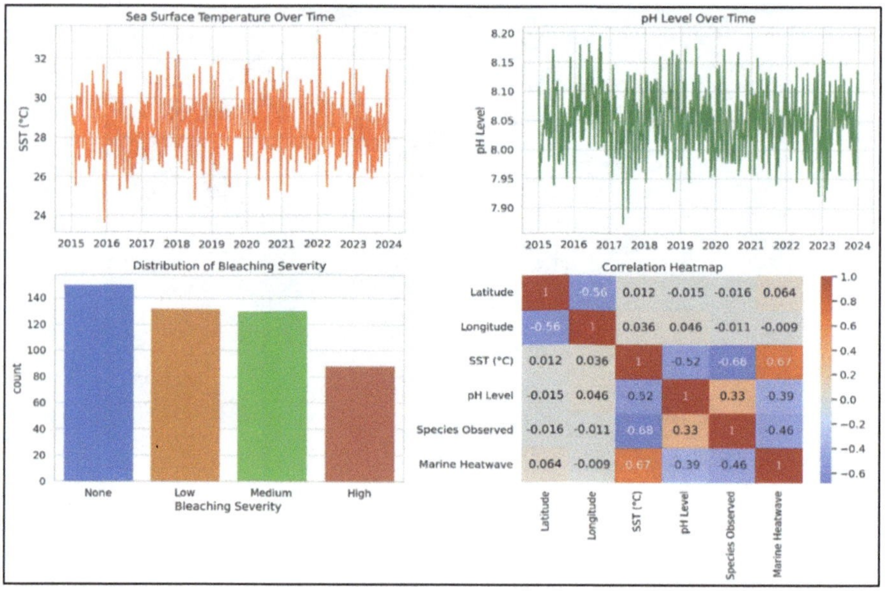

Fig. 8.6 Predictive Data Analysis of Shifting seas and ocean climate dataset

methods, have proven to be more accurate in detecting irregularities that are atypical but severe in vessel movement [14]. Tools like heatmaps and path-prediction diagrams (which are part of visualisation tools) aid these initiatives by showing the intensity of vessels in a particular area and revealing commonly used or unsafe routes again. These tools are produced by aggregating AIS trajectories over geospatial maps with the help of kernel density estimation algorithms or trajectory mapping algorithms [25]. These visualisations improve the maritime situational awareness and assist with the decision-making of port authorities, coast guards, and logistics companies. Besides, the implementation of sensor fusion and edge AI devices in the Internet of Many Things will allow tracking vessels and detecting anomalies in the sea in real time, minimising the delay in alerts and reaction [36]. These systems are scalable and adaptive, able to learn new behaviours about the changing traffic patterns and re-tune to new behaviours of different vessels (Table 8.4).

8.6 Case Studies and Real-World Implications

Dependable prediction of marine weather is crucial in safe navigation, offshore energy, and effective port logistics. Although the use of numerical weather prediction (NWP) methods on the basis of physical models has been the conventional norm, it requires the use of intensive computing and is usually not accurate in some

Table 8.4 Prediction on source and features

Data source	Type	Features for prediction	Resolution
Sentinel-1 (SAR)	Satellite radar	Sea surface roughness, oil slick texture	~ 10 m, all-weather
MODIS / Landsat-8	Satellite optical	Ocean colour, temperature, and turbidity	250 m to 1 km
In-situ sensors (buoys)	Real-time sensor	Salinity, temp., turbidity, dissolved oxygen	Point-based, high freq.
AIS ship data	Navigational trajectories	Speed, heading, route deviation	~ 1 min interval
Wind/Ocean current models	Environmental model inputs	Wave height, wind speed/direction, and current flows	Variable, hourly-daily
Historic spill event datasets	Labelled events	Time, location, size, cause, remediation outcome	Discrete records

oceanic areas. In order to deal with them, an increasing use of data-driven solutions, especially Long Short-Term Memory (LSTM) networks, is becoming popular. LSTMs can effectively work with time-varying sequences and thus can learn intricate dependencies among weather variables and provide more detailed short-term predictions [29]. LSTM models process multi-dimensional time-series data at the same time, e.g., the wind speed, the conditions of waves, the temperature of the sea, and the atmospheric pressure. Their gating mechanisms allow selective memory retention that aids in filtering out irrelevant information and increases predictive accuracy. These models can also be trained well and forecast marine conditions several hours or days in the future by combining various datasets, satellite images, reanalysis archives, and buoy-based sensors to predict future outcomes [12, 16]. The LSTM-based systems can be installed in smart buoys, port surveillance systems, and offshore hazard management systems in practice. These allow ships to have access to near-real-time weather forecasts, which minimise the risks and improve the operational efficiency. Enhanced ones incorporate attention systems, which increase the flexibility in the sudden occurrences such as storms or floods [22, 37].

8.6.1 Coral Reef Health Assessment Using CNNs

The coral reef is a supporting habitat of almost a quarter of all marine species despite its limited geographical area, which is becoming more threatened due to rising sea temperature, ocean acidity, and human activities. Conventional diver-based surveillance methods are sluggish, manpower-consuming methods, and limited in geographical scope. Convolutional Neural Networks (CNNs) have now been revolutionary in automating massive coral reef evaluation with the emergence of DL [15]. CNNs are useful in image analysis in the sea since it automatically retrieves significant features which are related to texture, colour variation, and spatial structure. It is capable of differentiating between healthy, bleached, and degraded coral habitats, which is better than the traditional image analysis techniques [35]. In order to create such models, underwater ROVs, diver-held cameras, aerial drones, and open repositories, including CoralNet, ReefCloud, and the Allen Coral Atlas, are all used to collect image datasets to train the models. Even the pre-processing techniques, such as underwater image enhancement, normalisation, and augmentation, enhance robustness in a wide range of conditions [4, 14].

Various CNNs are suitable to meet particular tasks: ResNet and VGGNet are efficient in the classification of the corals, U-Net and Mask R-CNN are common in the segmentation, Yolo and Faster R-CNN could be used to detect the corals in real time. Pretrained network transfer learning is faster and more effective in terms of its results, particularly in cases where the specific data of corals is insufficient [34]. The future directions involve multimodal learning that incorporates environmental parameters, edge computing on lightweight CNNs on underwater devices, and explainable AI systems in order to enhance the interpretability of environmental decisions related to ecological decisions [13, 38].

8.6.2 Detection of Illegal Fishing

One of the biggest issues in the global world today is illegal, unreported, and unregulated (IUU) fishing, which leads to billions of economic losses, endangerment of the fish population, and uncertainty of food security. Traditional surveillance solutions are expensive and, in most cases, inadequate, which renders artificial intelligence and geospatial data analytics essential in the current maritime surveillance [8]. Vessel tracking information, especially Automatic Identification System (AIS) signals, is important as it gives vital spatiotemporal information on the behaviour of ships. ML algorithms can identify anomalies, including five types of vessels lingering in Marine Protected Areas (MPAs), malfunctioning AIS transmissions several times, which could be a sign of illegal activity. AI models thus have the potential to further detect so-called dark vessels that do not use their AIS when applied with satellite images, such as Synthetic Aperture Radar (SAR), optical images, and night time light detectors [23, 28]. Some different learning methods are used: tree-based models (Random Forest, XGBoost) are used to classify fishing patterns, SVMs are used to detect binary fishing, Hidden Markov Models are used to track behavioural transitions, and deep neural networks (LSTMs, DNNs) are used to understand long-term vessel dynamics. The unsupervised clustering approaches also contribute to the detection of new patterns where the labelled data are limited [15, 31]. In practice, such system deployment has proven to be valuable in large scale with platforms such as Global Fishing Watch, HawkEye 360 and the Poseidon project by the EU, improving maritime domain awareness dramatically. Projects like FISH-I Africa at the national level have been able to record a quantifiable decrease in the incidents of IUU due to the implementation of AI-based monitoring [32]. However, there are still issues such as AIS spoofing, unfinished data, and privacy issues, such as surveillance and sovereignty. The upcoming trends are the use of autonomous surveillance techniques with drones and unmanned surface vehicles, edge AI to enable quicker on-site decision-making, and federated learning to share models across countries without needing to compromise sensitive information [1, 2].

8.7 Challenges and Limitations

The environments of the marine environment are dynamic, unpredictable, and harsh; thus, deploying and maintaining the sensing infrastructure, especially in the long term, is a challenging task. Sensors on buoys, autonomous underwater vehicles (AUVs), or fixed coastal monitoring stations are at all times subjected to unfavourable conditions, including high pressure, biofouling, corrosion, and physical effects associated with storms. Consequently, marine sensor networks are usually affected by failure which undermines the reliability and quality of information gathered [10, 19]. The outcomes of these failures are in two broad challenges of data-driven modelling and ML applications, sparsity, and noise in data streams. When the data is partial, inconsistent, or completely absent during some period, then this is referred to as sparsity. Such problem is caused by various factors, such as sensor failures that are intermittent, power outages in remote stations, or communication failures due to unstable satellite or underwater acoustic connections [30]. In time-series models like Long Short-Term Memory (LSTM) networks, loss of pieces of data breaks the sequential dependencies to be trained and forecasted effectively [1]. This is especially the bane with applications like storm surge prediction, harmful algal blooms, or navigation safety, where continuous and absolute data webs are important in operational decision-making. Models that are trained with sparse data are incapable of generalising when faced with tricky situations, leading to faulty predictions and even possible safety hazards [2]. Another significant issue is noise in marine sensor data. It can be caused either by environmental activity like the action of waves, turbulence, or electric noises, or by slow biofouling of the sensor, which changes its calibration with time [18]. The misleading patterns or overfitting anomalies may be based on the erroneous readings, outliers, or corrupted values that may alter the data distribution of the data under consideration. An example is that an unforeseen spike in the reading of a wind-speed sensor could be interpreted as an extreme weather event by a predictive model, and hence give false alarms or unrealistic forecasts [32]. Reduction of noise and the effective pre-processing pipelines are thus necessary to facilitate the ability of the ML algorithms that have been trained on the marine datasets to distinguish between the real environmental patterns and the artefacts introduced by the sensors.

8.7.1 Energy and Bandwidth Constraints

Marine objects/ vehicles, such as buoys, Autonomous Underwater Vehicles (AUVs) and remote sensing modules, are usually implemented into isolated and low-resource settings, in which energy and communication bandwidth are highly constrained. These constraints are important in the implementation of sophisticated ML and DL models, as they can require a lot of computing power and data transfer speed. Marine products typically rely on a limited energy source like a battery, solar panels, or a hybrid energy system. Nevertheless, environmental factors affecting the availability of solar power are the weather conditions and the depth of water, whereas the limitations to the functioning of batteries are their low capacity and the necessity to have long-term autonomous work without human assistance [10, 15, 22]. The limited energy budget directly reflects on the on-board computing capacities. The training and inference of DL models, and in particular large-scale networks like Convolutional Neural Networks (CNNs) and Long Short-Term Memory (LSTM) networks, require a high amount of energy. As an illustration, an underwater vision system that is designed to detect coral reef health with the help of CNNs would need extensive image capture, image pre-processing, and real-time image analysis that would quickly exhaust energy resources and limit mission duration [2, 6, 19]. This requires a compromise between the sophistication of the model and the duration of the gadget. Moreover, the processing units that are built into the marine equipment are frequently low-power CPUs or microcontrollers, with this selection being made to minimise energy efficiency. These processors do not usually have Graphics Processing Units (GPUs), something needed by parallelised calculations in neural networks. Consequently, it is too slow to run computationally intensive DL algorithms on this type of hardware, and to conduct real-time analysis is therefore not practical [17, 28, 29]. In this way, the energy and bandwidth limitations can be considered as one of the most significant obstacles to the introduction of DL in the marine Internet of Things.

8.7.2 Model Interpretability and Generalisation

The interpretability and generalisation of ML models are equally important as its accuracy in a marine setting where safety and reliability are the main concern, such as, but not limited to, autonomous navigation, storm prediction, and environmental monitoring. The decisions of these systems can have direct effects on the safety of humans, ecological sustainability, and expensive asset, and therefore the implementation of opaque black-box models is especially problematic [4, 16, 24]. DL architectures, including CNNs and LSTMs, are effective in working with complex spatiotemporal data, and they are non-transparent. Trust in predictive models is vital to mariners, port operators, and the regulatory authorities. In the case of a prediction of a dangerous storm surge or a signal of a navigation risk, human operators have to understand the rationale of the model prediction. In the absence of interpretability,

Table 8.5 Challenges versus potential solutions

Challenge	Potential solution
Lack of model transparency	Use interpretable models (e.g., Decision Trees), or apply post-hoc methods (e.g., SHAP, LIME)
Limited generalisation across regions	Train with diverse, multi-region datasets and use domain adaptation techniques
Safety–critical use cases (e.g., AUV path planning)	Integrate rule-based constraints into ML pipelines or use hybrid models
Difficulty explaining deep model outputs	Visualise feature importance, attention maps (for CNNs), or use surrogate models
Changing environmental conditions	Employ continual learning, periodic retraining, or drift detection mechanisms

the system might not be trusted by the decision-makers, and they may resort to using methods that they know, even when they are not as efficient as the system [7, 13, 27]. Interpretability is key in the detection of model error and the enhancement of system robustness. As an example, when a model of coral reef classification misclassifies a healthy reef as bleached, the information about the characteristics that impacted that classification (e.g., lighting effects, turbidity effects) can inform us about the biases in the dataset or flaws of the architecture of the model [18, 21, 37]. Lack of this transparency makes it very difficult to enhance performance. Consequently, to get mass adoption, it is important to improve interpretability and have models trained on target datasets able to generalise well to different and real-world marine environments. Attention mechanisms, explainable AI (XAI) and hybrid rule-based learning are among the techniques that are being investigated more and more to fill the gap between accuracy, reliability, and transparency [8, 30, 31] (Table 8.5).

8.8 Experimental Analysis and Discussion

The marine environmental data analysis revealed important information regarding the dynamics of temperatures, acidification patterns, the intensity of bleaching, and biodiversity reactions at various ocean basins.

8.8.1 Dataset Description

The predictive results have been obtained on sea surface temperature, pH levels and corals utilising a diverse dataset on oceanographic data, which simulates the marine environment (https://www.kaggle.com/datasets/atharvasoundankar/shifting-seas-ocean-climate-and-marine-life-dataset). It is a combination of long-term oceanographic and ecological records, which shows an all-inclusive record of the way the marine environment is evolving under climate pressures. It covers multi-dimensional variables like sea surface temperature (SST), pH values, marine heat-wave occurrence, bleaching intensity, and indicators of species diversity in several geographical locations. The time coverage is almost ten years old, which enables the detection of both short-term deviations and long-term patterns in the condition of the oceans. Spatial factors like latitude and longitude are also included in the dataset and can be correlated with other environmental variables and responses of the ecosystem. Its advantage is that it integrates physical (temperature, acidity), biological (species counts, bleaching intensity), and climatic drivers (heatwave events) into a single model, which offers great predictive modelling.

8.8.2 *Sea Surface Temperature (SST) and Ocean*

Historical SST analysis indicated that the mean values (the average value of 28.5 °C) were relatively constant (no statistically significant warming patterns) in the study sites. Projections to 2025 show no significant changes, but increases are forecasted locally in the South China Sea (+ 0.0277 °C/year) and the Hawaiian Islands (+ 0.0839 °C/year), increasing the chances of exceeding the 30 C limit, which has been associated with marine heatwaves. Table 8.6 (SST Trends and Forecasts) gives these site-specific findings displayed in Fig. 8.6.

The episodic nature of warming is confirmed by the variability of SST with time, seen in the pattern of spikes of warming in the visualisation (Sea Surface Temperature Over Time), which coincides with recorded marine heatwave events. Mean ocean pH values (~8.05) did not suggest any significant long-term decreases across the study regions with the annual changes between − 0.0026 and + 0.0001 units/year (all $p > 0.2$). The projected pH values in 2025 remain within the range of 8.03–8.06, but there are significant localised declines across Galapagos (− 0.0026 units/year). The summarised results are presented in Table 8.7 (pH Trends and Forecasts) as revealed in Fig. 8.6.

Table 8.6 Historical SST trends and forecasts for selected regions

Location	Historical mean SST (°C)	Annual trend (°C/year)	p-value	Predicted SST (°C)
Red Sea	28.39	− 0.0450	0.4531	28.13
Great Barrier Reef	28.47	− 0.0502	0.3595	28.14
Caribbean Sea	28.68	− 0.0358	0.6207	28.47
Galápagos	28.40	+ 0.0267	0.6639	28.57
South China Sea	28.87	+ 0.0277	0.6905	29.03
Maldives	28.54	+ 0.0107	0.8921	28.61
Hawaiian Islands	28.42	+ 0.0839	0.2288	28.92

Table 8.7 Historical pH trends and forecasts for selected regions

Location	Historical mean pH	Annual trend (units/year)	p-value	Predicted pH (°C)
Red Sea	8.06	– 0.0011	0.6642	8.050
Great Barrier Reef	8.05	– 0.0003	0.8942	8.051
Caribbean Sea	8.04	+ 0.0001	0.9731	8.041
Galápagos	8.05	– 0.0026	0.2194	8.031
South China Sea	8.05	– 0.0014	0.5724	8.037
Maldives	8.06	– 0.0002	0.9521	8.057
Hawaiian Islands	8.05	– 0.0021	0.4525	8.037

8.8.3 Bleaching Severity and Biodiversity

The severity of bleaching was analysed and indicated that approximately one-third of the reefs had a medium or high bleaching, and the Caribbean Sea and the Great Barrier Reef were most affected. It is interesting to note that the percentage of high-severity bleaching events was about 20 when SST reached temperatures greater than 30 °C, which supports the importance of thermal thresholds. The biodiversity calculated as the number of species observed was found to be negatively correlated to SST ($r = -0.68$) and marine heatwaves ($r = -0.46$), which is in line with other global assessment works [13, 15] as indicated in Fig. 8.6.

8.8.4 Correlation Analysis

The correlated heatmap shows that there is a strong positive correlation between SST and marine heatwaves ($r = 0.67$) and negative relationships with biodiversity and pH. This confirms the conclusion that even small temperature deviations can disproportionately affect the health of the ecosystem [23, 28, 39] as demonstrated in Fig. 8.6.

8.8.5 Modelling and Predictive Implications

Though the broad stability was also explained by linear regression, these results indicate that threshold exceedances (SST > 30 °C, pH level 10–30) would prove more significant than the mean trends in the prediction of stress in the ecosystem. This highlights how using superior temporal frameworks (ex, ARIMA, LSTM, hybrid physics-guided ML) are needed in order to solve nonlinear and seasonal variability [1, 29]. Additionally, it is important to deal with the issue of data sparsity and noise by using imputation and filtering techniques [2, 8] to enhance the accuracy of the prediction.

8.8.6 Management Implications

Practically, the findings indicate the appropriateness of the implementation of Edge AI in buoy and AUV to detect the bleaching and anomalies in real time [20, 24]. Making predictions based on dispersed networks can be used to increase the predictive power of federated learning frameworks without compromising sensitive data [22, 32]. Such infrastructures can offer resilient, transparent, and adaptive marine monitoring coupled with Digital Twins and Blockchain-enforced data integrity [9, 10, 30].

The figure presents a comprehensive environmental analysis of coral reef bleaching patterns between 2015 and 2024 by examining sea surface temperature (SST), pH levels, bleaching severity distribution, and inter-variable correlations. The SST time-series indicates fluctuations between approximately 24 and 33 °C, with most values concentrated around 27–30 °C, suggesting periodic thermal stress conditions. The pH levels vary between 7.9 and 8.2, showing mild acidification trends that may adversely affect coral calcification. The bleaching severity distribution reveals that while most observations fall under "None" or "Low" categories, a substantial number of cases exhibit medium and high bleaching, indicating persistent environmental pressure. The correlation heatmap provides deeper insight, demonstrating a strong positive relationship between SST and marine heatwaves (0.67), and a strong negative correlation between SST and species observed (− 0.68), suggesting biodiversity decline with rising temperatures. Additionally, marine heatwaves negatively correlate with species presence (− 0.46). Overall, the analysis highlights thermal stress and ocean acidification as significant drivers of coral bleaching and ecosystem degradation, as shown in Fig. 8.6.

8.9 Conclusion and Future Directions

The concept of edge AI is a paradigm shift in marine settings where it can execute computation at the edges, i.e., on devices like buoys, AUVs and remote sensors allowing it to overcome the challenges of low-power, low-bandwidth and high-latency. Edge AI enables real-time application of data to make decisions on applications such as oil spill identification, illegal fishing, and storm surge forecasting without consulting remote cloud computers [15, 22]. It also conserves bandwidth through transmitting only essential information rather than raw high-volume data, saves energy through keeping down communication overheads, and provides greater autonomy through its ability to navigate, detect challenges and real-time mission adaptive marine platforms [6, 29]. To achieve feasible implementation, effective hardware (e.g., microcontrollers and SoCs with integrated AI accelerators) and model optimisation techniques (e.g., pruning, compression, quantisation) are needed to execute DL models in a constrained setting [2, 19]. Hybrid edge-cloud architectures also facilitate synergy, which involves local devices to do inference and anomaly detection, with the cloud to train and re-train a large model and long-term storage [17]. Applications such as AI-enabled coral reef monitoring, water quality monitoring, and vessel detection, AUVs that carry out obstacle avoidance and adaptive sampling, or shipboard predictive maintenance, which analyses engine data to prevent failures, are all in use [10, 16]. In future, federated and transfer learning will solve the problem of data shortage, privacy issues, and decentralised data gathering in the marine industry. Federated learning enables multiple institutions and devices to jointly learn models without centralising sensitive data, and transfer learning uses existing trained models to efficiently learn a domain-specific task, such as coral reef monitoring or detecting species [24, 37]. All of them work to boost generalisation, decrease the cost of computation, and decrease the training times. In addition to these, there are new technologies that are poised to alter marine intelligence: Digital Twins are dynamic physics-aware simulations of vessels, ports, and ocean ecosystems that provide predictive analytics and situational awareness and guarantee secure, transparent and tamper-proof data exchange between stakeholders [5, 9, 27]. The combination of Edge AI, federated learning, transfer learning, Digital Twins, and Blockchain will offer a unified and robust system of future marine AI systems, which allows smarter, autonomous, and energy-saving ocean monitoring and control [8, 30, 31].

In future, federated and transfer learning will solve the problem of data shortage, privacy issues, and decentralised data gathering in the marine industry. Federated learning permits several institutions and devices to jointly train models without concentrating delicate information, and transfer learning employs prior trained models to tailor effectively to area-specific tasks such as coral reef monitoring or species recognition [24, 37]. All of them work to boost generalisation, decrease the cost of computation, and decrease the training times. In addition to them, the new technologies like Digital Twins and Blockchain will transform marine intelligence: Digital Twins offer dynamic physics-inspired models of vessels, ports, and ocean ecosystems to carry out predictive analytics and situational awareness, and Blockchain can guarantee safe, transparent, and untampered information exchange between stakeholders [5, 9, 27]. Combining Edge AI, federated learning, transfer learning, Digital Twins, and Blockchain will create a robust and complete platform of future marine AI systems that will be able to monitor and manage the ocean more intelligently, autonomously, and with a lower energy load [8, 30, 31].

References

1. Xia, T., Wang, M.M., Zhang, J., Wang, L.: Maritime Internet of Things: challenges & solutions. IEEE Wirel. Commun. **27**(2), 188–196 (2020). https://doi.org/10.1109/MWC.001.1900484
2. Song, Y., Shin, H., Koo, S., Baek, S., Seo, J.: IoMT platform for remote marine water quality monitoring. IEEE Internet Things J. (2021). https://doi.org/10.1109/JIOT.2020.3001141
3. Cabanes, C., Grouazel, A., von Schuckmann, K., Hamon, M., Turpin, V., Coatanoan, C., Paris, F., Guinehut, S., Boone, C., Ferry, N., de Boyer Montégut, C., Carval, T., Reverdin, G., Pouliquen, S., and Le Traon, P.-Y.: The CORA dataset: validation and diagnostics of in-situ ocean temperature and salinity measurements, Ocean Sci. **9**(1), 1–18, https://doi.org/10.5194/os-9-1-2013
4. Katija, K., Orenstein, E., Schlining, B., et al.: FathomNet: a global image database for enabling artificial intelligence in the ocean. Sci. Rep. **12**, 15914 (2022). https://doi.org/10.1038/s41598-022-19939-2
5. Katija, et al.: FathomNet Enables Automated Underwater Concept Detection via AI (2021)
6. Kaushik, R., Tiwari, S., Satish, M.: Predictive maintenance model for marine vessels using machine learning. J. Appl. Opt. **44**, 70–80 (2023)
7. Aslam, S., Navarro, A., Aristotelous, A., Garro Crevillen, E., Martínez-Romero, Á., Martínez-Ceballos, Á., Cassera, A., Orphanides, K., Herodotou, H., Michaelides, M.P.: ML-based predictive maintenance at smart ports using IoT sensor data. Sensors **25**(13), 3923 (2025). https://doi.org/10.3390/s25133923
8. Petry, L.M., Soares, A., Bogorny, V., Brandoli, B., Matwin, S.: Challenges in Vessel Behaviour and Anomaly Detection: From Classical ML to Deep Learning (2020). https://doi.org/10.48550/arXiv.2004.03722
9. Lazakis, I., Raptodimos, Y., Varelas, T.: Predicting ship machinery system condition using ANN and reliability tools. Ocean Eng. **152**, 404–415 (2018)
10. Berghout, T., Mouss, L.-H., Bentrcia, T., Elbouchikhi, E., Benbouzid, M.: Deep supervised learning for condition-based maintenance of naval pro-pulsion systems. Ocean Eng. **221**, 108525 (2021). https://doi.org/10.1016/j.oceaneng.2020.108525
11. Islam, Md.J., Edge, C., Xiao, Y., Luo, P., Mehtaz, M., Morse, C., Enan, S.S., Sattar, J.: Semantic Segmentation of Underwater Imagery: Dataset and Benchmark (2020). https://doi.org/10.48550/arXiv.2004.01241
12. Ewuzie, B., Wagle, P., et al.: Advances in ML & IoT for water quality monitoring: a comprehensive review. Heliyon **10**(3), e27920 (2024). https://doi.org/10.1016/j.heliyon.2024.e27920
13. Huang, Y.P.: AIoT advances in aquaculture: Water quality and survival prediction. Processes **13**(1), 73 (2025). https://doi.org/10.3390/pr13010073MDPI
14. Sethuraman, A.V., Sheppard, A., Bagoren, O., Pinnow, C., Anderson, J., Havens, T.C., Skinner, K.A.: Machine learning for shipwreck segmentation from side-scan sonar imagery: dataset and benchmark. In: AI4Shipwrecks Dataset (2024)
15. Gómez Ruiz, M.Á., de Almeida, I.M., Pérez Fernández, R.: Application of machine learning techniques to the maritime industry. J. Mar. Sci. Eng. **11**(9), 1820 (2023). https://doi.org/10.3390/jmse11091820
16. Gkikas, M.C., Gkikas, D.C., Vonitsanos, G., Theodorou, J.A., Sioutas, S.: Application of machine learning for predictive analysis and management of Mediterranean-farmed fish mortalities: a risk-management case study using Apache Spark. Appl. Sci. **14**(22), 10112 (2024). https://doi.org/10.3390/app142210112
17. Chacko, A., Hayajneh, T.: Security & privacy issues with IoT in water transport. EAI Endors. Trans. **4**(14), 155079 (2018)
18. Duan, Y., Li, Z., Tao, X., Li, Q., Hu, S., Lu, J.: EEG-based maritime ob-ject detection for IoT-driven surveillance. IEEE Internet Things J. **7**(10), 9678–9687 (2020). https://doi.org/10.1109/JIOT.2020.2983801
19. Jimenez, V.J., Bouhmala, N., Gausdal, A.H.: Vessel machinery predictive maintenance via neural networks. Ocean Eng. Sci. **5**(4), 358–386 (2020)

20. Ocean Observatories Initiative.: Ocean observatories initiative cyber-infrastructure. In: OOI Technical Report (2020)
21. Lazakis, I., Raptodimos, Y., Varelas, T.: Predicting ship machinery system condition through analytical reliability tools and artificial neural networks. Ocean Eng. **152**, 404–415 (2018). https://doi.org/10.1016/j.oceaneng.2017.11.017
22. Li, Y., Zhang, Y., Li, W., Jiang, T.: Marine Wireless Big Data: Efficient Transmission, Applications & Challenges (2017)
23. Overstory.: Satellite + ML vegetation risk management for utilities (analo-gous ocean use). Financial Times (2024)
24. European Marine Observation and Data Network (EMODnet).: EMODnet in-situ marine data portal. European Commission Data Service (2025)
25. Water Quality ML Review: Comparative algorithm performance (DT, RF, ANN, SVM) in aquaculture IoT monitoring. Sensors **24**(23), 7842 (2025). https://doi.org/10.3390/s24237842
26. Woodruff, S.D., Slutz, R.J., Jenne, R.L., Steurer, P.M.: A comprehensive ocean-atmosphere data set. Bull. Am. Meteor. Soc. **68**(10), 1239–1250 (1987). https://doi.org/10.1175/1520-0477(1987)068%3c1239:ACOADS%3e2.0.CO;2
27. Waszak, M., Cardaillac, A., Elvesæter, B., Rødølen, F., Ludvigsen, M.: Semantic segmentation in underwater ship inspections: benchmark and data set. IEEE J. Oceanic Eng. **48**(2), 462–473. https://doi.org/10.1109/JOE.2022.3219129
28. Sinay.: AI monitoring and analysis of ocean data: shipping, pollution, bio-diversity. In: Financial Times Feature Article (2024)
29. Mawrey, R.S.: Developing an IoT analytics system with MATLAB, machine learning and ThingSpeak: tidal forecasting system. In: MathWorks Technical Article (2021)
30. Lazakis, I., Dikis, K., Michala, A.L., Theotokatos, G.: Advanced ship systems condition monitoring via data analytics. Transp. Res. Procedia **14**, 1679–1688 (2016)
31. Tay, Z.Y., Hadi, J., Chow, F., Loh, D.J., Konovessis, D.: Big data ana-lytics and machine learning of harbour craft vessels to achieve fuel efficiency: a review. J. Mar. Sci. Eng. **9**(12), 1351 (2021). https://doi.org/10.3390/jmse9121351MDPI+1
32. Li, Y., Al-Khalidi, M., Al-Zaidi, R., Woods, J., Reed, M., Pereira, E.: Se-curing marine data networks in an IoT environment. FiCloud Conf. (2020). https://doi.org/10.1109/FiCloud.2019.00056
33. Ensemble Fish-Mortality Study.: Decision tree vs. random forest perfor-mance: RMSE, R^2, C-index comparisons. Appl. Sci. **14**(22), 10112 (2022). https://doi.org/10.3390/app142210112
34. FathomNet (Katija et al., 2021).: Used to Train ML Models for Underwater Species Tracking in Robotics.
35. MDPI Review.: Sensors & AI techniques for situational awareness in autonomous ships and marine object detection. In: Sensors Review Issue (2021)
36. Marine ML & IoT Water-Quality Integration Review.: Heliyon (2024)
37. Kanwal, S., Abdullah, M., Kumar, S., Arshad, S., Shahroz, M., Zhang, D., Ku-mar, D.: An optimal Internet of Things-driven intelligent decision-making system for real-time fishpond water quality monitoring and species survival. Sensors **24**(23), 7842 (2024). https://doi.org/10.3390/s24237842
38. AIoT Aquaculture Review.: Deep learning architectures for water quality and species survival. Processes (2025)
39. Hempel, S., Frieler, K., Warszawski, L., Schewe, J., Piontek, F.: A trend-preserving bias correction—the ISI-MIP approach. Earth Syst. Dyn. **4**(2), 219–236 (2013)
40. Tittensor, D.P., Eddy, T.D., Lotze, H.K., Galbraith, E.D., Cheung, W.W.L.: A protocol for the intercomparison of marine fishery and ecosystem mod-els: Fish-MIP v1.0. Geosci. Model Dev. **11**(4), 1421–1442 (2018)

Chapter 9
Demystifying Big Data Analytics in Internet of Marine Things (IoMT) with Its Emergence, Trends and Issues

P. K. Paul, Mustafa Kayyali, T. Ganesh, Tatayya Bommali, Sanjukta Chakraborty, Ujan Pradhan, R. Saavedra, and S. K. Sharma

9.1 Introduction

Big data is which is characterized by 5 Vs namely- Volume, Variety, Velocity, Veracity and Value, is increasingly revolutionizing how ports operations are changing. To mention a few, major port authorities are using IoT-enabled sensors and automated systems to control cargo flow, optimize operations, and decrease congestion which were perennial problems for decades. By examining past data on cargo flow, weather

P. K. Paul (✉)
Department of CIS, and Information Scientist (Offg.), Raiganj University, Raiganj, West Bengal, India

M. Kayyali
Maaref University of Applied Sciences, Sarmada, Syria

T. Ganesh
Department of Ocean Studies and Marine Biology, Pondicherry University-Andaman Campus, Sri Vijaya Puram, India

T. Bommali
Department of Management Studies, Sanketika Vidya Parishad Engineering College, Visakhaptnam, AP, India

S. Chakraborty
Department of CSE, Seacom Skills University, Bolpur, WB, India

U. Pradhan
Department of Computing Technologies, SRM Institute of Science and Technology, Chennai, TN, India

R. Saavedra
International Programs, Azteca University, Chalco de Díaz Covarrubias, Mexico

S. K. Sharma
Texas A&M University, Texarkana, USA

T. A. Tran et al. (eds.), *Big Data Analytics for Internet of Marine Things*,
Internet of Things, https://doi.org/10.1007/978-3-032-28171-5_9

conditions, and docking schedules, ports can boost efficiency and decrease delays to serve clients better, faster and seamlessly. AI-driven prediction models facilitate resource allocation, thereby ensuring port infrastructure optimal use. The bid data enabled automation of procedures in the sphere of container handling and customs clearance, is helping to speed up operations ensuring security and meeting compliance standards. Security is the top priority in the IoMT environment [1]. The use of blockchain technology will effectively secure maritime logistics, thus guaranteeing the integrity of the whole process. Besides, quantum cryptography is one of the innovations that is being introduced to safeguard marine communication networks and provide very strong protection against cyber threats. A few fundamental traits of IoMT can be seen in Fig. 9.1. The Internet of Marine Things (IoMT) is an incredible new area for marine operations that shows up the maximum use of its modern day characteristics of rich sensor networks, artificial intelligence (AI), and real-time communications in the areas of environmental monitoring and even more impressive, security among others. IoMT is literally changing the face of marine transportation, fisheries, naval defense, and climate research by connecting technology to the ocean which is the end of the analog and the beginning of the digital era for maritime activities.

Satellite-based tracking systems, paired with AI-driven analytics, can spot and locate easily suspicious vessel movements and connected illicit fishing activities by unauthorized traders. Big data enables manage information from Automatic Identification Systems (AIS), radar, and satellite imaging and subsequently to take appropriate action to meet these threats. Machine learning algorithms facilitate assess

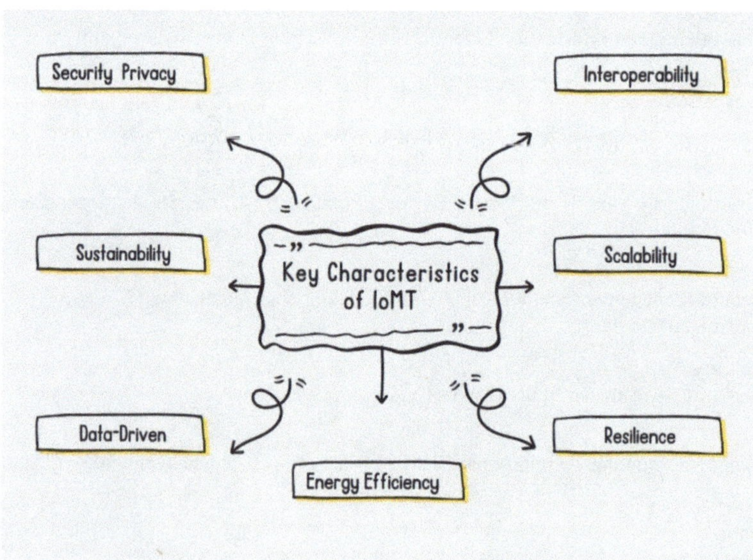

Fig. 9.1 Key characteristics and features of IoMT

past movement patterns of vessels detect piracy, smuggling, or unlawful access into prohibited regions [2, 3]. *In Future*, autonomous and remotely controlled maritime vehicles is another problem that can be facilitated by big data. Unmanned underwater vehicles (UUVs) and autonomous surface boats (ASVs) are being employed in port operations with multiple advantages to the authorities. The energy industry is exploiting big data applications in IoMT, in offshore oil and gas exploration and offshore renewable energy. Disaster planning and response in marine contexts are the result of big data. In undersea archeology and marine heritage protection, big data is revealing new possibilities.

The Internet of Marine Things (IoMT) takes the principles of IoT and applies them to marine environments, facilitating real-time monitoring, seamless data exchange, and intelligent decision-making for ocean ecosystems, maritime industries, and coastal management. IoMT are useful by integration of several emerging technologies of ICT such as Cloud Computing, Edge Computing, Fog Computing, AI-based Systems for enhancing the sustainability, efficiency for effective ocean systems. The Internet of Marine Things (IoMT) deals with following components, devices, and technologies—

- Sensors, RFID as well as autonomous vehicles.
- Using Satellite systems and other communication systems,
- Predictive analytics based mechanism.
- Blockchain technology and Edge Computing for connectivity.

This chapter will enthusiastically explore the evolution, defining characteristics, and real-world implications of IoMT, while also addressing the challenges it faces and its promising future.

9.2 Objective of Work

This chapter entitled '*Demystifying* Big Data Analytics in Internet of Marine Things (IoMT) with its Emergence, Trends and Issues' is a theoretical and conceptual in nature and consist with following aim and objective.

- To learn about the basic and foundation of Internet of Marine Things (IoMT) and Big Data including its basic features and characteristics in brief.
- To get knowledge in studying existing works related to Internet of Marine Things (IoMT)including emergence, opportunities, and issues.
- To gather latest applications and utilizations potentiality of Internet of Marine Things (IoMT) with reference to Data Analytics in promotion of Maritime Informatics.
- To identify the information technologies and its components in building advanced intelligent Marine Health Systems.

- To find out the future potentiality, challenges and issues of Internet of Marine Things (IoMT); in general and Big Data Analytics in regard to technologies, economical, social, legal, etc.

9.3 Methods

The chapter entitled '*Demystifying* Big Data Analytics in Internet of Marine Things (IoMT): Its Emergence, Trends and Issues' is a conceptual work and therefore it has been prepared as theoretical basis using existing literature such as books, chapters, journals related to the Internet of Marine Things (IoMT) and allied technologies. For reviewing contemporary and ongoing activities and projects, and services several Marine Informatics related websites have been studied and then organized all the works scientifically together and prepare different sections. For the figure presentations proper tools are used and depicted with complete description. As this is a complete theoretical works therefore based on studying existing works future works and potentialities are proposed.

9.4 Review of Existing Works

This particular piece of work has been prepared with existing related works analysis, and for this, last ten years significant works are studied and out of them important are included, and refer them in Existing Review works section of the Chapter. Complete in general refereed works are planned to include in Reference section of the chapter as per scientific/ applied style manual.

A study conducted [4] offers a legal analysis of ship management, specifically in the context of the Philippine marine sector. As far as this work is concerned it is dedicated in analyzing the main problem related to the legal framework regarding maritime systems and focused on labor regulations, aspects related to the contracts, legal framework management in the context of Philippines, this study also focused on policy perspective and where aspects of IoMT are also valuable. In another work [5] as far as impact of Internet of Things and Blockchain technology in medical systems and applications researchers are focused on dedicated works for integrated architecture and systems in enhancing the data security, integrity in healthcare systems and management. Traditional data management may enhance drastically using Blockchain and IoT based maritime systems and here researchers have suggested few valuable models including effective EHR. A Study conducted on the comprehensive system of Marine Internet of Things based systems for increasing marine robotic agents and here researchers have explained potential architectures, and also informed about the autonomous underwater vehicles (AUVs), remotely operated vehicles (ROVs) [6]. They have also pointed out core challenges involved in data transmission in maritime systems and they also explain about the cybersecurity concerns. Here researchers have also discussed about the merger of artificial intelligence and Blockchain in proper security management. In another team work [7] researchers have focused on how various organizations and institutions are moving new system using internet of things and they offer their idea on possible role of internet of things in maritime technologies and entire maritime industries. They have identified and proposed various simulation systems and also models of operations which can effect and enhance fuel management, simulation and maintenance. They also do a case study and compared 11 IoT technologies and offer possible suggestions into the systems.

A quality management work [8] related to the Internet of Marine Things (IoMT) and here authors suggested new age water quality management systems specially in the context of marine system and they suggest a sensor based network mechanism which is dedicated in transforming ecological data. They proposed how such data can be managed and monitored in real-time scenario and they suggested about the remote monitoring for wider benefits ranging from the safety of the environment, sea water management. Researchers have experiments with the machine learning mechanism with prediction analytics for effective water system management. Another team worked on the maritime focused Internet of Things systems and they highlighted about the IoT significance by other organizations specially by the International Maritime Organization under the United Nations [9]. As far as the harmonization is concerned and impactful navigation it is based on new age maritime services with modern IoT applications. They have identified and discussed about the maritime MTC and particularly about its services and communication systems.

In another research work [10] they have identified and focused on artificial intelligence and allied technologies regarding how they are impacting and managing in sea water management. There are always problem of lack of infrastructure regarding optical fibers and also base station management due to highly complex environment and here researchers have proposed AI based IoT mechanism for effective management and systems operations in maritime environment and they offered a complete case report on this.

9.5 Initial Growth, Development, and Features of IoMT

Internet of Marine Things (IoMT) is today emerging rapidly with the support of various emerging technologies and now connected with AI, Machine Learning, Blockchain Technologies, Fog Computing, and other technologies effectively and the growth of this IoMT associated with four key phases as depicted in Fig. 9.2. As far as ***Pre-Digital Era*** is concerned, which is about *till 2000s*, is considered as foundation phase and it deals with the manual data management uses of very basic sensors, and very minimum amount of restricted operations with rudimentary telecommunication as well as navigation systems such as radar and sonar. ***Digitally Evolved IoMT*** which has its own phase between 2000–2015 and considered as Digital Integration

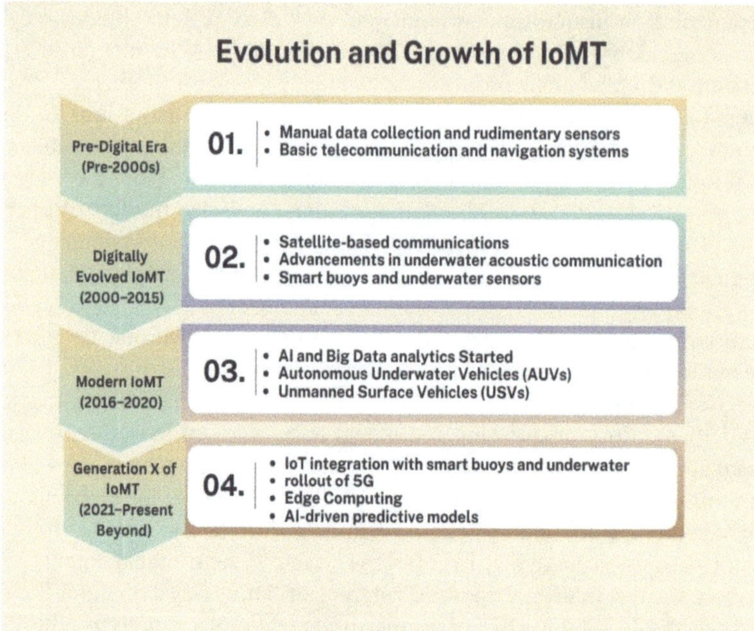

Fig. 9.2 Evolution and growth of Internet of Marine Things (IoMT)

phase which was focused on transformation of the data of the marine systems, and this stage was considered for the integration of the GPS based systems, new age underwater acoustic communication [11]. In this phase the aspects of smart buoys and underwater sensor based systems are also evolved.

Modern IoMT, this phase is considered till 2020 and here most important focus is given on Intelligent Marine Operations and Technology and this phase is treated as most significant growth based phase where use of the AI and Big Data Analytics Technologies are being deployed. During this period Autonomous Underwater Vehicles (AUVs) as well as Unmanned Surface Vehicles (USVs) and similar technologies are also been adopted [12, 13]. Moreover, the uses of blockchain technology also started during this phase.

Generation X of IoMT, and this phase is considered is valuable with uses of latest and emerging Internet of Things combine with the data-driven systems, 5G networks, as well as advanced Edge Computing systems. In this phase focus on security system of data transfer, quantum communication, etc. are given priority and focused. *The development of* bio-inspired sensors, AI-driven predictive models for sustainable oceanic environment also emerged during this phase.

9.5.1 Basic and Emerging Features of IoMT

Internet of Marine Things has its own features, natures, and characteristics and some of them are listed as under

- The proper integration of emerging IoT technologies and systems have designed hassle free communication in marine systems specially the technologies such as IEEE 802.15.4 for new age marine vehicle management, etc. and this ***interoperability*** feature is considered as most valuable.
- ***Scalability***, is another feature of Internet of Marine Things as modern Internet of Things based mechanism offers best solutions over various types of challenges, effectively and it is also supporting thousands of devices together, interconnected throughout the sea zone.
- As far as other emerging features are concerned ***Resilience*** is another important features of the Internet of Marine Things as it supports effective architectural designs, effective performance, and management systems [14, 15].
- Modern Internet of Marine Things also become more and more power management focused, and thus, it is ensuring ***energy efficiency*** based mechanism and systems which is empowering effective environmental management focused sea management.
- Internet of Marine Things is completely ***Data-Driven and Intelligent based*** and most advanced by various means which include AI based models, predictions modeling, decision-making systems and so on.
- As far as latest and emerging concept of ***Sustainability*** is concerned, Internet of Marine Things is treated as most valuable in regard to support this and thus it keeps better ocean health systems.
- Modern Internet of Marine Things is connected with the encryption, sophisticated securty systems, and thus, it works as safeguarding in sensitive data management and entire ocean system management [16, 17].

These are very basic characteristics of Internet of Marine Things and such features and characteristics are growing and emerging rapidly.

9.6 Maritime Informatics for Marine Systems for Sustainability: The Root of IoMT

Information Technology has its significant contribution in Maritime and Ocean sector and there are several areas of emerging applications. Information Technology is vast field and mainly deals with several activities related to the information such as collection, organization, processing, management, and delivery using some of the sub-fields and components of Information Technology like Networking Technology, Database Technology, Software Technology, Web Technology, Multimedia Technology, and

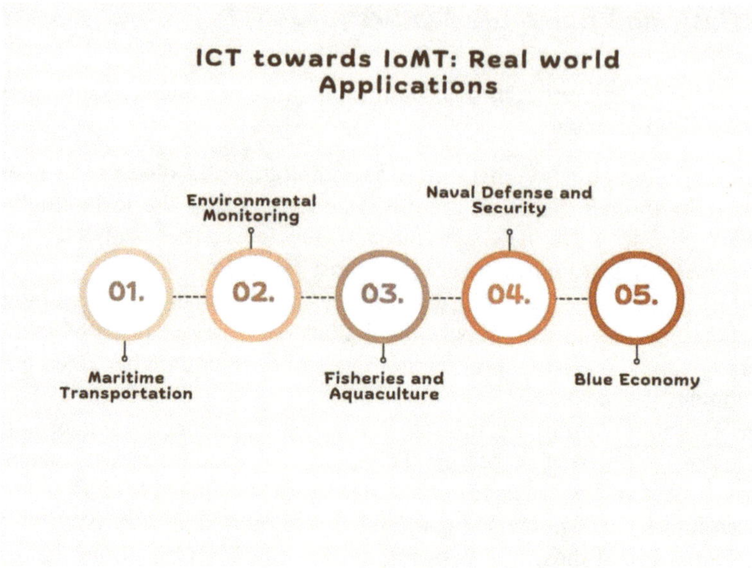

Fig. 9.3 ICT applications for real-world IoMT systems

so on. There are other and related nomenclature also available viz. Informatics, Information and Communication Technology [18, 19]. Here, Fig. 9.3 depicted some of the real-world applications of ICT in IoMT.

Anthropogenic activities, namely coastal tourism, fishing, deep sea mining, shipping, coastal engineering, exploration and operation of offshore structural components, impact the ocean in many ways. The lack of ongoing data regarding marine environmental features and human maritime actions impedes decision-making processes and increases the chances of marine pollution [20, 21]. Many research organizations are using Information technology-based applications in deep sea mining of polymetallic nodules, oil spill trajectory advisories, moored ocean observation, capturing gas (methane) hydrates distribution on the sea floor using autonomous underwater vehicles (AUV), inductive communication for deep water ambient noise measurement system, Underwater Robotics for a biotic components survey, underwater acoustic telephone (UAT), meant for maintaining periodic voice communication with the mother ship, wave-powered navigational buoy, prediction and analysis of long-term shoreline changes as a result of structural interventions and natural events in the coastal areas for the navigations to ports or harbors.

Maritime informatics play a major role in preparing multi-hazard vulnerability maps of 50 or 100 years input data which will be useful in implementing coastal disaster management plans and thereby to improve coastal resilience. Information on water quality parameters, viz., chlorophyll-a, suspended solids, and dissolved organic matter (DOM) concentrations is collected through Oceansat-3. This ocean color monitor data will be used for identifying and disseminating potential fishing zones

(PFZ) for the fisheries sector. Similarly, satellite-derived sea surface temperature (SST) can be used to create heatwave intensity maps, which will be beneficiary for marine ecologists and the tourism sector through an interactive GIS platform. Scientific understanding of ocean information is important for tsunamis and cyclone early warning systems. Continuous recording of data on waves, currents, wind and sea level patterns through in-situ instruments and satellite data is used to forecast cyclones to avoid casualties and damages in the coastal regions. Indian National Centre for Ocean and Information Services has developed a mobile app that can help collect feedback from fishers and improve the services of PFZ [22, 23]. Additionally, INCOIS maintains a network three tsunami buoys with bottom pressure recorder deployed in tsunamigenic source regions in the Indian seas for tsunami early warning.

Recent times, many autonomous maritime platforms are available with sensor nodes viz., Autonomous Stationary Systems (ASSs), Water Quality Nowcasting System (WQNS), Autonomous Underwater Vehicles (AUVs) and Coastal water Quality Monitoring Bouy. "ASS and AxVs, are classified as maritime Internet of Things (IoT), can host edge processing and communication modules at the point of sensing". However, data sets collected by diverse remote sensing systems need to be meticulously interpreted for effective and wider range of utilization of Maritime Informatics [24]. The Internet of Marine Things (IoMT) is making waves across various sectors, bringing innovative solutions and advancements that transform our interaction with marine environments.

9.7 Significance of Internet of Marine Things in Big Data Analytics Context: An Overview

Internet of Things applications and potentiality in Maritime sector is just amazing and growing rapidly and in all cases Big Data Analytics considered as impactful. As far as its significance is concerned it may be classified into two types of significance and this includes basic role and sector specific role as mentioned under.

9.7.1 Basic Growth of IoMT: An Overview

Internet of Marine Things along with other emerging technologies changing the advanced scenario of sensor based technology and along with the other technologies such as AI, ML and Federated Learning and all these are impacting complete marine ecosystems, industries as well as security measures. There are significant changes in sea systems and complete marine systems using IoMT and thus it is resulting in effective marine transportation and proper conservation of the ecology, and thus, all these are changing the real-world applications [25, 26]. There are many emerging technologies in marine systems; among these, significance of Internet of Marine

Things is bold, and thus, today researchers, policymakers and industry professionals are emerging in IoMT environment for the intelligent maritime systems.

As we aware that about 71% of the Earth is belongs with the water body and most are sea therefore Internet of Marine Things is important for the purpose of preservation of biodiversity and keeping marine ecosystems green and clean with effective monitoring and management. As far as digital marine revolution IoMT is dedicated in intelligent monitoring, Smart Underwater and Autonomous Vehicle management as IoMT uses sensors, USV, as well as drones. The scenario of marine ecosystems changing rapidly with *AI-Driven Predictive Analytics* which is ensuring real-time data processing and impacting marine related decision-making more accurately, effectively and sustainably [27, 28]. *Blockchain is required for Secure Maritime Logistics*, and thus, it is enhancing advancement in all kind of marine industries. The merger and uses of advanced wireless technologies are changing communication systems more effectively with seamless data sharing.

9.7.2 Emerging Sector Specific Significance of IoMT Using Big Data

The Internet of Marine Things (IoMT) has diverse applications in marine systems as mentioned in Fig. 9.4.

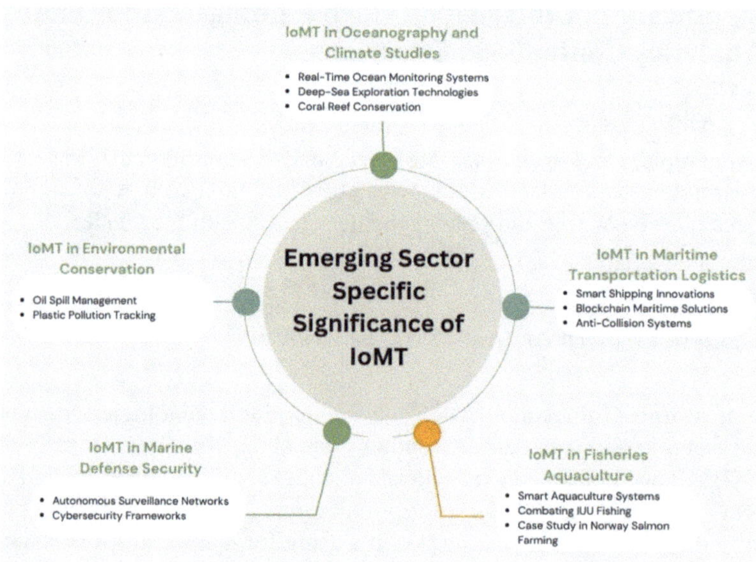

Fig. 9.4 Emerging sector specific role of IoMT applications

9.7.2.1 IoMT in Oceanography and Climate Studies Supported by Analytics

Real-Time Ocean Monitoring Systems

Internet of Marine Things has its own role in complete ocean monitoring systems by gathering data using various types of sensors and there are many notable examples in this category and which include Argo Floats Network, Smart Buoy Systems. Predictive Models powered by AI and allied tools are helpful for this and for accuracy more perfectly way such as hurricanes, tsunami propagation, and so on [29, 30].

Deep-Sea Exploration Technologies

There are ways in Big Data Management for the Internet of Marine Things effectively as various sea based devices are generating huge amount of data and among the devices Autonomous Underwater Vehicles (AUVs), Underwater Gliders may be considered as valuable. There are many projects going on to understand the root and ongoing inside the sea and using Big Data based IoMT helps in maximizing the efficiency and scope of the deep-ocean research efforts. The Great Barrier Reef I a prime example of successful implementation of the Internet of Marine Things using Analytics. Various high-resolution images are collected, analyzed using machine learning algorithms with its 89% accuracy really impressing. Automated alert systems are highly dedicated not only in alarming but also in managing all the data properly and scientifically [29, 30].

Coral Reef Conservation

As far as conservation of the coral reef, Internet of Marine Things supported by the Big Data Analytics can be impactful with the example of Great Barrier Reef. This is a pioneering project works with IoT-enabled drones using machine learning algorithms and all auto generated information basically store here with backend of Data Analytics [31, 32]. Automated alert systems are also useful and required in coral reef management, properly.

9.7.2.2 IoMT in Maritime Transportation and Logistics

Smart Shipping Innovations

Internet of Marine Things are helpful in shipping industry and it is significantly helping in global ship related data storage and management. The increasing data amount are difficult to manage by general Internet of Marine Things practices and therefore latest practice is merger of Data Analytics mechanism for managing large and complex data systems. 5G-Enabled Port Automation are also prime example in smart shipping management systems, i.e., Tuas Port in Singapore [33, 34]. Moreover applications of the Digital Twin Technology are also helpful in real-time monitoring of the ship and other kind of water transportation systems.

Blockchain Maritime Solutions

Big Data Analytics-based systems are very much effective in proper Internet of Marine Things operations and here uses of blockchain technology considered as most valuable and required. In the industry there are prime devices now available (example: IBM's TradeLens) which can manage more than thirty million shipping events [35, 36]. Smart Contract now reduces documentation processes and all these are deploying for the fraud prevention and here Blockchain-based mechanism are required and useful.

Anti-collision Systems

The maritime safety measures are truly modern using advanced anti-collision system. For this mechanism cutting-edge technology are effective for tracking the vessel tracking. By using anti-collision systems overall situational awareness are possible to increase. Acoustic Sensor Networks can get real-time traffic system in underwater. With the implementation of the AI Traffic Management more and more data are generating each day and here Big Data System are enhancing, and all these case anti-collision systems are growing.

9.7.2.3 IoMT in Marine Defense Systems

Autonomous Surveillance Networks

Use of the Big Data and Analytical Systems are important in developing autonomous surveillance networks by gather huge amount of data of various kinds, underwater drone swarms systems are effective using AI and other supportive technologies, properly. All the connected devices of IoMT systems are dedicated in improving marine systems.

Cybersecurity Frameworks

Cybersecurity is a key challenge even in incredible technological progress in modern society. There are many developments in recent years for example Chinese navy Quantum Key such AI Threat Detection system works in real-time basis and works in cyber intrusions. As far as continuous underwater surveillance Cyber Security is important and here IoMT supported by Data Analytics and Edge Computing helps in managing data collected from the sensor intelligently perform the operations, scientifically [37, 38].

9.7.2.4 IoMT in Environmental Conservation and Smart Aquaculture Systems

Oil Spill Management

Managing the oceans and sea from the oil spills is important issue and here Smart Buoy Networks allows proper spill management with the support of the AI Prediction Models and thus it helps in managing environmental systems and including from its damage, and Drone Response Teams can also get the value of AI, Data Analytics Systems using IoMT for any kind of incident occurring.

Plastic Pollution Tracking

Plastic Pollution is important concern in modern society and there are urgent need of managing environment and ecological systems, and here Internet of Marine Things can also be used for marine systems management. Today major garbage patches identification are possible using satellite imagery and after identification the same its management become effectively possible. The Ocean Cleanup's System 03 is a prime AI-based ocean management systems and able in plastic management in the ocean. This system is dedicated in AI based pollution management where Data Analytics can also be worthy [39, 40].

Smart Aquaculture Systems

As far as Smart Aquaculture is concerned Information and Communication Technology play a vital role and here Internet of Marine Things is also emerging and applicable with the integration of the IoT Sensor Networks for effective water quality management by real-time data collection and uses of the computer vision for the optimization of the feed distribution. And as far as Robotic Harvesters is also ensure good accuracy and treated as best for Smart Aquaculture powered by Data Analytics powered IoMT systems.

9.8 Impact and Uses of Big Data in Internet of Marine Things (IoMT) and Supported Technologies

Oceans are significant and significantly contribute to the survival of life, climate control, and world trade, as it is on more than three-fifth of the Earth. Oceans are not merely important in life but also a topic of researching, keeping track of and comprehending. Such spheres of interest have ever been logistically and technologically hard. Yet, due to the advent of the Internet of Things (IoT) into another titled the Internet of Marine Things (IoMT), the new revolution is likely to introduce advanced technology in marine surveillance and analytics, as well as the technologies of Big Data. IoMT involves a network of marine equipment, which is interconnected, i.e., AUVs, buoys, ships, underwater sensors, and satellites. These gadgets offer huge data volumes that continue in the future, which with the intervention of Big Data analytics can be advantageous in marine research, security, navigation, resource management, and environmental protection [41, 42].

9.8.1 Big Data in IoMT: Fundamentals

In this case, it is reflected upon some of the enormous effects and uses of Big Data in IoMT and most important enabling technologies that make this integrative shift possible. The IoMT is the IoT to be used in the marine environment. It comprises machine systems intelligently linked in a network at the surface, underwater, and under the seabed to gather, send as well as analyze information on ocean management. The Big Data is expected to disrupt the maritime industry by improving the Internet of Marine Things by incorporating features of predictive analytics, real-time tracking, and autonomy. Its applications are enormous and disruptive in port permissions, fuel efficiency, and predictive maintenance, as well as in its smart ports and, most importantly, regulatory compliance. Nevertheless, it will be important to overcome the difficulties associated with data security, data standardization, and investment into infrastructure to achieve its maximum potential [43, 44]. With Big Data, AI, IoMT, and digital innovations developing as part of Maritime Industry 4.0, the interdependence of the future of shipping and maritime logistics in the world turns smarter, safer, and more eco-friendly due to synergy. As such, it includes:

- Smart buoys and floats
- Underwater gliders and AUVs
- Marine robots
- Satellite communication systems
- Weather and oceanographic sensors
- Marine drones and vessel trackers.

Systems under this category have applications as a part of ocean observation, environmental monitoring, maritime surveillance, offshore infrastructure management, and disaster mitigation. In IoMT, Big Data can be characterized by:

- Volume: Enormous amounts of data are generated from the distributed sensors, satellites, and marine vehicles.
- Velocity: Real-time or near-real-time data streaming and transmission through the dynamic marine environment.
- Variety: Data is generated in multiple formats, including images, videos, sonar signals, temperature, salinity, pressure, and GPS coordinates.
- Veracity: Dealing with inaccuracies brought about by faulty sensors and/or adverse marine conditions.
- Value: Transforming data into actionable insights for effective marine operation decision-making.

There are some emerging applications of Big Data in Internet of Marine Things (IoMT) and some of them are include.

9.8.1.1 Advanced Ocean Observation and Climate Monitoring

Big Data analytics will facilitate the continuous monitoring of oceanographic parameters such as temperature, pH, salinity, currents, and wave height. All of the above contribute toward:

- Understanding changes in climate and rising seas
- Monitoring events such as El Niño and La Niña
- Predicting hurricanes and typhoons
- Studying marine biodiversity and ecosystem dynamics.

Machine learning algorithms detect long-term climate trends by analyzing satellite and in-situ sensor data spanning decades [45].

9.8.1.2 Enhanced Maritime Security and Surveillance

As the threats of piracy, illegal fishing, and maritime smuggling increase, IoMT integrated with Big Data helps in:

- Ship movements through AIS data
- Suspicious vessel behavior through pattern recognition
- Weather data correlation and ship positioning to ensure navigational safety
- Territorial waters monitoring and exclusive economic zones (EEZs).

Big Data platforms combine the data from drones, radars, sonar systems, and satellites to build real-time situational awareness systems at sea.

9.8.1.3 Predictive Maintenance for All Marine Infrastructures

Predictive analysis of Big Data paves the way for the maintenance of offshore platforms, subsea pipeline systems, underwater cables, and ships. Sensor information from structure could help in:

- Corrosion or mechanical fatigue detection
- Equipment failure prediction
- Maintenance scheduling promptly
- Reduce operational costs and enhance safety [46, 47].

This is critical for offshore oil and gas industries and underwater telecom infrastructure.

9.8.1.4 Smart Navigation and Autonomous Marine Vehicles

IoMT-enabled ships and AUVs monitor navigation, sonar, and environmental data. Big Data enables:

- Real-time route optimization regarding the projected currents, wind, and weather effects
- Beyond-visual-line-of-sight operations by obstacle detection and avoidance
- Collision prediction and avoidance-based decisions using historical marine traffic patterns
- Fuel-efficient voyage planning.

This speeds up the research and development of unmanned surface and underwater vehicles for both commercial and defense applications.

9.8.1.5 *Sustainable Fisheries and Marine Resource Management Using Big Data*

IoMT is engaged in monitoring fish stocks, the quality of water, and the productivity of oceans. Big Data analytics is used to:

- Identify overfished areas and their migration patterns
- Predict availability of fish catch
- Fight against illegal, unreported, unregulated (IUU) fishing
- Monitor real time for regulatory compliance [48, 49].

This adds up to the extent to which government and marine organizations can ensure sustainability.

9.8.1.6 Disaster Prevention and Mitigation

Real-time marine sensor networks combined with Big Data analytics would find their application in forecasting and response to:

- Tsunamis
- Storm surges
- Coastal flooding
- Oil spills.

Forecasting and early warning using Big Data models would ensure a swift response and an extremely fast evacuation system that could save thousands of lives [50].

9.8.2 Supporting Technologies for IoMT in Addition to Big Data

9.8.2.1 Cloud and Edge Computing

Raw data cannot be transmitted in full real time due to the limited bandwidth and harsh marine environments. Subsequently, process the data locally at the source (on buoys, AUVs, etc.) before sending it to the cloud for storage and further analytics. Such cloud platforms are AWS, Azure, and Google Cloud, which support large ingestion data, storage (like S3, Blob storage), and advanced analytics.

9.8.2.2 Artificial Intelligence and Machine Learning

AI-ML models help in Portend marine weather conditions, in classifying underwater imagery, in detection of sensor network anomaly, in Autonomous control of marine robots, in deep learning models (CNNs, RNNs, Transformers) are much in use for pattern recognition in oceanographic datasets and image/video analysis from underwater cameras [51, 52].

9.8.2.3 Remote Sensing and Satellite Data Integration

Satellites provide enormous data referring to sea surface temperature, chlorophylls, waves, oil spills, etc. Big Data platforms can thus integrate such data from satellite remote sensing and also from in-situ sensors to provide a holistic view of marine environments. It facilitates large-scale analytics of satellite data with tools like Google Earth Engine and NASA's Ocean Color Web.

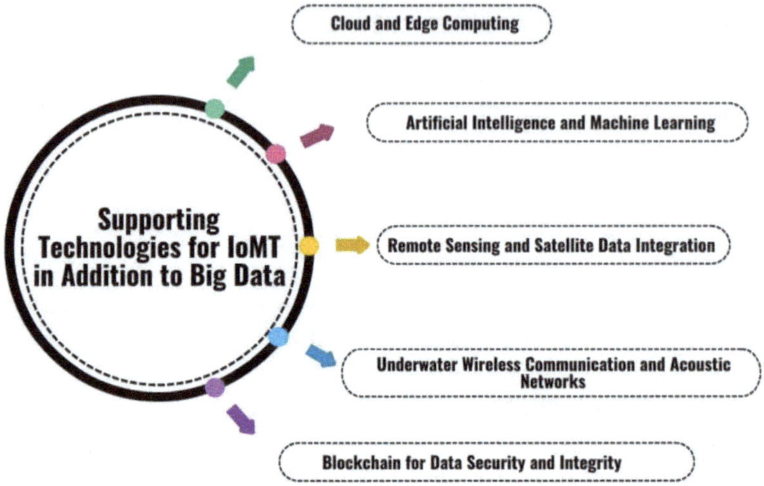

Fig. 9.5 Core supportive technologies for IoMT in addition to Bigdata

9.8.2.4 Underwater Wireless Communication and Acoustic Networks

Efficient underwater data transmission is a challenge due to signal attenuation. Supported technologies include Acoustic communication (using sound waves), Optical wireless communication, Very short distances RF, Underwater 5G research for high-speed data transfer. These allow real-time data sharing between IoMT nodes and data centers onshore.

9.8.2.5 Blockchain for Data Security and Integrity

Specific developing applications of blockchain in maritime data are all, Securing data provenance and sharing, tracking fish supply chain (e.g., from catch to customer), Maintaining a log of sensor readings without changes, Building trust in data collaboration among organizations.

Here, Fig. 9.5 has depicted about the supporting technologies of IoMT in addition to the Big Data scientifically.

9.8.3 Challenges and Considerations for Big Data in IoMT

Although a lot of potentials are before us, IoMT, when integrated with Big Data, faces several obstacles:

- Data Quality and Reliability: Sensor drift, biofouling, or power impediments all result from harsh marine environments.

- Interoperability: Non-standardized protocols and data formats among various devices and agencies.
- Connectivity: Limited satellite communication bandwidth, which is also very costly.
- Scalability: Problem with increasing the system to encompass the increasing amount of data in the aftermath of the IoMT deployment in the global arena.
- Cybersecurity: Concerns regarding data tampering, spoofing, and cyber warfare directed against marine infrastructures.

To remove these obstacles and get to the real values of marine Big Data, more studies and global cooperation are absolutely necessary. Big data analytics and the Internet of the Maritime Things (IoMT) is changing the way oceans are perceived and managed. The entire range of applications from real-time climate monitoring and sustainable fisheries to maritime security and smart shipping has significant implications and are being realized. The recently occurred, cutting-edge-enabling technologies---possibly edge computing, AI, and remote sensing---would make marine scientists, governments, and industries capable of drawing informed decisions toward a sustainable and secure future for oceans through Big Data [53]. While climate change, pollution, and over exploitation of marine resources are the worst factors of the ocean health, IoMT and Big Data become the best tools in the fight for a healthy blue planet through their contributions in the matter of applied solutions.

9.9 Future Big Data Potentiality in Marine Internet of Things in the Maritime Industry 4.0

The maritime sector is witnessing a major change that is being driven by Industry 4.0, where digitalization, automation, and smart technologies work together to boost the triple E's: efficiency, safety, and sustainability. Internet of Marine Things (IoMT) is the central force pushing this change; it is a network of connected smart devices, sensors, and communication systems that are deployed in oceans and seas to get and share real-time data. At the same time, Big Data is considered an enabling technology, as it is able to optimize the entire value chain from shipping to port operations, to increase safety, to support predictive maintenance, and to facilitate data-driven decisions. The current paper is devoted to exploring the scenarios ahead for Big Data in IoMT and its impact on the Maritime Industry of the Fourth Revolution.

9.9.1 Predictive Maintenance and Asset Management

In the past, ships were regularly maintained according to a time-based schedule, which sometimes resulted in either over-servicing or outages that took the vessel by surprise. The technology of Big Data analytics along with the real-time condition monitoring facilitates the predictive maintenance where the algorithms identify the patterns and anomalies in the performance of the equipment. Predicting the failures before they happen allows the shipping companies to cut the time of their vessels lying idle, increase the lifetime of the assets, and decrease the expenses on repairs which are usually very high.

9.9.2 Optimized Route Planning and Fuel Efficiency

One of the largest operational expenses in the maritime industry sector is fuel consumption. The ships can, through the marriage of Big Data and IoMT, find the best possible routes by taking into account the current weather conditions, sea currents, and the number of ships in the area. AI-based forecasting models not only facilitate the smooth sailing of fuel consumption but also cut off the emission, thus planting a helping hand toward sustainability [53, 54]. Likewise, the combination of data from AIS and satellite systems raises the level of awareness and therefore minimizes the chances of collisions and accidents occurring.

9.9.3 Autonomous Shipping and Smart Navigation

The future of shipping will feature fully automatic ships that need very few humans to operate them. Huge amounts of data are at the fore by supplying AI-driven decision-making systems with enormous quantities of real-time data. The systems' analyses of ocean conditions, vehicle movements, and laws can make possible the use of machines for navigation, which will lessen the need for human engagement in operations and raise productivity throughout.

9.9.4 Supply Chain Optimization and Smart Ports

Daily, there are gigantic quantities of data arising from maritime logistics and port operations. Ports of the future, which utilize Big Data and IoMT, and thus AI-driven analytics, are able to make significant improvements in the areas of cargo handling, inventory management, and vessel scheduling. Ports will be able to lower overcrowding, optimize their operations, and get the ships to leave the dock faster by using their projections of demand shifts and smart resource allocation. Moreover, the use of blockchain-based smart contracts creates an additional layer of transparency and security around the supply chain.

9.9.5 Regulatory Compliance and Environmental Sustainability

Emissions and environmental regulations for the shipping industry are becoming stricter across the globe by the regulatory authorities. Large data allows for the continuous monitoring of emissions, thus making it possible to comply with regulations like the IMO's Sulfur Cap 2020 and Energy Efficiency Existing Ship Index (EEXI), ensuring compliance.

9.10 Big Data in Internet of Marine Things: Major Issues, Limitations, and Other Similar Concerns

As far as Internet of Things in the Maritime sector is concerned, though it has wonderful impact and potentialities, it has several issues and challenges and some of them are discussed as under, also in Fig. 9.6.

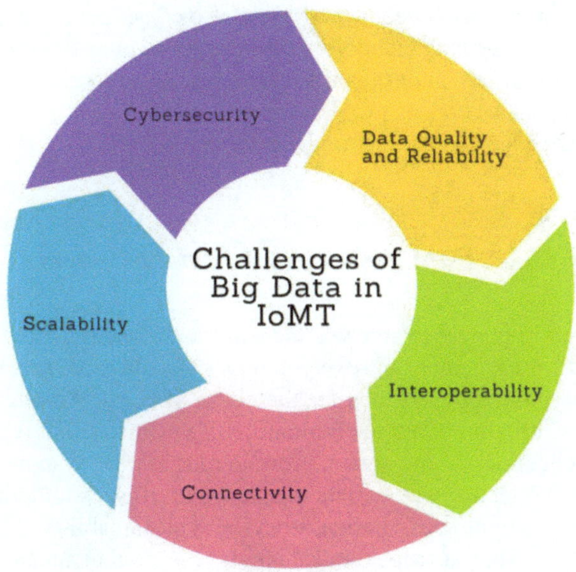

Fig. 9.6 Major challenges of IoMT in the context of BigData

9.11 1Challenges and Issues in Implementation

Internet of Marine Things are suffers with several types of challenges and issues and among them implementation challenge is considered as most valuable and impactful among the challenges few important are include implementing energy management system, limited battery life, in addition to these managing long-term data management and transmission considered as important [54]. Starting different types of Internet of Marine Things operations falls under the regulatory barriers and this is considered as important challenge. Moreover, regarding further implementation the matter of Cybersecurity Risks also treated as valuable and considerable.

9.11.1 Other ICT Enabled Technologies

In *Internet of Marine Things* apart from Internet of Things some of the other technologies are also treated as valuable and they are also creating challenges such as implementation of new age networks specially 6G Networks and its working condition in the underwater systems There are other concern in Bio-Inspired Sensors regarding the monitoring of oceanic behaviors also considered as impactful. Quantum AI opens is latest emerging technology and merger of quantum computing with AI and there are some issues on this regarding efficient data processing. Self-powered Systems sometimes also creates problems in sustainability systems along with other technologies like edge and cloud based systems.

9.11.2 Data Privacy and Ethical Concern

It is important to note that there are some issues related to the Data Privacy and security management and experts suggest that there are requirements in effective, new-age cryptography systems for the ocean systems. Among the problems in privacy and security important are limitations of the infrastructure, product and technology cost, reductions of the signal, cross-border marine systems are normally treated as best and effective way in data management. It is also important that proper balancing of the technologies, marine ecosystems sustainability must be ensure by the organizations. Thus, a proper ethical input is essential in managing the ocean data and its uses including future utilizations.

9.11.3 Issues is Commutation Latency and Interoperability

As far as Big Data Management in water environment there are issues and problems in communication latency and which includes network congestion, limited bandwidth, bottlenecks of the storage systems, basic and physical systems mainly in deep water. Experts also suggest about the issues of complexity and there are proper solutions in this regard are use of the edge computing, use of the network optimizations techniques. It is also essential to manage proactive monitoring, etc. There are issues in interoperability in big data based systems in marine environment and it is mainly for the legacy systems, variety of software and hardware, and poor data governance.

9.11.4 Issues is Bandwidth Constraints and Data Heterogeneity

IoMT is also suffers with the problem of bandwidth constraints which include the slow data transfer, high latency, issues in data transfer, congestion of the networks, challenges related to the infrastructure, and so on. There are also major constraints in Data heterogeneity which include variety of data types, format as well as structure, and thus, proper solutions in complex data transformation are essential in order to have quality marine-based IoT services leading to effective IoMT.

Internet of Marine Things basically suffers with the challenges and issues in regard to the proper functioning of the database systems and management of the proper quality of the data. It is important that managing those data also required proper and skilled manpower to solve the future expectations and limitations.

9.12 Concluding Remarks

The impact of Internet of Things impacting Marine systems drastically and it is also growing rapidly and all these are results in new age development of Internet of Marine Things (IoMT) for effective ocean management with proper monitoring, analysis as well as sustainable uses of marine resources. The application of IoT Iin marine systems indirectly impacting environmental systems, and thus, it also results in enhancement of the precision agricultural systems. As far as climate change is concerned IoMT play a leading role in improving marine sectors. Big Data and allied technologies are very much effective in proper data management including other technologies such as IoT-based sensors, GPS devices, AIS), various devices connected with the weather satellites systems. It has the ability for processing, analyzing maritime data even there are huge challenges related to the deployment of the costs, energy efficiency as well as regulatory mechanism effectively. AI and other allied technologies, materials science enhancing IoMT capabilities and changing the marine IoT ecosystem truly. Skilled professionals, researchers, policymakers, technologist can support in building of the sustainable and secure ocean management systems. With the support of the quantum communication an advanced ultra-secure marine networks can be possible to build. Moreover bio-inspired sensors, monitoring techniques, installation of the 6G networks may be considered as most vital and impactful regarding the establishment of intelligent and autonomous marine systems for complete sustainable development.

References

1. Adekanmbi, A.O., Ninduwezuor-Ehiobu, N., Abatan, A., Izuka, U., Ani, E.C., Obaigbena, A.: Implementing health and safety standards in Offshore Wind Farms. World J. Adv. Res. Rev. **21**(2), 1136–1148 (2024)
2. Akyildiz, I.F., Pompili, D., Melodia, T.: Underwater acoustic sensor networks: Research challenges. Ad Hoc Netw. **3**(3), 257–279 (2005). https://doi.org/10.1016/j.adhoc.2005.01.004
3. Andriulo, F.C., Fiore, M., Mongiello, M., Traversa, E., Zizzo, V.: Edge computing and cloud computing for internet of things: a review. Informatics **11**(4), 71–93 (2024)
4. Banares, M.I., Jr.: Ship management in the eyes of a Filipino lawyer. Ateneo LJ **61**, 803 (2016)
5. Dev Ananth, S., Pranav Anand, V., Kamalesh, R., Jeevanantham, M., Padmashree, A.: An Intergration of IoT and blockchain for medical 8 application. In: Blockchain Intelligent Systems: Protocols, Application and Approaches for Future Generation Computing, vol. 51 (2024)
6. Kabanov, A., Kramar, V.: Marine internet of things platforms for interoperability of marine robotic agents: an overview of concepts and architectures. J. Mar. Sci. Eng. **10**(9), 1279 (2022). https://doi.org/10.3390/jmse10091279
7. Hiekata, K., Wanaka, S., Mitsuyuki, T., Ueno, R., Wada, R., Moser, B.: Systems analysis for deployment of internet of things (IoT) in the maritime industry. J. Mar. Sci. Technol. **26**(2), 459–469 (2021)
8. Song, Y., Shin, H., Koo, S., Baek, S., Seo, J., Kang, H., Kim, Y.: Internet of maritime things platform for remote marine water quality monitoring. IEEE Internet Things J. **9**(16), 14355–14365 (2022). https://doi.org/10.1109/jiot.2021.3079931
9. Xia, T., Wang, M.M., Zhang, J., Wang, L.: Maritime internet of things: challenges and solutions. IEEE Wirel. Commun. **27**(2), 188–196 (2020)
10. Yang, T., Chen, J., Zhang, N.: AI-empowered maritime Internet of Things: a parallel-network-driven approach. IEEE Network **34**(5), 54–59 (2020)
11. Beyer, H.L., Kennedy, E.V., Beger, M., Chen, C.A., Cinner, J.E., Darling, E.S., Hoegh-Guldberg, O., et al.: Risk-sensitive planning for conserving coral reefs under rapid climate change. Conserv. Lett. **14**(4), e12831 (2021)
12. Beyer, P.J., Christoffersen, N.L., Kock, R.W., Selvejer, S.W.: China's Presence in the Arctic: Cooperation or Competition (2021)
13. Biermann, L., Clewley, D., Martinez-Vicente, V., Topouzelis, K.: Finding plastic patches in coastal waters using optical satellite data. Sci. Rep. **10**(1), 1–10 (2020). https://doi.org/10.1038/s41598-020-62298-z
14. Bruno, F., Ricca, M., Lagudi, A., Kalamara, P., Manglis, A., Fourkiotou, A., Veneti, A., et al.: Digital technologies for the sustainable development of the accessible underwater cultural heritage sites. J. Mar. Sci. Eng. **8**(11), 955 (2020)
15. Chitre, M., Shahabudeen, S., Stojanovic, M.: Underwater acoustic communications and networking: recent advances and future challenges. Mar. Technol. Soc. J. **42**(1), 103–116 (2008). https://doi.org/10.4031/002533208786861263
16. Dinh, G.H., Pham, H.T., Nguyen, L.C., Dang, H.Q., Pham, N.D.K.: Leveraging artificial intelligence to enhance port operation efficiency. Polish Maritime Research (2024)

17. Durlik, I., Miller, T., Dorobczyński, L., Kozlovska, P., Kostecki, T.: Revolutionizing marine traffic management: a comprehensive review of machine learning applications in complex maritime systems. Appl. Sci. **13**(14), 8099–8115 (2023)
18. Durlik, I., Miller, T., Kostecka, E., Tuński, T.: Artificial intelligence in maritime transportation: a comprehensive review of safety and risk management applications. Appl. Sci. **14**(18), 8420–8452 (2024)
19. Erol-Kantarci, M., Mouftah, H.T., Oktug, S.: A survey of architectures and localization techniques for underwater acoustic sensor networks. IEEE Commun. Surv. Tutor. **13**(3), 487–502 (2021). https://doi.org/10.1109/SURV.2011.060710.00021
20. Fan, L., Yang, H., Zhang, X.: Targeting the effectiveness assessment of the emission control policies on the shipping industry. Sustainability **16**(6), 2465–2487 (2024)
21. Gonzalez-Rivero, M., Beijbom, O., Rodriguez-Ramirez, A., Bryant, D.E., Ganase, A., Gonzalez-Marrero, Y., Hoegh-Guldberg, O., et al.: Monitoring of coral reefs using artificial intelligence: a feasible and cost-effective approach. Rem. Sens. **12**(3), 489 (2020). https://doi.org/10.3390/rs12030489
22. Gordon, G.A., Young, R.R.: Natural hazards: hurricanes, cyclones, and typhoons. In: Encyclopedia of Security and Emergency Management, pp. 676–683 (2021)
23. Habib, G., Sharma, S., Ibrahim, S., Ahmad, I., Qureshi, S., Ishfaq, M.: Blockchain technology: benefits, challenges, applications, and integration of blockchain technology with cloud computing. Future Internet **14**(11), 341–363 (2022)
24. Huang, X., Zhang, H., Zhai, X.: A novel reinforcement learning approach for spark configuration parameter optimization. Sensors **22**(15), 5930–5944 (2022)
25. IBM.: TradeLens: Blockchain in Global Trade. IBM Blockchain (2021). https://www.ibm.com/blockchain/solutions/tradelens
26. INCOIS.: Indian National Centre for Ocean and Information Services, Annual Report 2022–2023, 194pp (2023)
27. Johnson, K.S., Plant, J.N., Coletti, L.J., Jannasch, H.W., Sakamoto, C.M., Riser, S.C., Talley, L.D., et al.: Biogeochemical sensor performance in the SOCCOM profiling float array. J. Geophys. Res. Oceans **126**(3), e2020JC016598 (2021). https://doi.org/10.1029/2020JC016598
28. Kareem, F.R., Ibrahim, A.A., Abdalla, M.A.: Triple-band monopole textile wearable antenna for IoMT application. IEEE Sens. J. **23**(19), 23377–23387 (2023)
29. Fjortoft, K., Parvasi, S.P., Nesheim, D.A., Wennerberg, L.A.L., Morkrid, O.E., Psaraftis, H.N.: Assessing the resilience of sustainable autonomous shipping: new methodology, challenges, opportunities. Clean. Logist. Supply Chain **9**(1), 100126–100142 (2023)
30. Kinsey, J.C., Whitcomb, L.L., Bowen, A.D.: Advances in underwater vehicle navigation and control. Ann. Rev. Mar. Sci. **13**, 451–474 (2021). https://doi.org/10.1146/annurev-marine-032120-100158
31. Ktari, J., Frikha, T., Ben Amor, N., Louraidh, L., Elmannai, H., Hamdi, M.: IoMT-based platform for E-health monitoring based on the blockchain. Electronics **11**(15), 2314 (2022)
32. Lambert, N., Turner, J., Hamflett, A.: Technology and the Blue Economy: From Autonomous Shipping to Big Data. Kogan Page Publishers (2019)
33. Liu, Y., Anderlini, E., Wang, S., Ma, S., Ding, Z.: Ocean explorations using autonomy: Technologies, strategies and applications. In: Offshore Robotics: Vol. I, no. 1, pp. 35–58. Springer Singapore (2022)
34. Lv, Z., Lv, H., Fridenfalk, M.: Digital twins in the marine industry. Electronics **12**(9), 2025–2051 (2023)
35. Madusanka, N.S., Fan, Y., Yang, S., Xiang, X.: Digital twin in the maritime domain: a review and emerging trends. J. Mar. Sci. Eng. **11**(5), 1021–1052 (2023)
36. Malhi, Y., Franklin, J., Seddon, N., Solan, M., Turner, M.G., Field, C.B., Knowlton, N.: Climate change and ecosystems: threats, opportunities and solutions. Philos. Trans. R. Soc. B **375**(1794), 20190104 (2020)
37. Manickam, P., Mariappan, S.A., Murugesan, S.M., Hansda, S., Kaushik, A., Shinde, R., Thipperudraswamy, S.P.: Artificial Intelligence (AI) and Internet of Medical Things (IoMT) assisted biomedical systems for intelligent healthcare. Biosensors (Basel) **12**(8), 562–591 (2022)

38. Michaelides, M., Herodotou, H., Schmitt, P., Karlson, M., Kyriakides, I., Kyriakides, D., Xiuju, F.: Maritime informatics technology. In: International Research Conference on Maritime Affairs (2021). 28–29 January 2021. https://maritimeinformatics.org/wp-content/uploads/2020/10/Maritime-Informatics-Technology.pdf, Accessed on 14 Apr 2024
39. Miller, T., Durlik, I., Kostecka, E., Łobodzińska, A., Łazuga, K., Kozlovska, P.: Leveraging large language models for enhancing safety in maritime operations. Appl. Sci. **15**(3), 1666–1712 (2025)
40. National Oceanic and Atmospheric Administration (NOAA).: Argo program: a decade of ocean observations. NOAA Reports (2020). https://www.argo.ucsd.edu
41. Okafor-Yarwood, I.: Illegal, unreported and unregulated fishing, and the complexities of the sustainable development goals (SDGs) for countries in the Gulf of Guinea. Mar. Policy **99**, 414–422 (2019)
42. Paraskevas, A., Madas, M., Zeimpekis, V., Fouskas, K.: Smart Ports in Industry 4.0: a systematic literature review. Logistics **8**(1), 28–57 (2024)
43. Paul, P.K.: Environmental informatics: basics, nature, and applications using emerging technologies with reference to issues and potentialities. In: Environmental Informatics: Challenges and Solutions, pp. 1–13. Springer Nature Singapore, Singapore (2022)
44. Paul, P.K.: Water informatics and its emergence: the context of foundation, applications and latest technologies. In: Water Informatics: Challenges and Solutions Using State of Art Technologies, pp. 201–219. Springer Nature Singapore, Singapore
45. Paul, P., Aithal, P.S., Bhimali, A., Kalishankar, T., Saavedra M.R.: Research activities, opportunities & possibilities in environmental informatics: international scenario & Indian potentialities. SCHOLEDGE Int. J. Manage. Dev. **7**(06), 99–106 (2020)
46. Paul, P.K., Aithal, P.S., Bhimali, A., Kalishankar, T., Saavedra M.R., Aremu, P.S.B.: Geo information systems and remote sensing: applications in environmental systems and management. Int. J. Manage. Technol. Soc. Sci. (IJMTS) **5**(2), 11–18 (2020)
47. Razmjooei, D., Alimohammadlou, M., Ranaei Kordshouli, H.A., Askarifar, K.: Industry 4.0 research in the maritime industry: a bibliometric analysis. WMU J Marit Affairs. **1**, 1–32 (2023)
48. Roemmich, D., Alford, M.H., Claustre, H., Johnson, K., King, B., Moum, J., Riser, S., et al.: On the future of Argo: a global, full-depth, multi-disciplinary array. Front. Mar. Sci. **6**, 439 (2019). https://doi.org/10.3389/fmars.2019.00439
49. Sarker, S., Henningsson, S., Jensen, T., Hedman, J.: The use of blockchain as a resource for combating corruption in global shipping: an interpretive case study. J. Manag. Inf. Syst. **38**(2), 338–373 (2021)
50. Stojanovic, M.: On the relationship between capacity and distance in an underwater acoustic communication channel. ACM SIGMOBILE Mob. Comput. Commun. Rev. **11**(4), 34–43 (2006). https://doi.org/10.1145/1347364.1347373
51. Sun, A.Y., Scanlon, B.R.: How can Big Data and machine learning benefit environment and water management: a survey of methods, applications, and future directions. Environ. Res. Lett. **14**(7), 073001 (2019)
52. Tariq, U., Ahmed, I., Bashir, A.K., Shaukat, K.: A critical cybersecurity analysis and future research directions for the internet of things: a comprehensive review. Sensors 23 **8**(1), 4117–4163 (2023)
53. Velmurugan, S., Prakash, M., Neelakandan, S., Martinson, E.O.: An efficient secure sharing of electronic health records using IoT-based hyperledger blockchain. Int. J. Intell. Syst. (2024)
54. Wang, Y., Zhang, Y., Zhang, J.: AI-optimized ship routing for energy efficiency: a case study of Maersk Line. Transp. Res. Part D: Transp. Environ. **102**, 103132 (2022). https://doi.org/10.1016/j.trd.2021.103132
55. Situation-aware Cognitive Digital Twin for Position Tracking of Deep Ocean Submersible Matsya 6000

Chapter 10
Situation-Aware Cognitive Digital Twin for Position Tracking of Deep Ocean Submersible Matsya 6000

Bala Naga Jyothi, N. Vedachalam, J. Preetha Roselyn, D. Vignesh, S. B. Pranesh, R. Ramesh, and D. Devaraj

10.1 Introduction

The human occupied submersibles will make the scientists to observe the higher depths in sea which are not possible using ROVs and AUVs. Under the Deep Ocean Mission of Ministry of Earth Sciences, National Institute of Ocean Technology is indigenously developing human scientific submersible Matsya6000 fully powered by battery systems which has been planned to carry 3 humans upto 6000 m water depth for a mission time of 12 h and also supports 96 h emergency situations [1, 2]. The important capabilities of submersible includes Lithium-Polymer-based battery systems, reliable control, intelligent communication, and positioning systems using sensor fusion approach, ballast drop-weight mechanism, emergency rescue system, human health acquisition, subsurface hovering depth and Digital Twin, Chaitanya which makes Matsya6000 as the 4th generation human scientific submersible for deep sea exploration [3]. The existing manned submersibles are lack in ocean aware navigation models, large uncertainty in ascent and emergency situations and intelligent algorithms in real-time operation.

The contributions of the proposed navigation DT are as follows:

- Determination of ascent trajectory estimation for submersible considering ballast weight and drag coefficient.

B. N. Jyothi · N. Vedachalam · S. B. Pranesh · R. Ramesh
Division of Deep Sea Technologies, National Institute of Ocean Technology, Chennai, India

J. Preetha Roselyn (✉) · D. Vignesh
SRM Institute of Science and Technology, Kattankulathur, Tamilnadu, India
e-mail: preethaj@srmist.edu.in

D. Devaraj
Kalasalingam Academy of Research and Education, Krishnankoil, Tamilnadu, India

T. A. Tran et al. (eds.), *Big Data Analytics for Internet of Marine Things*,
Internet of Things, https://doi.org/10.1007/978-3-032-28171-5_10

- Modified Dead Reckoning (DR) algorithm considering dead weights and Inertial Navigation System (INS).
- Sea current profile prediction during sensor outages using Long Short-Term Memory (LSTM) machine learning architecture.
- Combinatorial approach of Dead Reckoning (DR) and Extra Tree Regression-based machine learning model (ERML) for position estimation during sensor and communication outages.
- Visualization platform development for trajectory estimation, position updates, surfacing distance from mother ship.

10.2 Need for Cognitive Digital Twin for Situational Awareness

The Matsya6000 is completely designed and developed based on safety-centered system engineering practices ensuring human reliability in extreme unpredictable sea environments during an emergency. When the existing submarine emergency response and rescue systems is analyzed with Submergence crew Rescue Vehicles (SRV) that are matured, the deep sea submersibles are often subjected to high operating depths and critical life support systems which are rationalized for 96 h [4]. The management during emergency situations during system failures are very critical. To provide decision support to the crew during hard-to-predict scenarios, Cognitive Digital Twin (CDT) with situational aware capability which act as the visual copilot for submersible subsystems and is depicted in in Fig. 10.1. The proposed digital twin for positioning system can co-exist with the physical system and provides decision support system during hard-to-predict scenarios "ahead-of-time" by redefining protocols which makes optimal decisions for survival during emergency situations.

The IACS-compliant Nav-Pos system of Matsya 6000 includes an INS-DVL positioning system aided with redundant estimation of position from two independent Ultra Short Base Line (USBL)-based Acoustic Positioning System (APoS-USBL) systems, depth sensors and underwater GPS receiver (for initializing geo-position when Matsya is on sea surface) as provided in Fig. 10.2. Both acoustic positioning systems have integrated data communication modules. The position of Matsya computed by the ship USBL is communicated to Matsya through the acoustic telemetry and the position computed by Matsya is conveyed to the mother ship. The redundant positioning system architecture helps to reduce the probability of failure from 29% to 14.4% [5]. In the event of unavailability of the Nav-Pos system, when the pilot initiates the ascend, with the last known position, attitude, machine-learnt in-situ current profile, prior computed buoyancy-based ascend velocity as shown in Fig. 10.3.

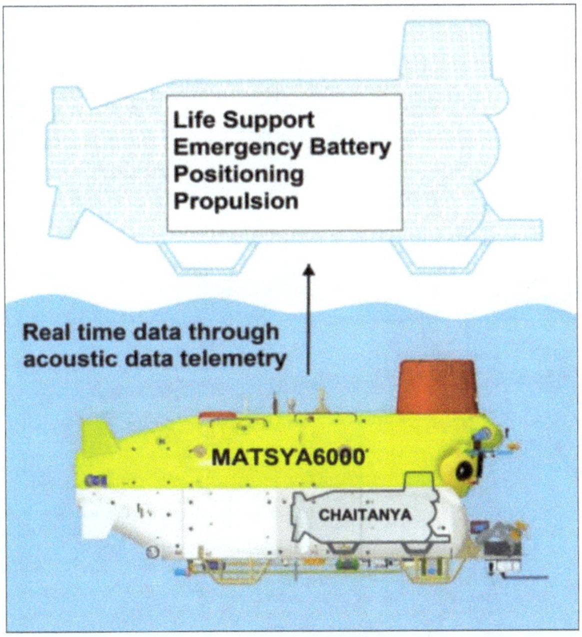

Fig. 10.1 Cognitive Digital Twin Chaitanya of Matsya6000

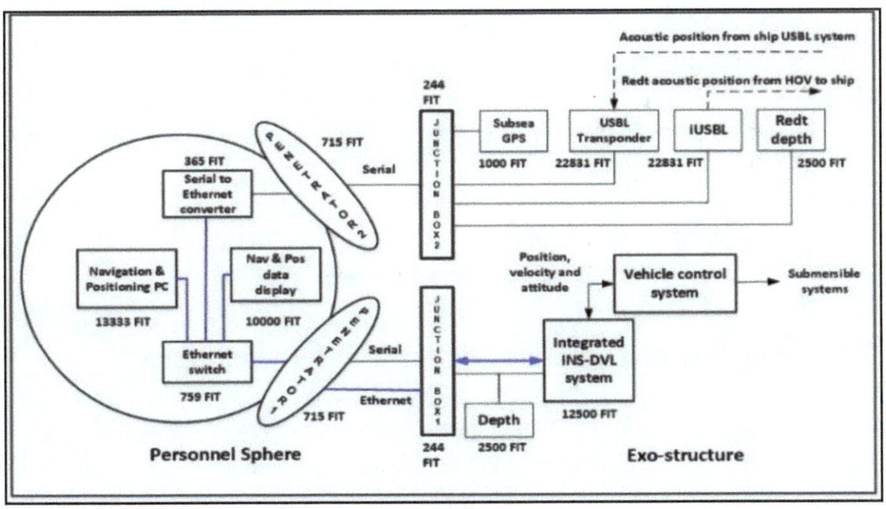

Fig. 10.2 Architecture of Matsya's Nav-Pos system [5]

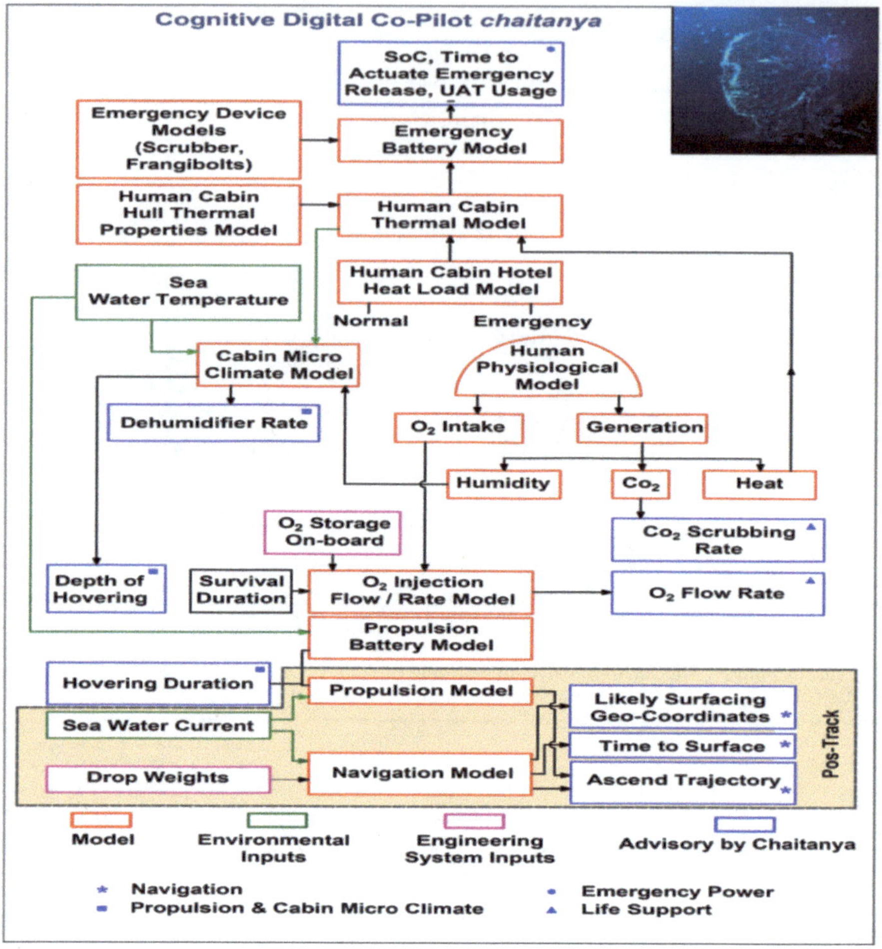

Fig. 10.3 Architecture of CDT Chaitanya showing Pos-Track

10.3 Overview of Cognitive Digital Copilot Chaitanya

The proposed module of the CDT Chaitanya shown in Fig. 10.3 comprises of multiple complex coupled models involving interactions between human systems and equipment under different operating conditions in varying sea environmental conditions. The CDT Chaitanya Pos-Track predicts in-advance the likely surfacing trajectory, surfacing position, surfacing time and offset from the deployment vessel, enabling appropriate position estimation of rescue assets for early recovery of the human crew. The proposed Chaitanya DT shall be operational in Matsya6000, as well as in Mission Control Center on-board deployment ship, are being validated for stable and consistent performance [6].

The power system module of emergency system is based on precise physical model of the lead acid battery unit which integrates machine-learnt profile of physical thermal conditions of the personnel sphere at various ocean depths. The proposed system provides an operating protocol for effectively using the energy for various load operations including critical operations. The life support system module (based on gas diffusion model and machine-learnt O_2 consumption and CO_2 generation model) supports optimizing O_2 consumption for 96 h and also ahead-of-time, till the crew is rescued. The residual energy in the propulsion battery units and the period of operation in subsurface and microclimate model which can be predicted much ahead-of-time, which provides decision support the optimal parking depth which is optimal compromise between propulsion system for position-keeping overcoming water currents and maintaining condensing human cabin microclimate for crew comfort, without using energy-intensive dehumidification units [6].

The capacity of the emergency batteries housed inside the human cabin of Matsya6000 are sized to cater for 96 h. A well-defined Emergency Operating Protocol (EOP) for effectively rationing the available energy to the essential life-critical systems and communication equipment during 96 h shall be in place. Any deviation from the EOP consuming increased energy than envisaged affects the energy availability during the remaining emergency period. However, during emergency, the possibility of the crew deviating the EOP could not be avoided. Under such circumstances, a DT for the Emergency Power System (EPS) shall help to redefine the EOP (hard-to-predict scenario) such that the residual energy in the emergency batteries is enough to operate the emergency systems reliably during the remaining emergency period. The life support system in MATSYA 6000 is dependent on machine learning-based learnt profile of oxygen consumption analysis, when the rescue operation is extended after 96 h. Depending on the residual energy available in the propulsion battery unit and the duration to remain in subsurface the propulsion and microclimate modules computes prediction unit, and recommends the suitable hovering depth for ensuring maximum hovering duration depth.

10.4 Description of Chaitanya Pos-Track Module

The architecture describing the inputs and outputs of the Pos-Track module is shown in Fig. 10.4. Central to the Post-Track is the Dead Reckoning Algorithm (DRA, Block-F), that is based on the ocean environmental situation-aware concept, in which movement of Matsya in surge, sway and heading operation Degrees of Freedom (DoF) is driven by water currents. During the ascend, the position of Matsya is continuously computed based on water current velocity vectors in surge and sway DoF (X- and Y-axis), which is casted prior to the deployment. In Block-A, the depth-wise velocity vectors (V_x and V_y) are interpolated (to a resolution of 1 m), validated and machine-learnt based on identified Long Short-Term Memory (LSTM) technique. The minimum water velocity for Matsya (to overcome the initial inertia and start forward movement) is experimentally identified (Block-B) from the maneuvering

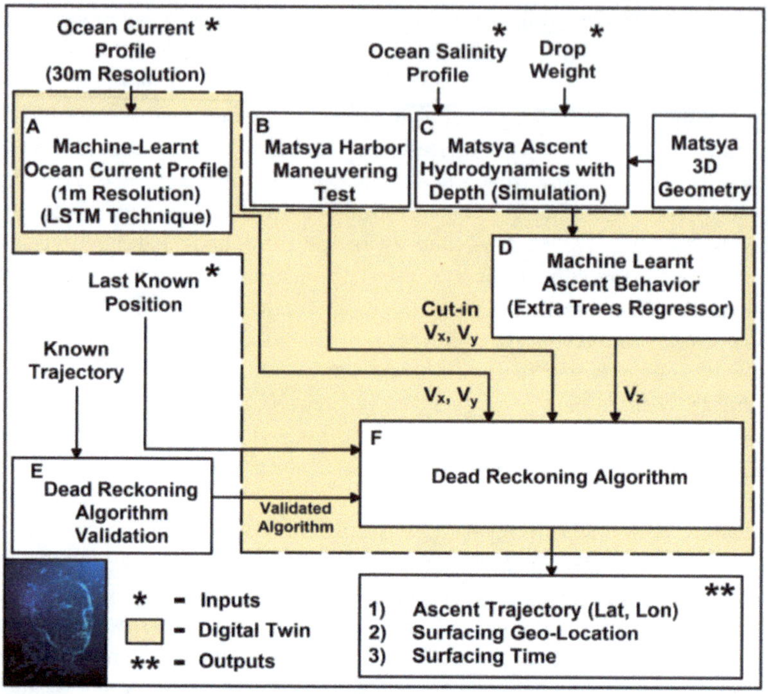

Fig. 10.4 Architecture of CDT Chaitanya Pos-Track module

test (cut-in water velocity to move Matsya from rest). With ocean salinity profile, mass of the drop weights and the 3-D geometry of Matsya6000 as inputs, Block-C computes the ascend hydrodynamic behavior with ocean depth, that are machine-learnt in Block-D using Extra Trees Regressor (ETS) ML algorithm. It serves depth-wise vertical velocity (V_z) to DRA. The DRA which is coded and validated with a known terrestrial trajectory (Block-E) is used in Block-F. The DRA outputs include the likely ascend trajectory, surfacing geo-coordinate and time to surface. These outputs enable the mission managers to precisely position the rescue assets for quick recovery system of human crew, in case of emergency ascend. Figure 10.5 depicts the architecture diagram of proposed position tracking module of Matsya 6000.

10.5 Machine Learning of Ocean Current Profile

The ocean water current profile which is depth against velocity profile in Central Indian Ocean basin where the water depth is 5500 m is logged during a scientific cruise in Research Vessel of NIOT. The ADCP has the provision to measure and log the 3-axis water current velocities and directions, at 50 m depth intervals [5]. In order to interpolate and learn the depth-series data using machine learning model,

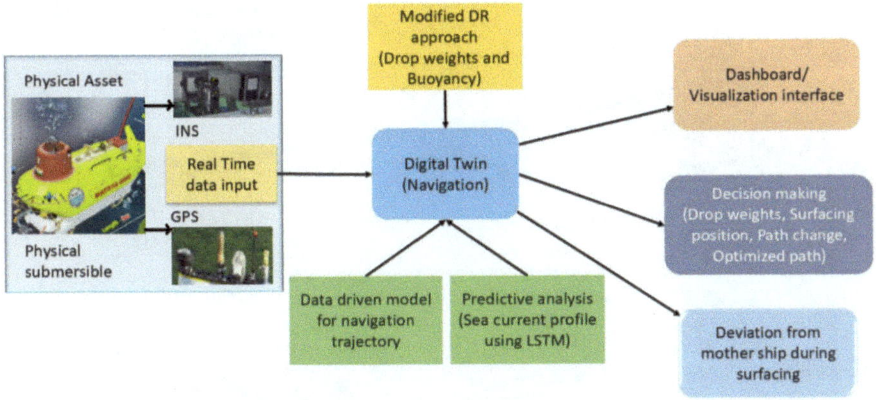

Fig. 10.5 Proposed position tracking module of manned submersible

RNN based LSTM machine learning technique is identified which has the capability to handle sequential data and learn long-term dependencies. LSTM is a proven technique which is used to model and predict ocean currents [7]. The machine-learnt V_x and V_y vectors using RNN LSTM technique having a depth-wise resolution of 1 m is plotted in Fig. 10.6. The effectiveness of the ML interpolation and learning process is analyzed using Mean Absolute Error, and the results of V_y is plotted in Fig. 10.7. This shows that the predicted and actual velocity vectors is close within 0.01 m/s.

Fig. 10.6 Learnt ocean current profile with a resolution of 1m

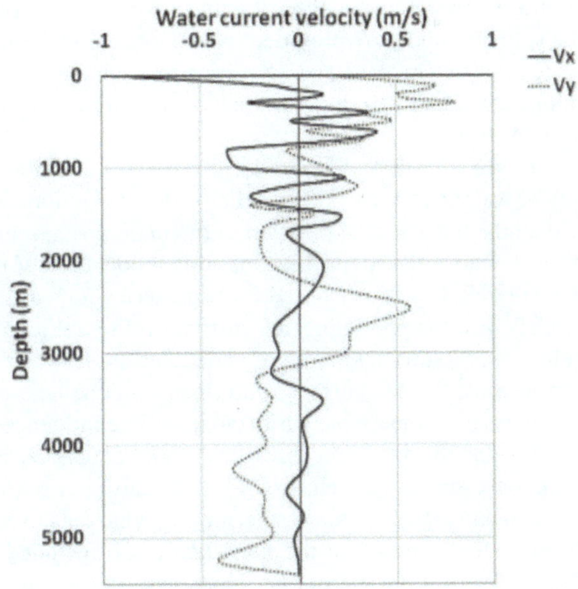

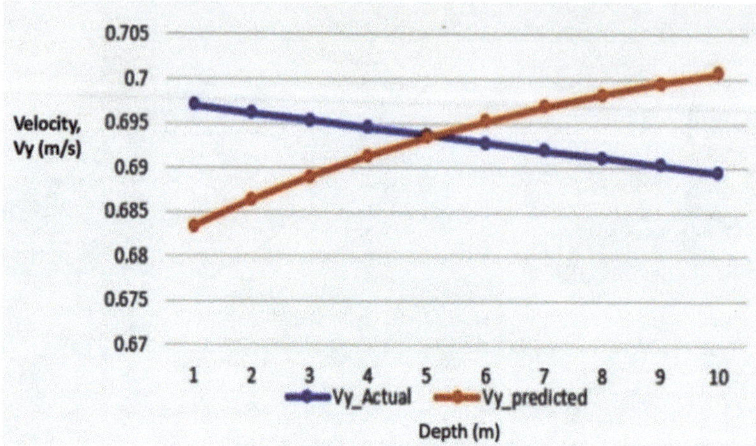

Fig. 10.7 Machine learning performance assessment of V_y in m/s

The position tracking module comprises of position, approach model and outcomes of the estimation model. The state variables include initial geo-coordinates, horizontal drift placement and position updates of the system. The computational approach consists of position track module, ascent velocity under different buoyancy conditions, numerical integration of vehicle ascent dynamics under different currents, outcomes provides predicted vehicle trajectory, updated position estimates, and drift characteristics during ascent.

The DRA computes the velocity of Matsya in surge and sway DoFs, for which the velocity inputs are provided from the machine-learnt current profile with a depth resolution of 1m. However, in order to navigate MATSYA from static position, a minimum velocity or cut-in velocity is required to overcome the initial inertia of the system.

The sea current profile provides crucial information about the direction and intensity of water currents. This data helps the navigation system account for drift effects and adjust the essential position of the underwater vehicle accordingly. Long Short-Term Memory-based machine learning model (LSTM) is a type of Recurrent Neural Network (RNN) that is suited for time-series forecasting and sequential time-series models, such as predicting sea current profiles. With a set of features such as water velocity, pressure, temperature and Euler angles over a specific period, the LSTM can forecast the sea current conditions over the next time frame, providing insight into how the submersible's trajectory will be influenced by underwater currents.

The LSTM layer processes the input sequences, retaining critical information about past sea current behavior while discarding irrelevant details using three gating mechanisms: Forget Gate—determining what data to remove from the memory cell; Input Gate—controlling the new data to be included to the memory cell; Output Gate—determining the output dependent on the current state of the memory cell. These gates help the LSTM layer maintain a balance between preserving long-term dependencies and adapting to new inputs. After processing the sequence with the LSTM layer, the output is passed through a fully connected layer networks. The dense layer outputs a vector of six elements, representing the predicted future values for each feature (e.g., velocity, pressure, temperature, etc.) at the next timestep. This allows for multivariate prediction, which is crucial for understanding the overall sea current profile. The output consists of predicted values for the next time step, which are used to adjust the submersible's trajectory in real time. With the prediction of next 10 rows of data, the model provides short-term forecasts that guide navigation decisions. This approach enhances the submersible's ability to operate autonomously in complex environments and helps ensure safe and energy-efficient navigation, especially in large depths where manual adjustments may not be feasible. By providing real-time sea current predictions, the model improves mission planning and operational efficiency in underwater explorations.

Table 10.1 provides the sample of estimated data profile using proposed ML. Table 10.2 provides the comparison of proposed model against other conventional forecasting ML models. It is understood that proposed RNN LSTM is more accurate than other ML models in predicting the ocean current profile.

Table 10.1 Sample of estimated sea current profile using LSTM-based Ml model

V_X	V_Y	V_Z	Roll	Pitch	Heading
– 0.1673	0.68336	0.50639	– 0.0002	– 0.0003	153.922
– 0.1633	0.68638	0.5055	– 0.0003	– 0.0007	154.587
– 0.16	0.689	0.50564	– 0.0004	– 0.0007	155.18
– 0.1571	0.69133	0.5058	– 0.0006	– 0.0008	155.704
– 0.1544	0.69339	0.50594	– 0.0007	– 0.001	156.168
– 0.152	0.69521	0.50607	– 0.0007	– 0.0011	156.577
– 0.1498	0.69683	0.50617	– 0.0007	– 0.0011	156.938
– 0.1478	0.69826	0.50625	– 0.0007	– 0.0012	157.256
– 0.1459	0.69954	0.5063	– 0.0006	– 0.0012	157.536
– 0.1442	0.70068	0.50634	– 0.0005	– 0.0013	157.784

Table 10.2 Comparison of ocean current profile estimation using proposed RNN LSTM

Method	MSE	RMSE	MAE
ARIMA	2.95	1.71	2.1
GRU	2.12	1.46	1.85
LSTM	1.86	1.36	0.96
Proposed RNN LSTM	1.05	1.024	0.65

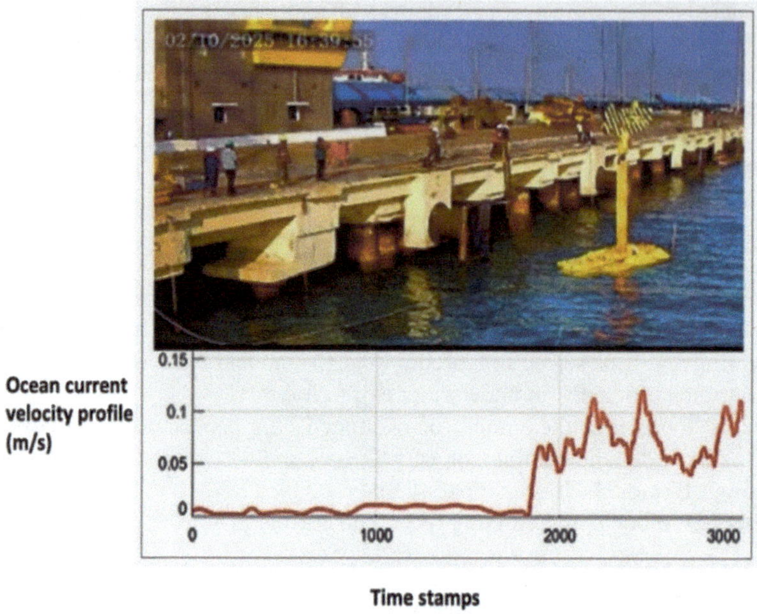

Fig. 10.8 Identification of cut-in velocity from maneuvering test

10.5.1 Determination of Cut-In Velocity by Maneuvring Tests

The cut-in velocity is determined experimentally using the maneuvering test in the harbor as provided in Fig. 10.8. It is identified that the minimum water velocity of 0.05 m/s is essential to override the inertia and makes Matsya to follow the water current.

10.6 Computing and Machine Learning Depth-Dependent Ascend Hydrodynamics

The ascend velocity (V_z) of Matsya as in (10.1) is an input to the DRA, that depends on weight of Matsya (W), displacement (Δ), depth-dependent sea water density (ρ), drag coefficient in heave DoF (C_D), and projected area (A_P) [8].

$$V_{\text{ascend}} = \sqrt{\frac{\Delta - W}{\frac{1}{2} C_{DA\rho} A_p}} \quad (10.1)$$

Based on the above relationship, the descend and ascend velocity of Matsya for a range of upward buoyancy are computed and depicted in Fig. 10.9.

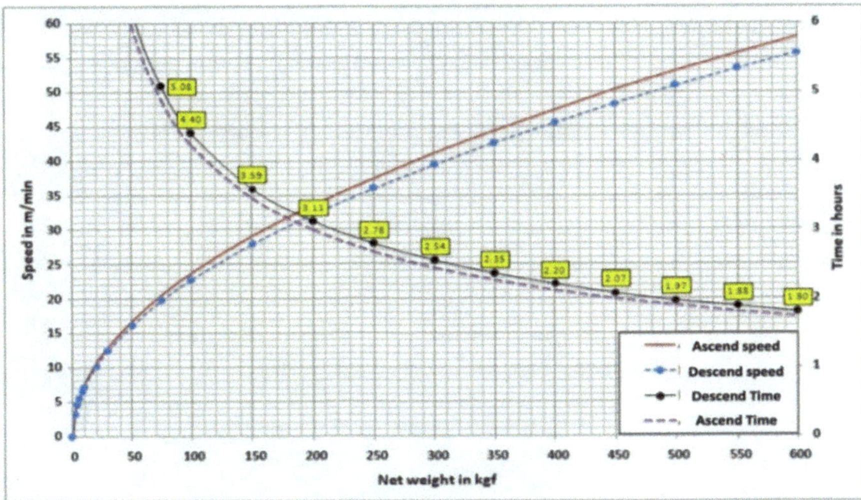

Fig. 10.9 Descend and ascend speed of Matsya with buoyancy

The results obtained through calculations are verified using Computational Fluid Dynamics (CFD). A standard RANSE-based CFD package is used for solving the flow around Matsya and estimate V_z, as well as visualizing flow parameters over her hull surface. Hexahedral meshes were used for meshing and six prism layers are generated adjacent to hull surfaces to capture boundary layer flow accurately. During the analysis, motion in other DoFs is constrained. Numerical simulation was carried out for using STAR-CCM + software to find out the terminal velocity of Matsya during various ascend scenarios. As an example, the flow field and velocity around Matsya in longitudinal plane (XZ) plane is shown in Fig. 10.9 [10]. This ensures that the ascent trajectory estimation from the seabed to the surface is performed with precision by calculating the net force, terminal velocity and time of ascent at each depth. This approach also accounts for variations in buoyancy due to changing depth, allowing for accurate predictions in scenarios involving different seawater densities and pressure conditions.

The ascend velocities with depth, for various values of buoyancy (drop weights) conditions are computed and shown as envelopes in Fig. 10.10. It can be seen that, with an upward buoyancy of 100 kg, ascend velocity which is 0.38 m/s at 6000 m, reduces to 0 m/s at 2850 m, and thereafter remains at the same depth. With an upward buoyancy of 500 kg, the average ascend velocity is 0.8 m/s. Buoyancy of 300 kg is required to ascend with an average velocity of 0.5 m/s. The ascend velocity of Matsya under various upward buoyancy conditions are machine-learnt using Extra Tree Regressor (ETR) ML technique. ETR builds multiple decision trees on random subsets of data and random feature splits to make predictions. It introduces more randomness in the tree construction process, potentially leading to faster training and higher variance in predictions [9]. The performance of the proposed model in

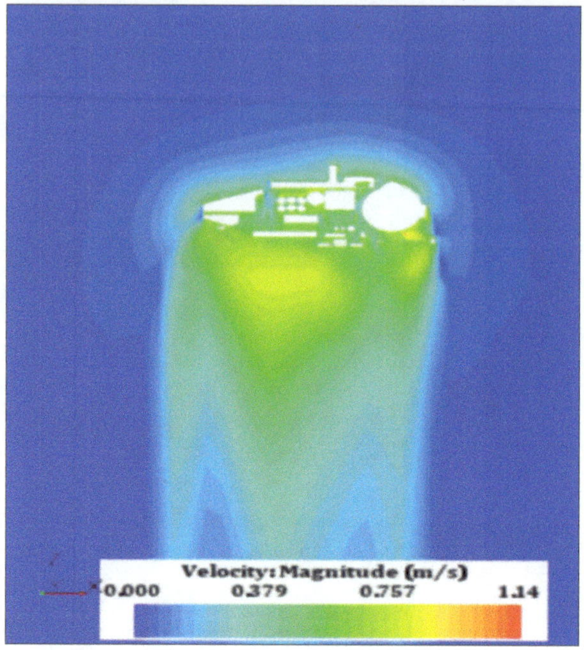

Fig. 10.10 Matsya ascent velocity profile for 300 kg buoyancy

this module is compared with random forest and decision tree models and the results are depicted in Fig. 10.11.

The proposed model provides better accuracy than conventional random forest and decision tree models by providing 0.025 MAE and 0.075 RMSE which is very high in conventional models (Fig. 10.12).

10.7 Dead Reckoning Algorithm Development and Validation

The machine-learnt input velocity vectors (V_x, V_y, and V_z) and based on body frame (BF) coordinates. For calculating the geo-position of Matsya respect to the navigation frame, measured body frame velocity is translated into navigation frame as provided in Fig. 10.13 [11–13].

The Direct Cosine Matrix (DCM) as in (10.2 and 10.3) is used to convert Matsya's BF velocity to NED velocity [14, 15].

$$V_n = C_b^n V_b \tag{10.2}$$

where C_b^e is the transformation matrix.

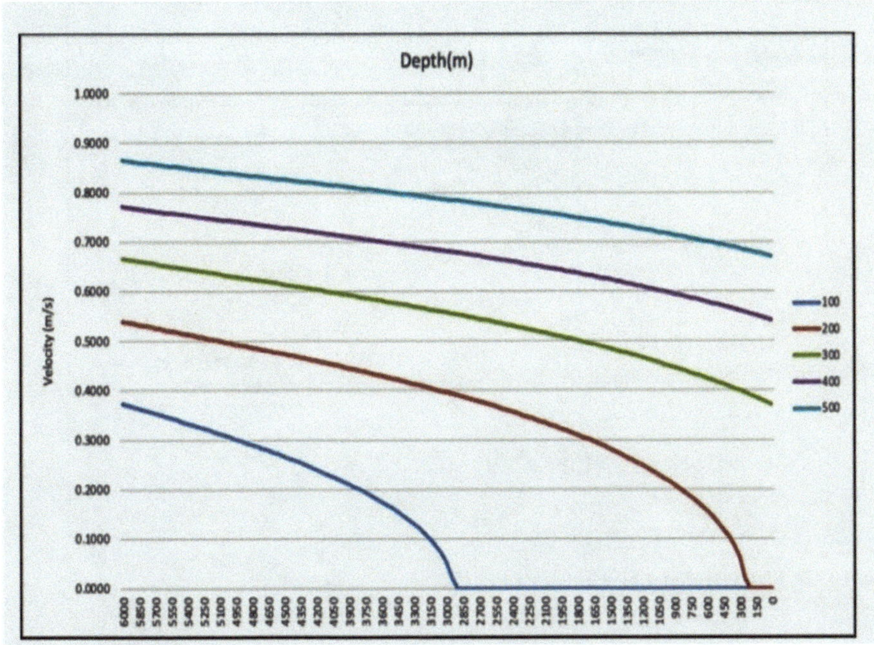

Fig. 10.11 Matsya ascent velocity profile for various buoyancy

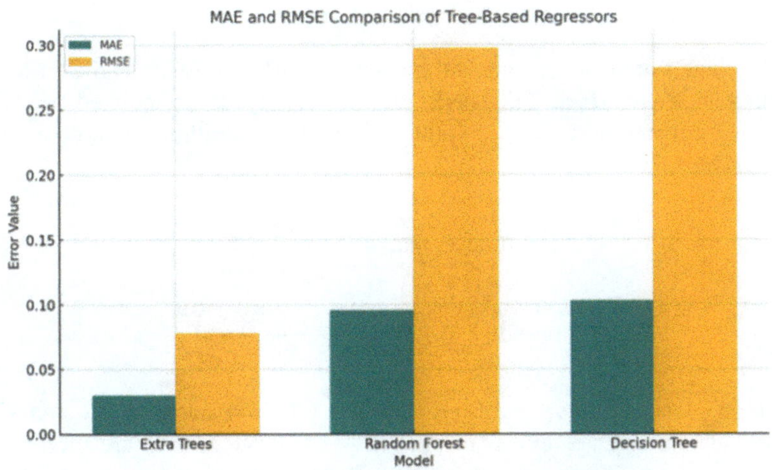

Fig. 10.12 Comparison of proposed ETR model against other tree-based ML regression models

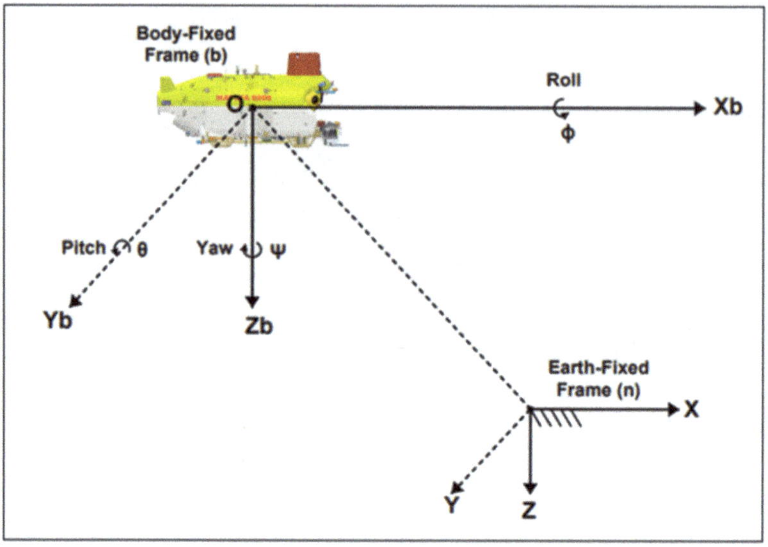

Fig. 10.13 Matsya BF and EF representation [5]

$$C_b^n = \begin{bmatrix} c\psi c\theta & -s\psi c\theta + c\psi s\theta s\phi & s\psi s\phi + c\psi c\phi s\theta \\ s\psi c\theta & c\psi c\phi + s\phi s\theta s\psi & -c\psi s\phi + s\theta s\psi c\phi \\ -s\theta & c\theta s\phi & c\theta c\phi \end{bmatrix} \tag{10.3}$$

where *s*, *c*, and *t* are sine, cosine, and tangent operation of Euler angles of Matsya. V_b is BF velocity matrix and V_n is NED frame velocity. The calculated NED frame velocity is estimated over a time to obtain the updated positions in geo-coordinates as in Eqs. (10.4, 10.5).

$$\text{Latitude } L = L + \left(\frac{v_N * \Delta t}{R_N}\right) \tag{10.4}$$

$$\text{Longitude } \lambda = \lambda + \left(\frac{v_E * \Delta t^* \sec L}{R_E}\right) \tag{10.5}$$

where L and λ are Latitude and Longitude per WGS 84 standard. V_N and V_E are Northing and Easting velocities (in m/s) of Matsya; R_N and R_E are meridian and transverse radius curvature, *e* is eccentricity. Based on the deployment ship position, the computed distance is recursively added to prior position to estimate the geo-referenced position as in (10.6 and 10.7).

$$R_N = \left(\frac{R_E\left(1 - e^2\right)}{\left(1 - e^2 \sin^2 L\right)^{\frac{3}{2}}}\right) \tag{10.6}$$

$$R_E = \left(\frac{R_E}{\left(1 - e^2 \sin^2 L\right)^{\frac{1}{2}}} \right) \tag{10.7}$$

The in-house developed DRA is validated for position estimation performance based on a terrestrial test, in which the DRA along with other validation systems are moved in stable land vehicle. During the test in which velocity inputs to DRA (under validation) is provided from a Commercially-off-the-shelf navigation-grade Inertial Navigation System. The outcome of the tests is summarized in Table 10.3. Results indicate that the DRA has position estimation accuracy of 0.7% of the distance traveled as provided in Figs. 10.13 and 10.14.

The DRA developed in MATLAB and implemented in Pos-Track is illustrated in Fig. 10.15. Based on the last known water depth, geo-position and heading of Matsya, DRA computes "ahead-of-time" (hence described as Cognitive Digital Twin) the ascend trajectory, surfacing position and real time of surfacing. During each iteration, with 1 m depth resolution (bin), from the last known depth, DRA fetches corresponding V_x, V_y, and V_z water velocity vectors of the respective bin, from the respective machine-learnt databases. If V_x and V_y are greater than the cut-in velocity,

Table 10.3 Performance assessment of DRA

Configuration	Distance traveled	Accuracy of DRA
DRA with velocity inputs from navigation-grade INS	678.1 m	0.7% of distance traveled
GPS-aided INS	683.3 m	

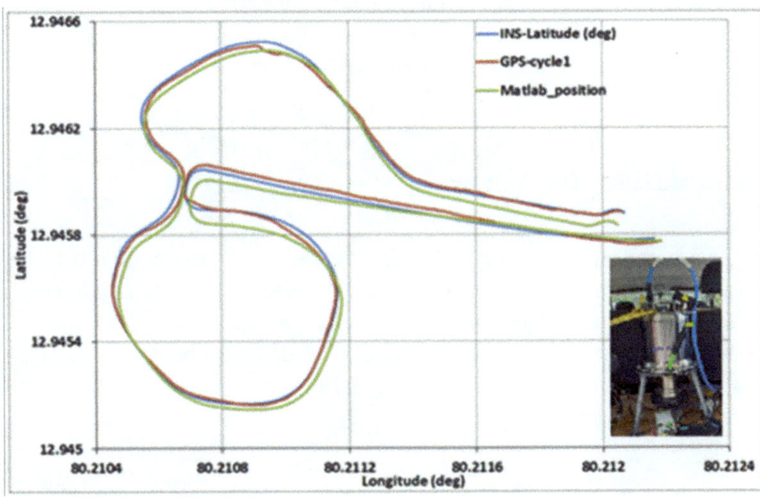

Fig. 10.14 Validation of DRA based on terrestrial tests

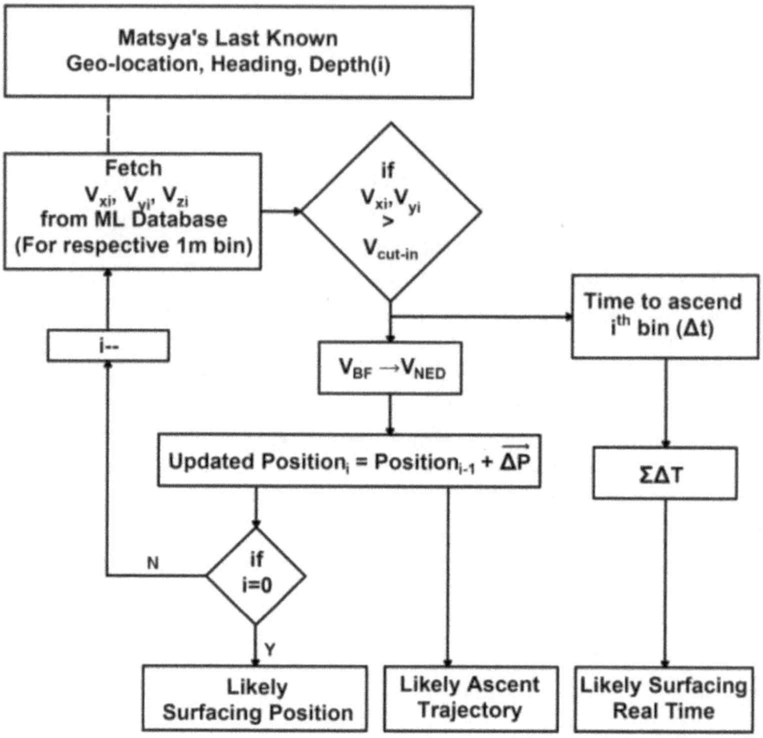

Fig. 10.15 DRA for obtaining the Pos-Track parameters

the velocity is used for iteratively updating the geo-location of Matsya, which finally provides the likely ascend trajectory and the surfacing geo-location. Based on the bin-wise V_z, the DRA computes the likely time to surface.

10.8 Simulations for Various Ascend Scenarios

Simulations are carried out using CDT Chaitanya Pos-Track module for analyzing Matsya's ascend scenarios for a range of likely upward buoyancy conditions. Inputs include the machine-learnt Central Indian Ocean water current profile and hydrodynamic behavior of Matsya6000. The simulation results with buoyancy of 300 kg and 500 kg as inputs are shown in Fig. 10.16a and b, respectively. The results are also summarized in Table 10.4.

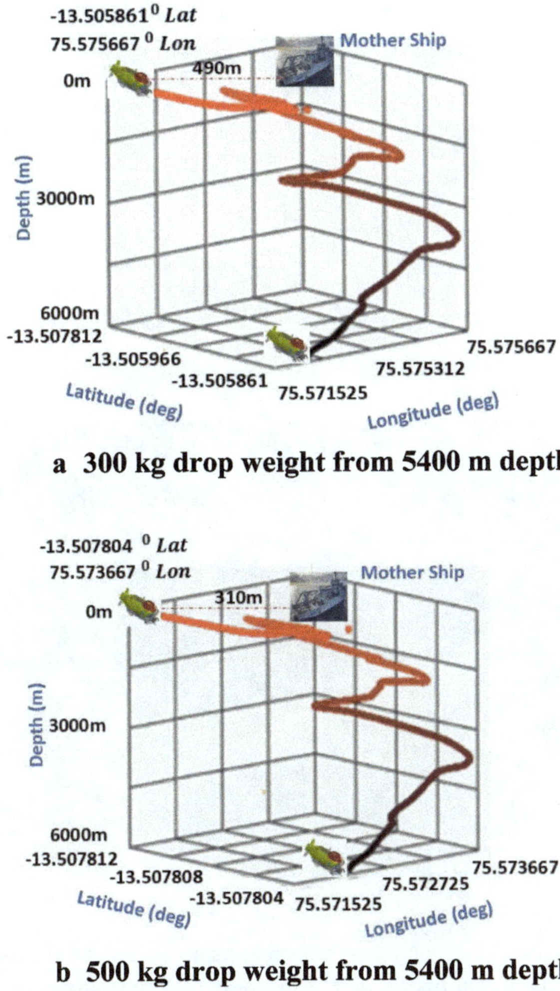

Fig. 10.16 Pos-Track simulations results for ascend with 300 kg and 500 kg buoyancies

10.9 Deployment of Chaitanya Pos-Track CDT in Matysa6000

The developed CDT Chaitanya Pos-Track module is deployed in the human cabin of Matsya6000 as provided in Fig. 10.17.

A user-friendly GUI is developed in LabView software which supports the mission personnel during the ascend, following non-availability of positioning information or sensor outages as shown in Fig. 10.18. Based on the last known information, the mission personnel can provide the status of the drop weights (buoyancy) as inputs, while all other parameters are machine-learnt. This front panel of GUI provides the

Table 10.4 Summary of simulation results Of CDT Pos-Track

Inputs/Outcome	Buoyancy	
	300 kg	500 kg
Inputs: Last known parameters Geo-position: – 13.505861^0 Lat, 75.571525^0 Lon Depth: 5400 m		
Average ascend velocity	0.5 m/s	0.8 m/s
Time to surface	3 h: 0 min	1 h: 52 min
Surfacing geo-location	Lat: – 13.507812^0 Lon: 75.575667^0	Lat: – 11.99853^0 Lon: 80.00399^0
Deviation from deployment ship	490 m	310 m

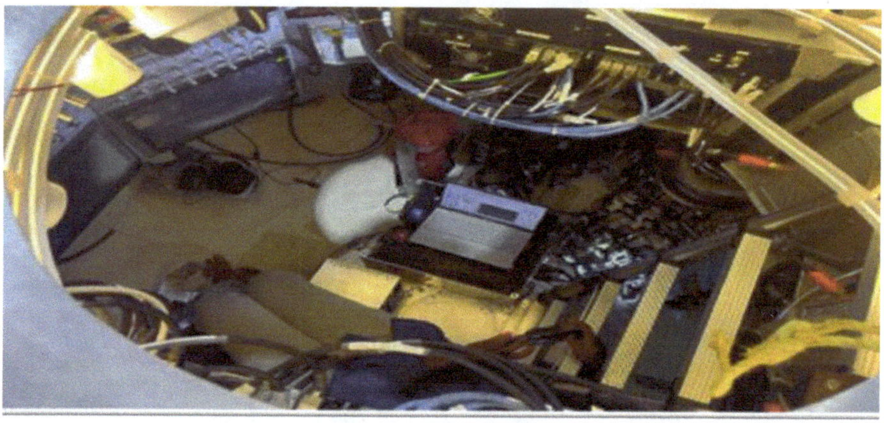

Fig. 10.17 Deployment of CDT in human cabin of Matsya6000

understanding of position track of the vessel at every 200 m depth, Time to reach the surface level during ascent operation by dropping different drop weights in different depths, distance of vessel from mother ship while ascending.

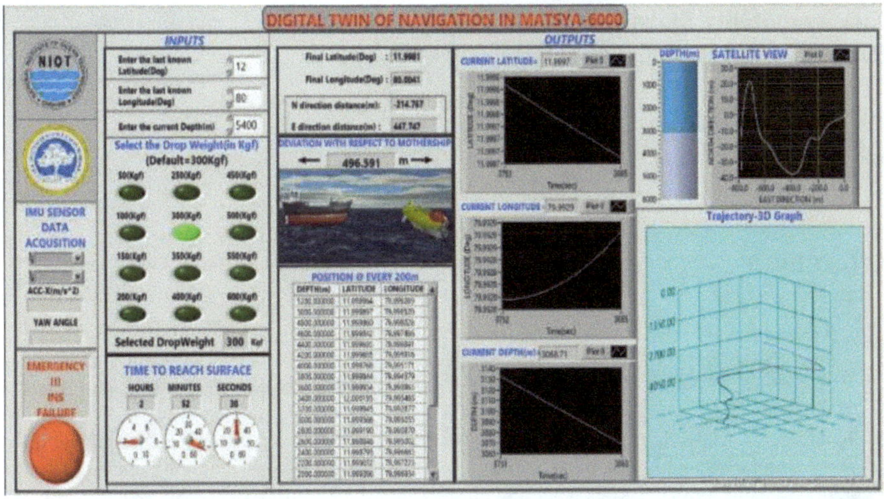

Fig. 10.18 Visualization front end of navigation DT of manned submersible

10.10 Conclusion

In order to ensure the reliability and safety of the 4th generation Matsya6000, certain standards like IEC15288 are strictly followed for system engineering. By implementing probabilistic failure analysis implementation using failure data obtained in field results as inputs, Probability of Failure in 1 year of operation in positioning system is reduced from 29% to 14.4%. This corresponds to a Mean Time to Fail duration of approximately 6 years. However, during a deep sea operations, during unavailability of positioning information or sensor outages leads to incapacitations of the crew, as Matsya is not tethered, hostile deep sea condition, fear of fault aggravation and unpredictable time to surface. In order to support the crew and mission during positioning system outages ahead-of-time, this work provides the solution for the development and validation of Cognitive Digital Twin Chaitanya Position Track module based machine-learnt ocean current profile, hydrodynamic behavior of Matsya6000, and field-tested dead-reckon position estimation algorithm. The proposed module of this paper can be extended to deploy for ensuring safety in human scientific submersibles with intelligence on decision support system during sensory outages/system failures leading to unavailability of position information, for which the system will predict the surfacing position, deviation from mother ship and estimation of position every 1 m depth profile.

Funding Declaration
The authors gratefully acknowledge the support extended by the NIOT, Ministry of Earth Sciences, Government of India, in undertaking the project work titled "Digital twin assisted operational decision support system for deep ocean manned scientific submersible", MoES/DOM/GIA/PAMC/01/2023 under the Deep Ocean Mission.

Data Availability Statement
There is no data availability for this manuscript.

Competing Interest
There is no conflicting interest in this manuscript.

References

1. Ravichandran, M.: India's deep ocean mission. Curr. Sci. **128**, 227–228 (2025)
2. Matsya-6000: India's Fourth-Generation Deep-Ocean Submersible successfully Completes Wet Testing. 2104039 (2025). https://www.pib.gov.in/PressReleaseIframePage.aspx?PRID=2104039
3. Palaniappan, M., Bala Naga Jyothi, V., Chowdhury, T., Sathianarayanan, D., Vedachalam, N.: A method for evaluating safety reliability of descend/ascend drop-weight system and mission abort decision for deep-ocean scientific human submersibles. Sci. Rep. **14**, 19266 (2024)
4. ISMERLO (International Submarine Escape and Rescue Liaison Office). https://ismerlo.org/. Accessed: Oct 2024.
5. Jyothi, V.B.N., Ramesh, R., Vedachalam, N., Ramadass, G.A.: Assessment of the technological maturity of manned submersible navigation positioning systems. Mar. Technol. Soc. J. **55**, 129–137 (2021)
6. Yang, L., Zhao, S., Wang, X., Shen, P., Zhang, T.: Deep-sea underwater cooperative operation of manned/unmanned submersible and surface vehicles for different application scenarios. J. Mar. Sci. Eng. **10**, 909 (2022)
7. Immas, A., Do, N., Alam, M.R.: Real-time in situ prediction of ocean currents. Ocean Eng. **228**, 108922 (2021)
8. Pranesh, S.B., Rajput, N.S., Sathianarayanan, D., et al.: CFD analysis of the hull form of a manned submersible for minimizing resistance. J. Ocean Eng. Mar. Energy **9**, 125–143 (2023)
9. Wang, Z., Mu, L., Miao, H., Shang, Y., Yin, H., Dong, M.: An innovative application of machine learning in prediction of the syngas properties of biomass chemical looping gasification based on extra trees regression algorithm. Energy **275**, 127438 (2023)
10. Bala Naga Jyothi, V., Chowdhury, T., Pranesh, S.B. et al.: Marine journal of the institute of marine engineers (India), Marine India engineers review, digital twin for India's indigenous deep-ocean human scientific submersible Matsya6000 evolution of next-generation bio-inspired AI-enabled homing guidance system for autonomous underwater vehicles-Part A. Mar. Eng. Rev. **19**, 8–24 (2025)
11. Ramesh, R., Jyothi, V.B.N., Vedachalam, N., Ramadass, G.A., Atmanand, M.A.: Development and performance validation of a navigation system for an underwater vehicle. The J. Navigat. **69**(5), 1097–1113 (2016)
12. Stovner, B.N.: Aided inertial navigation of underwater vehicles. PhD Dissertation, Norway, Dept of Marine Technology, Norwegian University of Science and Technology (2018)
13. Li, S., Zhao, Y., Chen, Y., Ben, Y., Wang, Z.: A novel GPS-aided robust calibration method for SINS/DVL integrated navigation system. IEEE Sens. J. **25**(9), 15851–15862 (2025)
14. Jia, C., Wen, Q., Zou, Z., Nie, Z.: Underwater real-time heading determination method by using a dual-beacon design of ultra-short baseline (USBL) acoustic positioning. IEEE Sens. J. **25**(6), 10163–10173 (2025)
15. Ben, Y., Wang, J., Zhang, L., Li, Q.: A novel initial position matching method utilizing multibeam bathymetry for underwater vehicles. IEEE Trans. Instrum. Meas. **74**, 1–8 (2025)

Chapter 11
Data Acquisition and Preprocessing in the Internet of Marine Things

Kamalika Bhattacharjya, Debashis De, Sapta Rathi Roy, Tiasha Dutta, Minakshi Sarkar, Rupsa Chatterjee, and Samarjit Roy

11.1 Introduction

Internet of Marine Things (IoMT) is crucial for real-time marine surveillance and decision-making. A crucial sub-component of IoMT is the Internet of Underwater Things (IoUT) network, which comprises underwater devices such as autonomous underwater vehicles (AUVs), remotely operated vehicles (ROVs), and seabed sensor nodes [1, 2]. The IoMT and IoUT systems capture data autonomously and enable real-time, critical decisions. The IoUT architecture for IoMT applications is present in Fig. 11.1. IoUT enables the acquisition of real-time underwater data that aerial networks cannot capture because water absorbs radio waves, which work as the primary communication medium for aerial networks. However, the underwater environment introduces unique challenges: high propagation delay, limited bandwidth, variable channel conditions, security, and high energy consumption [3, 4]. In this context, data acquisition and preprocessing become essential to ensure that collected data is clean, meaningful, and analyzable. Efficient preprocessing minimizes data transmission costs, improves information quality and energy consumption that significantly

K. Bhattacharjya (✉) · D. De
Centre of Mobile Cloud Computing, Department of Computer Science and Engineering, Maulana Abul Kalam Azad University of Technology, West Bengal, Nadia, West Bengal, India
e-mail: kamalika.bhattacharjya@makautwb.ac.in

D. De
e-mail: debashis.de@makautwb.ac.in

S. R. Roy · T. Dutta · M. Sarkar · R. Chatterjee
Department of Information Technology, Maulana Abul Kalam Azad University of Technology, West Bengal, Nadia, West Bengal, India

S. Roy
Computing and Information Technology, EIU, Becamex IDC Corp, Binh Duong, Ho Chi Minh City, Vietnam
e-mail: samarjit.roy@ieee.org

T. A. Tran et al. (eds.), *Big Data Analytics for Internet of Marine Things*, Internet of Things, https://doi.org/10.1007/978-3-032-28171-5_11

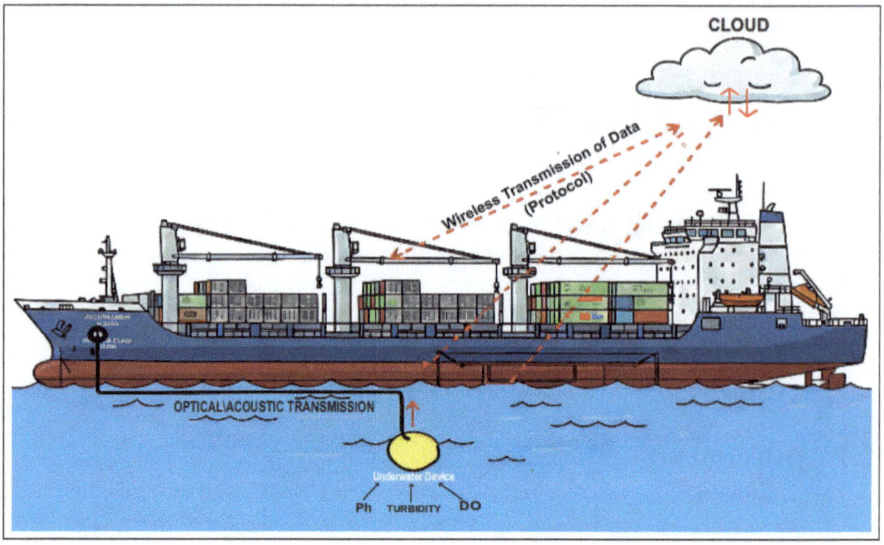

Fig. 11.1 Internet of underwater things network architecture for Internet of Marine Things application

enhances the model performance [5]. IoMT applications include bioflocks, environmental monitoring, coral health monitoring, military surveillance, and natural disasters [6]. Underwater sensors capture key physical data, including dissolved oxygen (DO), pH, turbidity, and temperature, in real time, filtering and validating the data before feeding it into the model [7, 8]. In real time, data is cleaned and normalized to support stable performance and efficient decision-making in autonomous underwater management systems and low-cost, customized hardware components. After collecting data from the IoUT monitoring system, the most important task is data preprocessing. This is performed to maintain the quality of the underwater data. The first few preprocessing steps include data collection, data cleaning, handling missing values, feature extraction, and data splitting. After that, several changes have been implemented across different data types. In tabular data, data imbalance is a significant issue. To mitigate class imbalance, various data-balancing techniques, such as SMOTE, are employed. The next preprocessing technique is feature scaling, which ensures equal variable contributions in the data. For image data, preprocessing techniques differ from those for tabular data. First verification and conversion are required to be performed on each image. Some geometric transformations and photometric transformations should be applied as necessary. For video data, preprocessing techniques are similar to those for images, but many other techniques have also been employed, such as geometric and photometric transformations. Pixel normalization and temporal frame sampling are conducted. Spatial preprocessing is also applied to individual frames. All these data preprocessing steps are performed to ensure high-quality data for model training and testing. These are essential for a mission-critical intelligent IoMT system for efficient data analysis.

11.2 Literature Survey

Our survey critically examines current developments and intractable issues in data acquisition and preprocessing within an Internet of Marine Things ecosystem. Marine IoMT is applied to resources because they are dynamic and, as such, complex to explore, monitor, and recover from disasters [9]. The volume, velocity, and variety of information received by sensors, hydrophones, AUVs, and cameras used in IoUT systems pose challenges for "Big Marine Data," which requires refined processing methods beyond traditional ones [1]. The challenges among limitations of IoMT include energy storage, latency, bandwidth, and communication [10, 11]. These problems are important to address in decision-making processes in the marine environment because raw data often contain missing values or outliers, which can compromise the integrity of the analytical process [12]. This means that proper data preprocessing methods are necessary to convert raw, usually noisy, high-dimensional data into a useful format for processing [1]. Communication approaches such as optical, magnetic induction, and acoustic signals are essential for integrating IoMT systems, despite current limitations in reliability and bandwidth [13]. In particular, acoustic communication, though widespread, has high latency, is limited by bandwidth, and has high bit error rates due to signal attenuation and the effects of marine currents [14, 15]. These environmental influences may also destroy data collection tools, which adds to inaccuracy and inadequate data [16, 17]. Although machine learning (ML) technology can analyze large datasets efficiently to solve underwater problems compared to conventional methods [18]. Additionally, built-in ML-enhanced autonomous sensors are essential to minimize maintenance costs and achieve greater operational choice for mission-critical IoUT applications [1].

To address network constraints, data-sharing limitations, and privacy preservation in the IoMT system, federated learning (FL) has become an effective solution for collaborative training and learning in IoMT environment. In an FL environment, IoMT nodes independently train local models using their local data and periodically share model parameters with the server, which aggregates model updates and generates a new global model [19]. This decentralized system allows clients to keep raw data with them instead of transferring the data to a centralized repository [17] to ensure privacy, security, and communication efficiency for mission-critical IoMT applications. The distributed nature of FL is particularly suitable for large-scale IoMT deployments involving sensors, autonomous vehicles, and marine monitoring platforms located across different geographic regions. By exchanging only model updates instead of complete datasets, the framework significantly reduces communication overhead and protects sensitive environmental information [20–22]. This feature is important in underwater communication systems where bandwidth is limited, and transmission costs are high [23]. In practical applications such as marine debris detection or water quality monitoring, multiple organizations–including research institutions, port authorities, and environmental agencies–can contribute to a shared learning model while maintaining control over their own data resources [24]. The iterative nature of FL also allows models to continuously improve as new observations become available, which is beneficial for real-time environmental monitoring [17].

Recent studies have also explored the integration of automated deep learning (DL) techniques to improve model development and interpretation for water quality monitoring and marine environmental analysis [1, 25]. These methods help overcome the generalization challenges commonly encountered in traditional big-data-driven approaches. Reinforcement learning has further been considered to support adaptive decision-making in complex underwater environments where communication channels are dynamic and uncertain [17]. In this direction, federated reinforcement learning allows multiple nodes to collaboratively optimize their decision by sharing encrypted learning parameters rather than raw data, which preserves privacy and improves learning efficiency [26].

Despite these advantages, several challenges remain when applying FL in underwater environments. Acoustic communication channels often suffer from high latency, signal attenuation, and intermittent connectivity, which can disrupt the timely exchange of model updates [17, 27]. In addition, IoMT devices typically operate under strict resource constraints, including limited processing power, memory capacity, and battery life. These limitations make it difficult to implement computationally intensive privacy-preserving mechanisms such as differential privacy or complex encryption methods [28]. Another challenge arises from the heterogeneous nature of data collected by distributed sensors, which is often non-IID and may affect model convergence and performance [28].

To address these issues, researchers have proposed several improvements to FL frameworks. Adaptive strategies, including federated meta-learning, have been explored to improve model generalization across varying marine conditions [29]. Techniques such as gradient clipping, dynamic privacy budgeting, and secure multi-party computation can also help maintain privacy while allowing resource-constrained devices to participate in collaborative learning [28]. In addition, communication-efficient approaches such as sparse parameter updates and model quantization are being investigated to reduce transmission costs and energy consumption in bandwidth-limited underwater networks [30]. Lightweight ML models are essential to ensure efficient training and timely parameter updates in IoUT environments [31].

Overall, FL provides collaborative intelligence in IoMT systems by supporting decentralized training while preserving local data privacy. It offers a practical solution for large-scale marine monitoring and environmental assessment applications.

11.3 Data Acquisition for Internet of Marine Things

Data acquisition in the IoMT system refers to the process of gathering data from the underwater environment through devices such as sensors and AUVs, which play a pivotal role in environmental monitoring, national defense, and other real-time applications [32]. It offers valuable insights into marine ecosystems, supports resource management, and encourages technological advancement in IoMT.

11.3.1 Sensors and Processing Units of Internet of Marine Things

A sensor is a device that measures physical quantities of environmental changes and converts them into electrical signals for processing [13]. Sensors collect different physical phenomena such as pH, temperature, and turbidity for underwater applications [33]. For example, temperature, DO, humidity, pressure, gas, and light sensors are used for natural disaster prediction, smart surveillance, and water quality monitoring. Accelerometers, gyroscopes, GPS, and proximity sensors are crucial for autonomous navigation and security systems [34]. Table 11.1 presents the sensor details.

Processing units in the IoUT are the components, like microcontrollers (MCUs), system-on-a-chip (SoC), and edge computing nodes, that handle data collection, analysis, and decision-making. These units process sensor data to control actuators, communicate with other devices for energy-efficient tasks, such as control processing and machine learning in IoMT. Table 11.2 presents the details of the processing units in IoMT.

Table 11.1 Sensors with detailed measurements

Components	Technical specification	Range
Temperature sensor	Category	NTC (Negative temperature coefficient)
	Resistance at 25°C (R25)	Typically 1 kΩ, 10 kΩ, 100 kΩ (model dependent)
	B-constant (Beta value)	3000 K–4500 K (material dependent)
	Operating temperature range	−55°C to +125°C (some models up to +150°C)
	Tolerance (on R25)	±1% to ±5%
Turbidity sensor	Operating voltage	5V DC
	Measurement range	0–1000 NTU (nephelometric turbidity units)
	Operating temperature	5°C–90°C
	Accuracy	±5% of full-scale value (typical)
PH sensor	Range	0–14 pH
	Measuring temperature	0–60 Ω
	Working voltage	5V DC
	Response time	Less than 2 min
Dissolved oxygen (DO) sensor	Density range	Range: 0 20 mg/L
	Pressure range	0 50PSI
	Response time	Less than 2 min
	Voltage	5v

Table 11.2 Details of processing units in IoUT

Components	Technical specifications	Details
Raspberry Pi zero 2W	CPU	1 GHz quad-core 64-bit arm cortex-A53 CPU
	RAM	512 MB SDRAM
	Wifi	2.4 GHz 802.11 b/g/n wireless LAN
	Bluetooth	Bluetooth 4.2, bluetooth low energy (BLE), onboard antenna
	Camera connector type	CSI-2 camera connector
	Pin	HAT-compatible 40-pin header footprint (unpopulated)
	Power port	Micro USB power(5V)
	Encoder/decoder	H.264, MPEG-4 decode (1080p30); H.264 encode (1080p30)
	Graphics	OpenGL ES 1.1, 2.0 graphics
	Dimension	65 mm × 30 mm
Arduino nano	Microcontroller	ATmega328
	Operating voltage	5 V
	Flash memory	32 KB of which 2 KB used by bootloader
	SRAM	2 KB
	Digital I/O pins	Digital I/O pins
	PCB size	18 × 45 mm

11.3.2 Working Principle of Internet of Marine Things

The sensors and working flow of the data acquisition process are presented in this section. Here, Diagram 11.2 shows how data flows through each layer, enabling it for further processing.

(i) **Input (Sensor)**: Initially, data is collected from multiple sensors. Some popular IoUT sensors are pH, temperature, acoustic, and turbidity.
(ii) **Primary Processing Unit**: The data collected by these multiple sensors is subsequently fed into a processing unit. Arduino and ESP32 are favorable primary processing units.
(iii) **Advanced Processing Unit**: More powerful and energy-efficient processing units than the primary processing unit act as an advanced processing unit. The Arduino then acts as an intermediary, passing the information it receives to a Raspberry Pi Zero 2W for subsequent processing. Raspberry Pi Zero 2W is an advanced processing unit.

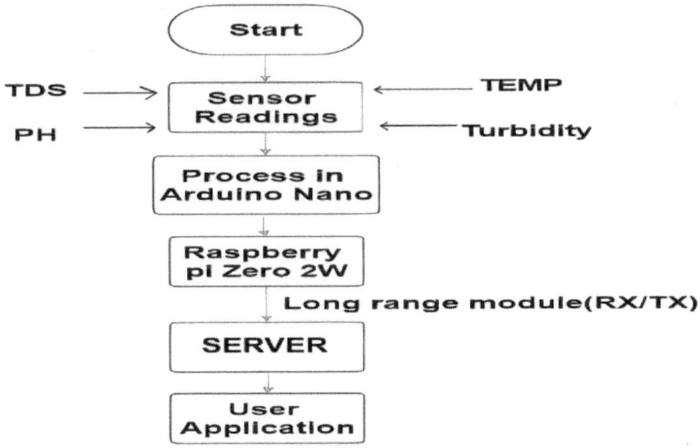

Fig. 11.2 Working flow diagram for IoUT system

(iv) **Power System**: To manage the power supply, the entire system is sustained by lithium-ion batteries (18650), and the power supply is managed using a Battery Management System (BMS).

The IoMT system is fundamentally divided into two main subsystems: (i) Data Module and (ii) Battery Module, as presented in Fig. 11.3. The Battery Module provides power to the Data Module via a "Power Connection." The Data Module is

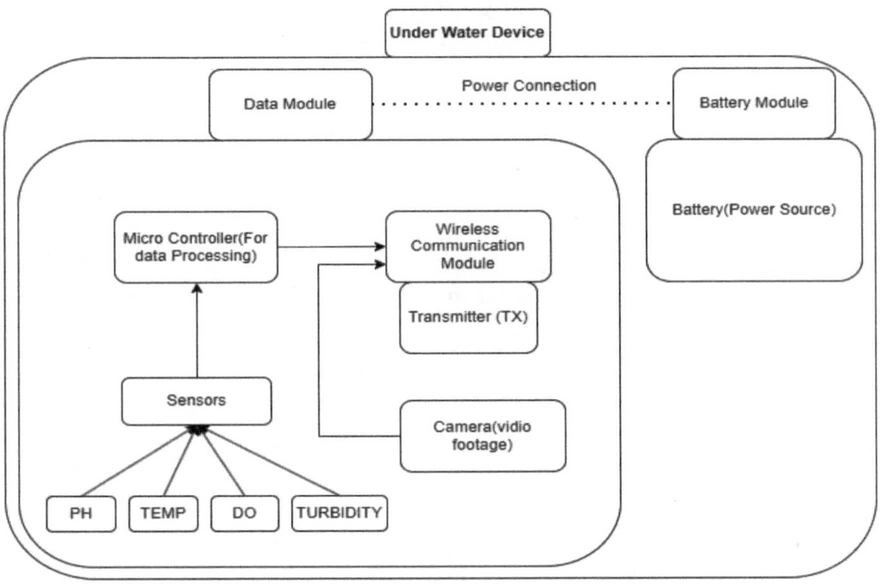

Fig. 11.3 Internal working structure of Internet of Marine Things device

responsible for collecting, processing, and transmitting sensor data. The data acquisition chain originates from a "Sensors" block, which interfaces with four distinct sensors: pH, temperature, acoustic, and turbidity. Sensor data is transferred to the microcontroller for initial processing or aggregation [35]. Camera sensors capture the video footage. Finally, the processing unit passes data to the Wireless Communication Module, which uses the transmitter to send the data to the end node. The battery module powers the entire system.

11.3.3 Autonomous Underwater Vehicle (AUV) for Internet of Marine Things

An Autonomous Underwater Vehicle (AUV) is an underwater robot, or "drone," that works entirely on its own without a human operator or a physical connection to a surface ship [11]. It is programmed for a specific mission and carries its own onboard power and navigation sensors. AUVs are classified according to design and movement, which are needed according to missions:

(i) Torpedo-Shaped (Cruising AUVs): This is the most common design. These AUVs are moved by thrusters and are suitable for speed and long-distance traveling. They are ideal for mapping the seafloor in "mowing the lawn" patterns.
(ii) Underwater Gliders: These gliders are designed for ultra-high endurance. Instead of propellers, they make use of buoyancy to sink and rise and then make use of wings to turn the vertical movements into gliding motions. They are slow and take up little energy, making them suitable for long-term monitoring of the environment.
(iii) Hovering AUVs: These are designed for high maneuverability and can stay in one position, similar to how a drone can stay in one position in the air. They are equipped with many propellers, allowing for precise movement in all directions, and are thus suitable for inspecting objects such as pipelines, ships, and bridge pillars.
(iv) Bio-mimetic AUVs: These are experimental and less common types of AUVs, where they are designed to mimic the form and movements of marine animals, such as fish and manta rays.

11.3.4 Protocols in Internet of Marine Things

A protocol is a set of rules that defines how data is sent and received over an IoUT network [5]. It makes sure that devices can communicate correctly and understand each other's messages. Underwater tools do not use normal Wi-Fi or radio waves well, so they use special rules that work with sound, light, or magnetic signals. The types of protocols are:

(i) ***JANUS Protocol*** —It's the Standard of NATO in a way that different underwater items can share their data through sound.
(ii) ***Aqua-Net***—A study rule for setting up underwater sensor links.
(iii) ***H2O Talk Rule***—Used for fast data in shallow water with light (optical).
(iv) ***UWAN Protocol***—Underwater Wireless Acoustic Network—supports data communication among mobile underwater nodes.
(v) ***SEA Swarm/SUN Protocols***—They belong to the set of experimental protocols for IoT and robotics under the water environment.

11.4 Preprocessing Pipelining for Internet of Marine Things

The information preprocessing pipeline in the IoMT plays a pivotal role in converting raw marine sensor information into reliable and meaningful information for downstream analytics. Marine sensing environments are inherently harsh and dynamic, leading to data that is often noisy, incomplete, and heterogeneous. As IoMT systems integrate data from underwater sensors, autonomous vehicles, and surface platforms, preprocessing becomes essential to ensure data quality, consistency, and efficient utilization of limited communication and energy resources. Raw marine information collected in IoMT deployments encompass a wide range of physical, chemical, acoustic, and visual parameters. These data streams are generated at varying sampling rates and are frequently affected by environmental disturbances such as water turbulence, acoustic multipath effects, corrosion, and biofouling. In addition, scattered connectivity and packet loss in underwater communication further degrade data completeness. Consequently, raw marine data cannot be directly used for analysis and decision-making without systematic preprocessing. The first stage of the preprocessing pipeline focuses on data cleaning, which aims to remove errors and inconsistencies introduced during sensing and transmission. Noise filtering techniques are applied to smooth sensor readings and suppress random fluctuations caused by environmental interference. Outlier detection mechanisms identify abnormal values that may arise from sensor drift, temporary faults, or extreme environmental events. Missing data handling techniques are then employed to reconstruct incomplete datasets using interpolation or predictive models, thereby improving continuity and reliability in long-term marine observations. Following data cleaning, the pipeline proceeds to data transformation, where heterogeneous sensor data is standardized and arranged. Normalization techniques rescale sensor measurements into comparable ranges, preventing dominance of specific parameters in analytical models. Time synchronization is critical in IoMT systems, as distributed marine sensors often operate with unsynchronized clocks and experience variable communication delays. Correcting temporal misalignment enables accurate correlation and fusion of multi-source data. Spatial referencing further enhances data usability by associating measurements with precise geographic coordinates and depth levels, allowing spatial analysis of marine environment. The data reduction stage addresses the demanding bandwidth and energy constraints of marine networks. Instead of transmitting large volumes of raw data, IoMT systems extract informative features that capture essential patterns while reducing data size. Aggregation techniques combine data over time to eliminate redundancy, and compression methods further minimize communication overhead. By reducing data volume at edge or fog nodes, the system achieves lower latency, extended network lifetime, and improved scalability. Once preprocessed, the refined data is forwarded to higher-level analytics and application layers. These analytics support a wide range of marine applications, including environmental monitoring, ecosystem health assessment, anomaly detection, and offshore infrastructure management. The effectiveness of these applications is directly dependent on the quality of preprocessing, as inaccurate or inconsistent data can propagate errors throughout the system and compromise decision-making. In modern IoMT architectures, the preprocessing pipeline is increasingly distributed across edge, fog, and cloud layers. Edge nodes perform initial cleaning and reduction close to the data source, fog nodes handle synchronization and aggregation, and cloud platforms conduct large-scale analytics and model generation. This hierarchical preprocessing approach improves energy efficiency, reduces communication costs, and supports sustainable marine monitoring aligned with long-term ocean observation objectives (Fig. 11.4).

Fig. 11.4 Information preprocessing pipeline in the context of Internet of Marine Things schema

11.5 Data Preprocessing for Big Data Analytics in Internet of Marine Things

Data preprocessing for the IoMT is a critical phase for ensuring the accuracy and reliability of marine data collected from harsh environments. It involves several key steps to transform raw, often noisy, sensor data into a format suitable for analysis and machine learning models.

11.5.1 Data Extraction and Organization in Internet of Marine Things

The initial step involves collecting raw data from the source and structuring it so that it can be used by an intelligent system. The organization method is highly dependent on the data modality. The diagram of Fig. 11.5 shows that data integration and organization are the most important and initial steps in data preprocessing techniques.

(i) ***Tabular/Sensor Data***: Data is typically extracted from .csv, .json, or databases (SQL) into a DataFrame (e.g., using Pandas). A sample is represented by each row, and a feature is represented by each column.
(ii) ***Image Data***: Datasets are often extracted from compressed files. A common organization is the folder-based classification format, where images are grouped into sub-folders named after their corresponding class [22].
(iii) ***Video Data***: Raw video files are organized in a similar folder-based classification format as images. Those files are grouped by their action or content class.

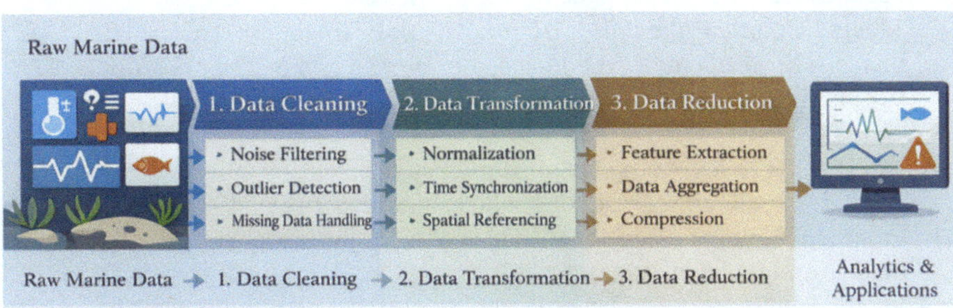

Fig. 11.5 Flow diagram of preprocessing techniques on different types of marine data

11.5.2 Handling Missing Values in Internet of Marine Things

Raw datasets are frequently incomplete or contain errors. Data cleaning is the most important and effective way to clear any kind of noise, which impacts the model performance in various datasets, as shown in

(i) Detection: Datasets are inspected for null, Not a Number (NaN), or placeholder entries.
$df.isnull().sum()$
(ii) Imputation Strategy: Tabular Data (Numerical Features): Missing values can be replaced with a statistical measure of the feature's column, such as the median or the mean.
$df = df.fillna(df.mean(numeric_{only} = True))$
Tabular Data (Categorical Features): Usually, the mode of the feature column is used to substitute missing values.
$df = df.fillna(df.mode().iloc[0])$
Image or Video Data: Missing data usually means a corrupt or non-existent file. Imputation is not feasible; the invalid sample is simply removed from the dataset.

11.5.3 Feature and Target Separation in Internet of Marine Things

For any supervised learning task, the input data (features, X) must be explicitly separated from the desired output (target, y). In the diagram of Fig. 11.5, feature extraction is a crucial part for all kinds of datasets.

(i) Features (X): The measurable characteristics used for prediction.
Tabular: Columns like "pH," "Hardness," and "Total Dissolved Solids."
Image: The pixel data of the image, represented as a numerical array.
Video: The sequence of image frames.

(ii) Target (y): The label or value to be predicted.
Tabular: A column representing a binary indicator (e.g., "1" for potable, "0" for non-potable) or a continuous value (e.g., pollution level).
Image: A class label (e.g., "$plastic_waste$").
Video: A class label for the entire video (e.g., "$anomalous_event$").

11.5.4 Handling Class Imbalance (Tabular Data) in Internet of Marine Things

In a marine pollution detection application, one class is far more common than another, which creates bias by ignoring the minority class and reducing model performance. To reduce bias and improve model performance, it is important to handle class imbalance. Handling class imbalance is a crucial phase for an efficient and intelligent model to manage big data in IoMT.

11.5.4.1 Resampling Techniques

(i) ***Synthetic Minority Oversampling Technique (SMOTE)***: Synthetic new. Interpolating between the samples in the majority class to provide minority class samples [36].

(ii) ***Borderline-SMOTE***: This is a modification that only creates artificial samples close to the class decision boundary [36].

(iii) ***SMOTEENN/SMOTETomek***: Hybrid combinations of SMOTE are referred to as SMOTEENN / SMOTETomek. Cleaning (undersampling) sampling to eliminate noisy samples [36].

(iv) ***Adaptive Synthetic Sampling (ADASYN)***: This is a variant of SMOTE which is more advanced and more concerned with the production of artificial samples of the minority data points that are harder to learn. This is a class learning mechanism that is best where this adaptive mechanism is concerned needed [36].

(v) ***Cluster-Based Oversampling***: The minority samples are put in clusters with the help of such algorithms as K-Means, and artificial data is created inside each cluster. This is useful in ensuring that the synthetic samples are diverse and is useful in preventing overfitting [36].

(vi) ***NearMiss***: A well-known method of undersampling which picks the majority of the samples. Those are nearest to a sample of minority classes, to assist the model in concentrating on the difficult boundary areas and enhancing discrimination among classes [36].

(vii) ***Tomek Links***: A data cleaning algorithm that passes over pairs of samples and removes those that are quite near to one another. The elimination of borderline samples contributes to cleanliness. The overlapping of classes enhances the separability of datasets.

(viii) ***Edited Nearest Neighbor (ENN)***: According to this method, samples that are different from most of their closest neighbors are excluded, practically scrubbing off the noise and enriching class purity. It may follow SMOTE to balance the oversampled dataset [36].

(ix) ***Cluster Centroids***: This methodology substitutes the conventional clustering approach with the random undersampling. It arbitrarily selects representative samples from the mainstream class, preserving the elemental characteristics of the information while lessening overall size of the dataset [36].

(x) ***Balanced Bagging Classifier***: A resampling-based ensemble technique where each bag (subset) of data is balanced before training a classifier. It combines the benefits of bagging and data balancing for better overall stability [36].

(xi) ***Random Oversampling Examples (ROSE)***: A smoothed bootstrap approach that creates synthetic samples by generating new examples around both minority and majority instances, helping the model learn a more continuous decision boundary [36].

Why Is SMOTE More Effective than the Other Balance-Handling Techniques? The SMOTE is useful for tabular data because, before applying SMOTE, the training set often suffers from class imbalance. The majority class dominates, and the classifier can be biased toward it. But, after applying SMOTE, new synthetic samples are generated for the minority class using interpolation between existing samples:

$$x_i' = x_i + \lambda(x_j - x_i), \tag{11.1}$$

where x_i is the minority sample, x_j is one of k nearest neighbors of x_i, λ = random scaler drawn uniformly between 0 and 1, and this produces synthetic points along the line segment joining x_i and x_j.

11.5.5 Data Splitting (Tabular/Image/Video Data) in Internet of Marine Things

Before any preprocessing that "learns" from the data (like scaling or oversampling), the dataset must be split. This prevents data leakage, where information from the test set influences the model.

1. ***Training Set***: The portion of data is used to train the model and adjust its parameters. Example: The largest portion of the dataset, at most 70%−−80% of the dataset, is used to train the model and is shown in Fig. 11.6.
2. ***Validation Set***: This portion of data is used to tune hyperparameters and select the best model configuration. Example: The smallest portion of the dataset, at most 10% of the dataset, is used to tune model hyperparameters and make decisions, as shown in Fig. 11.6.
3. ***Test Set***: This portion is used to evaluate the final model's performance on unseen data. Example: At most 20% of the dataset is used only once to evaluate the final, trained model's performance on unseen data, which is shown in Fig. 11.6.

$$\begin{aligned} &X_{train}, X_{temp}, y_{train}, y_{temp} \\ &= train_test_split(X, y, test_size = 0.2, random_state = 42), \end{aligned} \tag{11.2}$$

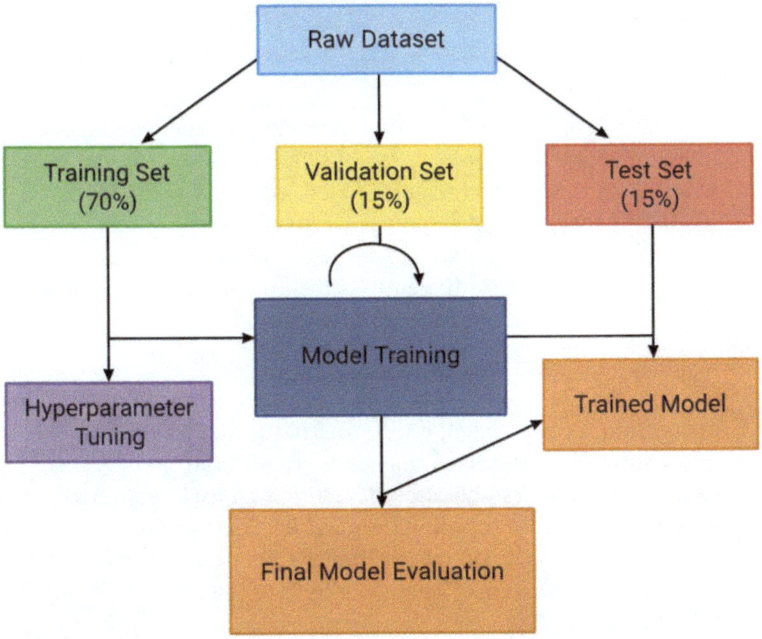

Fig. 11.6 Data splitting workflow

$$X_{val}, X_{test}, y_{val}, y_{test} \\ = train_test_split(X_temp, y_temp, test_size = 0.5, random_state = 42), \tag{11.3}$$

where X is feature and y is label. X_{train} is training input features, y_{train} is training labels, X_{temp} is temporary input set, X_{test} is test input features, y_{test} is test labels.

11.5.6 Feature Scaling (Tabular Data, Image Data, Video Data) in Internet of Marine Things

Feature scaling ensures that all numerical variables contribute equally during training, which is vital for many models. It ensures all variables contribute proportionally to the learning process.

(i) *StandardScaler*: This feature scaling method transforms data to have a mean of 0 and a standard deviation (variance) of 1, known as Z-score normalization. This method is suitable for models that assume normally distributed features, such as linear regression, logistic regression, and SVMs [51].

$$X' = \frac{(X - \mu)}{\sigma}, \tag{11.4}$$

where
X = original value, X' = scaled value, μ = mean of the feature, σ = standard deviation of the feature.

(ii) *Min MaxScaler*: Rescales all features to a fixed range, typically [0, 1]. Model-Specific Importance: This is particularly critical for neural networks like Long Short-Term Memory (LSTMs). LSTMs use activation functions (like sigmoid and tanh) that operate most efficiently in this [0, 1] or [−1, 1] range. Unscaled data can lead to unstable gradients and failed training.

(iii) *Data Verification and Conversion*

(a) Verification: Check that all image files have valid and corresponding label files. Corrupt or unreadable images are removed from the dataset.
(b) Data Conversion: In some cases, data formats must be converted. For example, data in an object detection format (like YOLO, with bounding box text files) might be converted into a simpler folder-based classification format by cropping the objects of interest or by labeling the entire image based on its contents.

(iv) *Geometric Transformations* These operations alter the spatial properties of the image.

(a) Resizing: This is the most critical step. All images must be resized to a uniform dimension (e.g., 224 × 224 pixels or 512 × 512 pixels) to be fed into the model in batches. This matches the input dimension requirements of pre-trained Convolutional Neural Network (CNN) models.

$$(x', y') = (s_x.x, s_y.y). \tag{11.5}$$

Equation (11.5) is a transformation that scales or dilates a point (x, y) by factors s_x and s_y in the x and y directions, respectively.
Code: $img_resized = transforms.Resize((224, 224))(img)$

(v) Cropping:

(a) Center Crop: Used during validation/testing to extract the central portion of the image.

$$(x', y') = \left(x + \frac{W - w}{2}, y\frac{H - h}{2}\right),$$

where
(x, y) = pixel in cropped region; (x', y') = mapped coordinate in original; W, H = original size; w, h = crop size.

(b) Random Crop: Used during training to randomly select a patch of the image, forcing the model to learn from different parts.

$$(x', y') = (x + x_0, y + y_0). \tag{11.6}$$

Code:
$img_cropped = transforms.CenterCrop(224)(img)$

(vi) Flipping: Random horizontal and vertical flipping. Vertical flipping is shown in Fig. 11.7.

(a) For horizontal flipping,

$$(x', y') = (w - 1 - x, y). \tag{11.7}$$

(b) For vertical flipping,

$$(x', y') = (x, H - 1 - y), \tag{11.8}$$

where original coordinate, (x', y') = flipped coordinate, W=width, H=height.
Code:
Horizontal flip: $img_flipped = transforms.functional.hflip(img)$
Vertical flip: $img_flipped = transforms.functional.vflip(img)$

(c) Rotation: Randomly rotating the image by a small angle.

$$\begin{pmatrix} x' \\ y' \end{pmatrix} = \begin{pmatrix} c_x \\ c_y \end{pmatrix} + \begin{pmatrix} \cos\theta & -\sin\theta \\ \sin\theta & \cos\theta \end{pmatrix} \left(\begin{pmatrix} x \\ y \end{pmatrix} - \begin{pmatrix} c_x \\ c_y \end{pmatrix} \right)$$

Where:
(x, y) = original coordinate, (x', y') = rotated coordinate;
θ = rotation angle (radians);
$c_x = \frac{W-1}{2}$, $c_y = \frac{H-1}{2}$ = image center.

Code: $img_rotated = transforms.functional.rotate(img, 45)$

(vii) Photometric Transformations These operations alter the pixel values to simulate different environmental conditions.

(a) **Brightness and Contrast Adjustment**: Randomly varying the brightness and contrast of the image to make the model robust to different lighting conditions. Brightness and contrast adjustment are shown in Fig. 11.7.
(b) **Hue and Saturation Control**: Randomly changing the color properties.
(c) **Grayscaling**: Converting the image from 3 channels (RGB) to 1 channel (greyscale), which can be useful if color is not an important feature. Grayscaling is shown in Fig. 11.7.

Fig. 11.7 Augmentation of a single image

(viii) **Data Augmentation** Data augmentation is the strategic application of geometric and color transformations randomly during the training process only. Each time the model sees an image from the training set, it sees a slightly different version (e.g., flipped, rotated, and with different brightness) [37]. This process artificially expands the dataset, and it's one of the most effective techniques to reduce overfitting and improve model generalization.

(ix) **Temporal Frame Sampling** A single video can contain thousands of frames. It is computationally prohibitive and often unnecessary to process all of them [38].

   (a) **Libraries**: OpenCV or FFmpeg is used to read video files and extract frames.
   (b) **Sampling Strategies**:
      (A) **Uniform Sampling**: Select N frames evenly spaced throughout the video's duration. Uniform sampling is shown in Fig. 11.8.
      (B) **Keyframe Extraction**: Only extract frames where a significant change in the visual content occurs.
      (C) **Dense Sampling**: Extract short, overlapping clips from the video.

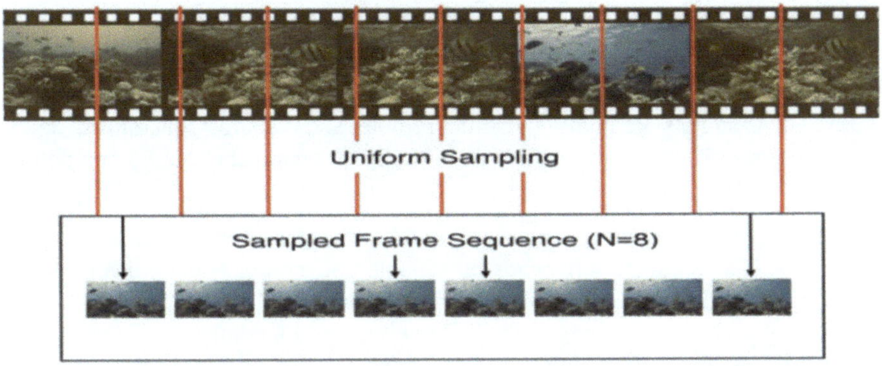

Fig. 11.8 Frame sampling

11.5.7 Normalization Technique in Internet of Marine Things

(a) **Structured Data (Tabular Form)**: For tabular form, normalization is used to fill in the limited range, usually it is [0,1].

  (A) **MinMaxScaler**: This scaler is one of the most important for NN models like LSTMs, because of the usage of the activation function like sigmoid and tanh, which work more efficiently where the input range lies between [0, 1] and [-1, 1]. When the data is not being normalized, the gradient becomes very high or vanishes, and ultimately, the model fails to train.

(b) **Pixeled (Image) Data**: When the inputs are in the form of images, then the values (pixels) need to be normalized before the training.

(c) **Video Data**: For video, the same method is used as for images, so the method is applied to multiple frames of the video frame sequence. The model is converged properly if it gets normalized and decreases the amount of flickering in between the frames. For this, two approaches are:

  (A) **Scaling**: Divide each pixel value by 255. Then intensities will fall to the range of [0, 1].

  (B) **Standardization**: Used to extract features and to match the format of the original model training data. Pixel values are subtracted from the mean dataset and divided by the SD.

11.6 Future Research Directions in Communication, Computing, Sustainability, and Security for IoMT

Marine environments are highly dynamic, remote, and resource-constrained, so the IoMT system must adopt sustainable and intelligent approaches across computing, communication, and energy management domains.

(A) **Challenge 1: How do communication technologies support IoMT?**
Marine data acquisition depends on reliable underwater communication technologies such as acoustic, optical, and radio frequency (RF) [39]. Acoustic communication is used for long-range and low-bandwidth communication, whereas optical communication supports short-range but high-speed data transfer. For surface-to-satellite communication, RF is used. Future research will focus on integrating 5G/6G technologies, enabling seamless, high-speed, and low-latency communication for remote IoMT monitoring [40]. Additionally, developing an intelligent algorithm that automatically switches between acoustic, acoustic-to-optical, and RF links depends on environmental conditions such as depth, noise level, turbidity, and mobility [41]. Energy-efficient communication is also a research challenge in IoMT. Incorporation of the low-power protocols enable energy aware routing and sleep scheduling mechanisms to extend lifetime of the anticipated IoMT system [42, 43].

(B) **Challenge 2: What role does computing play in IoMT system?**
Marine data acquisition and preprocessing generate massive volumes of heterogeneous data [44]. Instead of transferring raw data to centralized servers, future research emphasizes intelligent data preprocessing, analysis, and compression at the edge before transmission. The adoption of edge-fog-cloud architecture will be crucial for handling large-scale marine data efficiently. Furthermore, lightweight preprocessing techniques and energy-aware intelligent algorithms at the edge layer will help reduce latency, computational burden, and communication overhead in IoMT systems [45, 46]. Integration of artificial intelligence and ML will allow IoMT system to become more autonomous and adaptive for real-time decision-making in marine environment [47, 48]. Incorporating quantum computing in IoMT system is also a future research direction to solve complex marine problem optimal route finding and resourse allocation [49].

(C) **Challenge 3: Why is sustainability important in IoMT?**
The IoMT system must be sustainable and energy-efficient as more marine sensors opt for limited battery power [34]. Three important sustainability aspects are environmental, social and economics sustainability. Efficient battery management significantly extends the operational lifetime of IoMT system. In the future, the research should focus on low-power, long-range communication protocols to extend battery lifespan. Additionally, an intelligent optimization mechanism that dynamically manages data transmission rates and sensor activity can maximize energy efficiency and prolong the IoMT system lifetime [43]. This reduces

the waste in IoMT system and supports environmental sustainability. The cost-efficient deployment, resource management and maintenance of IoMT system supports social and economic sustainability.

(D) **Challenge 4: How can privacy be preserved in IoMT?**
The marine environment contains sensitive information making privacy preservation crucial [50, 51]. Future research will focus on a federated learning approach, which enables collaborative model training without sharing raw data. Federated learning enhances security by keeping data localized at the edge and reducing the risk of data exposure [52]. Moreover, secure aggregation and encryption mechanisms should be integrated to strengthen privacy in the IoMT system. Information security is a challenge in IoMT. Incorporating multiobject authentication [53], cross-domain authentication [54], and blockchain mechanism [55] are the future direction to provide security in IoMT.

11.7 Conclusion

Effective data acquisition and preprocessing are the fundamental and most crucial parts of an intelligent Internet of Marine Things. Data acquisition is performed in IoMT through sensors, AUVs, and an IoUT network. Raw marine data often suffers from missing values and noise. Therefore, preprocessing techniques such as data cleaning, missing value handling, and feature extraction are crucial for a high-quality real-time intelligent system. An efficient data acquisition and preprocessing ensures the reliability and accuracy of an intelligent IoMT system that supports automated decision-making in complex marine environments. In this chapter, we discuss the data acquisition technique in the marine scenario. Besides these, we also studied different preprocessing techniques for tabular data, image, and video data in the IoMT environment. In this chapter, we also studied the challenges with research directions of IoMT scenario.

References

1. Jahanbakht, M., Xiang, W., Hanzo, L., Azghadi, M.R.: Internet of underwater things and big marine data analytics-a comprehensive survey. IEEE Commun. Surv. Tutorials **23**(2), 904–956 (2021)
2. Masek, A., Rudi, M., Sekeri, S.: Water quality monitoring for goldfish aquarium using IoT. Int. J. **8**(6), 1177–1180 (2020)
3. Bhattacharjya, K., De, D.: Bfl: Blockchain-federated learning for privacy preservation in internet of underwater things. In: Global Conference on Applications of Artificial, pp. 236–245. Springer (2024)
4. Theocharidis, T., Kavallieratou, E.: Underwater communication technologies: a review. Telecommun. Syst. **88**(2), 54 (2025)
5. Bhattacharjya, K., Alam, S., De, D.: Performance analysis of dymo, zrp and aodv routing protocols in a multi hop grid based underwater wireless sensor network. In: International

Conference on Computational Intelligence, Communications, and Business Analytics, pp. 465–476. Springer (2018)
6. Al Mamun, M.R., Ashik-E-Rabbani, M., Haque, M.M., Upoma, S.M.: IoT-based real-time biofloc monitoring and controlling system. Smart Agric. Technol. **9**, 100598 (2024)
7. Bhattacharjya, K., De, D.: Federated learning-based privacy-preserving internet of underwater things: a vision, architecture, computing, taxonomy, and future directions: K. bhattacharjya and d. de. J. Supercomputing **81**(8), 870 (2025)
8. Zandi, R., Dehghani, M.J., Kavoosi, V.: Asynchronous tracking of sensor nodes in an internet of underwater thing (IoUT) system. Telecommun. Syst. **88**(2), 64 (2025)
9. Dai, M., Li, Y., Li, P., Wu, Y., Qian, L., Lin, B., Su, Z.: A survey on integrated sensing, communication, and computing networks for smart oceans. J. Sens. Actuator Netw. **11**(4), 70 (2022)
10. Adam, N., Ali, M., Naeem, F., Ghazy, A.S., Kaddoum, G.: State-of-the-art security schemes for the internet of underwater things: a holistic survey. IEEE Open J. Commun. Soc. (2024)
11. Kumar, P., Jha, S., Singh, R.K., et al.: Internet of underwater things (IoUT): a systematic review research. J. Recent Innov. Comput. Sci. Technol. **2**(1), 28–47 (2025)
12. Hu, L., Shu, Y.: Enhancing decision-making with data science in the internet of things environments. Int. J. Adv. Comput. Sci. Appl. **14**(9), (2023)
13. Mohsan, S.A.H., Li, Y., Sadiq, M., Liang, J., Khan, M.A.: Recent advances, future trends, applications and challenges of internet of underwater things (IoUT): a comprehensive review. J. Marine Sci. Eng. **11**(1), 124 (2023)
14. Ding, Y., Wang, X., Xu, J., Xie, G., Liu, W., Li, Y.: Multi-objective-optimization multi-AUV assisted data collection framework for IoUT based on offline reinforcement learning. In: 2025 IEEE Wireless Communications and Networking Conference (WCNC), pp. 1–6. IEEE (2025)
15. Lima, K., Nguyen, N.T., Heldal, R., Kristensen, L.M., Oyetoyan, T.D., Pelliccione, P., Knauss, E.: A data-flow oriented software architecture for heterogeneous marine data streams. In: 2024 IEEE 21st International Conference on Software Architecture (ICSA), pp. 146–157. IEEE (2024)
16. Lan, H., Lv, Y., Jin, J., Li, J., Sun, D., Yang, Z.: Acoustical observation with multiple wave gliders for internet of underwater things. IEEE Internet Things J. **8**(4), 2814–2825 (2020)
17. Victor, N., Alazab, M., Bhattacharya, S., Magnusson, S., Maddikunta, P.K.R., Ramana, K., Gadekallu, T.R.: Federated learning for IoUT: concepts, applications, challenges and opportunities (2022). arXiv:2207.13976 arXiv preprint
18. Omeke, K.G., Abubakar, A.I., Zhang, L., Abbasi, Q.H., Imran, M.A.: How reinforcement learning is helping to solve internet-of-underwater-things problems. IEEE Internet Things Mag. **5**(4), 24–29 (2023)
19. Celik, A., Saeed, N., Shihada, B., Al-Naffouri, T.Y., Alouini, M.S.: A software-defined opto-acoustic network architecture for internet of underwater things. IEEE Commun. Mag. **58**(4), 88–94 (2020)
20. Niu, H., Li, X., Zhang, Y., Xu, J.: Advances and applications of machine learning in underwater acoustics. Intell. Marine Technol. Syst. **1**(1), 8 (2023)
21. Sumanth, N., Heggere, S.B., Prathiksha, E., Shetty, C.A., Nikhitha, H., Soundarya, B.: Federated Learning: Healthcare, Security, Challenges, and Threats (2023)
22. Zetas, M., Spantideas, S., Giannopoulos, A., Nomikos, N., Trakadas, P.: Empowering 6G maritime communications with distributed intelligence and over-the-air model sharing. Front. Commun. Networks **4**, 1280602 (2024)
23. Gou, Y., Zhang, T., Liu, J., Qi, Z., Zheng, D.: An efficient and adaptive framework for achieving underwater high-performance maintenance networks. In: International Conference on Internet of Things, Communication and Intelligent Technology, pp. 409–418. Springer (2024)
24. Hossam, Y., Nagdy, H., Adel, R., Mehanny, S., Fahim, I.S., Gomaa, O., Keshavarz, T., Al-Kabbany, A.: Emerging AI Solutions for Hazardous pet Waste in Marine Environments: A Review of Underexplored Paradigms, (2025)
25. Aouedi, O., Vu, T.H., Sacco, A., Nguyen, D.C., Piamrat, K., Marchetto, G., Pham, Q.V.: A survey on intelligent internet of things: applications, security, privacy, and future directions. IEEE Commun. Surv. Tutorials (2024)

26. Omeke, K.G., Mollel, M., Shah, S.T., Zhang, L., Abbasi, Q.H., Imran, M.A.: Toward a sustainable internet of underwater things based on auvs, swipt, and reinforcement learning. IEEE Internet Things J. **11**(5), 7640–7651 (2023)
27. Li, K., Zhang, Z., Pourkabirian, A., Ni, W., Dressler, F., Akan, O.B.: Towards Resilient Federated Learning in Cyberedge Networks: Recent Advances and Future Trends, (2025). arXiv:2504.01240 arXiv preprint
28. Farooqi, S.A., Rahman, A.A., Saad, A.: Advanced privacy-utility optimization techniques in federated learning with differential privacy for IoMT—a review. Int. J. Interact. Mobile Technol. **19**(19), (2025)
29. Zhao, H., Ji, F., Li, Q., Guan, Q., Wang, S., Wen, M.: Federated meta-learning enhanced acoustic radio cooperative framework for ocean of things. IEEE J. Sel. Top. Sign. Proces. **16**(3), 474–486 (2022)
30. Kaur, I., Jadhav, A.J.: Federated learning in IoT: a survey from a resource-constrained perspective. In: 2023 International Conference on Artificial Intelligence Robotics, Signal and Image Processing (AIRoSIP), pp. 376–381. IEEE (2023)
31. Bhattacharya, P., Verma, A., Tanwar, S.: Federated Learning for Internet of Medical Things: Concepts, Paradigms, and Solutions, CRC Press (2023)
32. Bhattacharjya, K., Alam, S., De, D.: CUWSN: energy efficient routing protocol selection for cluster based underwater wireless sensor network. Microsyst. Technol. **28**(2), 543–559 (2022)
33. Bhattacharjya, K., De, D.: IoUT: modelling and simulation of edge-drone-based software-defined smart internet of underwater things. Simul. Model. Pract. Theory **109**, 102304 (2021)
34. Menon, V.G., Midhunchakkaravarthy, D., Sujith, A., John, S., Li, X., Khosravi, M.R.: Towards energy-efficient and delay-optimized opportunistic routing in underwater acoustic sensor networks for IoUT platforms: an overview and new suggestions. Comput. Intell. Neurosci. **2022**(1), 7061617 (2022)
35. Nellore, K., Polasi, P.K.: An improved underwater wireless sensor network communication using internet of things and signal to noise ratio analysis. Trans. Emerg. Telecommun. Technol. **33**(9), e4560 (2022)
36. Mujahid, M., Kına, E., Rustam, F., Villar, M.G., Alvarado, E.S., De La Torre Diez, I., Ashraf, I.: Data oversampling and imbalanced datasets: an investigation of performance for machine learning and feature engineering. J. Big Data **11**(1), 87 (2024)
37. Ben Tamou, A., Benzinou, A., Nasreddine, K.: Targeted data augmentation and hierarchical classification with deep learning for fish species identification in underwater images. J. Imaging **8**(8), 214 (2022)
38. Rajasekar, M., Celine Kavida, A., Anto Bennet, M.: A pattern analysis based underwater video segmentation system for target object detection. Multidimension. Syst. Signal Process. **31**(4), 1579–1602 (2020)
39. Menaka, D., Gauni, S., Manimegalai, C., Kalimuthu, K.: Vision of IoUT: advances and future trends in optical wireless communication. J. Opt. **50**(3), 439–452 (2021)
40. Wei, T., Feng, W., Chen, Y., Wang, C.X., Ge, N., Lu, J.: Hybrid satellite-terrestrial communication networks for the maritime internet of things: key technologies, opportunities, and challenges. IEEE Internet Things J. **8**(11), 8910–8934 (2021)
41. Saleem, K., Wang, L., Almogren, A., Ntizikira, E., Rehman, A.U., Bharany, S., Hussen, S.: Cognitive intelligence routing protocol for disaster management and underwater communication system in underwater acoustic network. Sci. Rep. **15**(1), 10004 (2025)
42. Bharany, S., Sharma, S., Alsharabi, N., Tag Eldin, E., Ghamry, N.A.: Energy-efficient clustering protocol for underwater wireless sensor networks using optimized glowworm swarm optimization. Front. Mar. Sci. **10**, 1117787 (2023)
43. Diwan, S.A.: A multi-layered energy efficient approach for performance aware internet of ocean things. Int. J. Interact. Mobile Technol. **16**(17), (2022)
44. Ahmad, W., Ali, F., Ullah, Y., Khan, M.A., Khan, M., Khan, S., Khan, M.A., Ali, A., Almansour, S.: IoUT and collaborative cloud computing based UAV channel modeling for agriculture communication. IEEE Trans. Consum. Electron. (2025)

45. Jung, S., Jeong, S., Kang, J., Kang, J.: Marine IoT systems with space-air-sea integrated networks: hybrid LEO and UAV edge computing. IEEE Internet Things J. **10**(23), 20498–20510 (2023)
46. Lin, B., Zhang, M., Jiang, Y., Qi, S., Yang, M., Wen, Z., Na, Z.: Harsh weather-oriented edge intelligence empowered maritime communication-computing converged network resource allocation. IEEE Internet Things J. (2026)
47. Ananthi, M., Lakshmana Kumar, R., Muthu, B., Punitha, P.: Adaptive marine intelligence and sensing architecture for autonomous underwater ecosystem monitoring using AI and IoT integration. Intell. Data Anal. **30**(1), 162–181 (2026)
48. Yang, T., Chen, J., Zhang, N.: AI-empowered maritime internet of things: a parallel-network-driven approach. IEEE Network **34**(5), 54–59 (2020)
49. Wang, T., Dai, M., Guo, H., Qian, L., Niyato, D., Wu, Y.: When quantum empowerment meets marine edge computing: challenges, novel paradigms, and enabling technologies. IEEE Network (2025)
50. Gaur, R., Prakash, S.: Privacy prevention and nodes optimization, detection of IoUT based on artificial intelligence. Wireless Pers. Commun. **138**(1), 67–97 (2024)
51. Victor, N., Chengoden, R., Alazab, M., Bhattacharya, S., Magnusson, S., Maddikunta, P.K.R., Ramana, K., Gadekallu, T.R.: Federated learning for IoUT: concepts, applications, challenges and future directions. IEEE Internet Things Magazine **5**(4), 36–41 (2023)
52. Qin, Z., Ye, J., Meng, J., Lu, B., Wang, L.: Privacy-preserving blockchain-based federated learning for marine internet of things. IEEE Trans. Comput. Soc. Syst. **9**(1), 159–173 (2021)
53. Ye, J., Cao, X., Guo, Z.: Secure marine environment communication: a multiobject authentication protocol based on secret sharing. Int. J. Intell. Syst. **2023**(1), 1814053 (2023)
54. Gong, B., Zheng, G., Waqas, M., Tu, S., Chen, S.: LCDMA: lightweight cross-domain mutual identity authentication scheme for internet of things. IEEE Internet Things J. **10**(14), 12590–12602 (2023)
55. Hu, B., Wang, X., Wang, J., Liu, Y., Li, K., Tsang, K.F.: Hybrid-blockchain-based cross-domain authentication for internet of marine things. Comput. Stan. Interfaces 104140 (2026)

Chapter 12
Design and Development of a Drone-Rover Hybrid for Marine Monitoring and Disaster Aid

V. Vignesh Babu, J. Preetha Roselyn, Prabha Sundaravadivel, Tien Anh Tran, Spyros Hirdaris, Zhang Mingyang, C. H. Vasanthkumar, Kudzaishe S. Chitsenga, V. Varun, and Nithyashree Pillai

V. V. Babu · J. P. Roselyn (✉) · C. H. Vasanthkumar · K. S. Chitsenga · V. Varun · N. Pillai
SRM Institute of Science and Technology, Kattankulathur, Tamilnadu, India
e-mail: preethaj@srmist.edu.in

K. S. Chitsenga
e-mail: kc6567@srmist.edu.in

V. Varun
e-mail: vv0015@srmist.edu.in

N. Pillai
e-mail: np6060@srmist.edu.in

P. Sundaravadivel
University of Texas, Tyler, USA
e-mail: prabhas2@srmist.edu.in

T. A. Tran
Department of Marine Engineering, Vietnam Maritime University, Haiphong, Vietnam

Department of Electrical Engineering, University of Malta, Msida, Malta

S. Hirdaris
American Bureau of Shipping, Global Ship Systems Centre, Athens, Greece

Z. Mingyang
School of Ocean and Civil Engineering, Shanghai Jiao Tong University, Shanghai, P.R. China

Key Laboratory of Marine Intelligent Equipment and System, Shanghai Jiao Tong University, Shanghai, P.R. China

T. A. Tran et al. (eds.), *Big Data Analytics for Internet of Marine Things*,
Internet of Things, https://doi.org/10.1007/978-3-032-28171-5_12

12.1 Introduction

Coastal regions are among the most dynamic and environmentally sensitive zones on the planet. They support critical ecosystems and help with economic activity [1]. These regions are increasingly threatened by climate change, driven sea level rise, extreme weather events, marine pollution, and the proliferation of hazardous marine organisms, all of which pose risks to both ecological stability and human safety [2, 3]. Effective monitoring and timely intervention in coastal environments therefore remain major scientific and engineering challenges.

Conventional coastal monitoring approaches rely heavily on static sensor stations, satellite observations, or manual inspection, each of which suffers from limited spatial resolution, delayed response, or high operational cost [4]. In recent years, the development and use of unmanned aerial vehicles gained attention. This is because of their ability to rapidly survey large areas and collect high-resolution visual and environmental data. However, aerial platforms are constrained by limited flight endurance, payload capacity, and vulnerability to adverse weather conditions [5]. These limitations restrict their effectiveness for prolonged monitoring and physical intervention tasks.

Ground-based robotic rovers offer complementary advantages, including extended operational duration, terrain interaction capability, and the ability to perform close-range inspection or object manipulation [6, 7]. Nevertheless, rovers face significant challenges in coastal environments, such as uneven terrain, soft sand, debris fields, and partially submerged areas that limit accessibility and mobility [8]. As a result, systems relying solely on either aerial or ground platforms provide incomplete situational coverage and reduced effectiveness during time-critical scenarios such as disaster response or search and rescue operations [9, 10].

Hybrid robotic systems that combine aerial and terrestrial platforms have emerged as a promising direction to address these limitations [11, 12]. By integrating the wide-area sensing capability of aerial vehicles with the endurance and physical interaction capability of ground robots, such systems can offer enhanced operational flexibility and resilience. Recent studies have demonstrated hybrid aerial-ground vehicles for applications including environmental surveillance, agricultural monitoring, and emergency response [13]. Despite these advances, existing hybrid systems often focus on mobility integration alone, with limited emphasis on intelligent perception, predictive environmental analysis, and coordinated autonomous decision-making tailored to complex coastal conditions.

Advances in artificial intelligence enabled robotic platforms to move beyond passive data collection toward intelligent environmental interpretation and prediction [14, 15]. Machine learning techniques such as deep neural networks have shown strong potential for visual object recognition in complex natural environments, while recurrent neural networks have been successfully applied to time series forecasting of oceanographic variables [16, 17]. An example of successful implementation is the NVIDIA Jetson Nano [18]. However, the integration of these techniques into unified, real-time, multi-domain robotic systems for coastal monitoring remains relatively unexplored [19, 20].

This work addresses these challenges by presenting an integrated drone–rover hybrid system designed specifically for autonomous coastal monitoring, environmental assessment, and disaster response. The proposed platform combines aerial surveillance, ground-based intervention, and onboard artificial intelligence within a single coordinated framework. The system supports real-time perception, predictive analysis of coastal conditions, and autonomous task execution. Accordingly, applications such as search and rescue assistance, shoreline waste detection, and environmental data logging are enabled. Because of the interlinks of operations between aerial and terrestrial agents, the proposed system enhances situational awareness, reliability, and scalability in dynamic coastal environments.

The main contributions of this work include the design of modular aerial-terrestrial robotic architecture, the integration of intelligent perception and environmental prediction capabilities, and the demonstration of coordinated autonomous operation for coastal monitoring tasks. Together, these contributions advance the state of the art in multi-domain robotic systems for environmental monitoring and disaster mitigation.

12.2 Drone Development

12.2.1 Drone Motor Specifications and Calculation:

Drone functions such as the aerial unit of the proposed hybrid platform are designed for high-altitude surveillance, aerial data acquisition, wave height measurement, and environmental monitoring over coastal regions. The system plays a pivotal dual role in terms of data collection and the physical deployment of the ground-based rover. This is because it includes a detachable coupling mechanism that allows the drone to airlift and release the rover in hard-to-reach coastal zones and at a later stage it may reattach for retrieval. Such a modular design enhances the system's adaptability in dynamic or inaccessible environments, enabling seamless air-to-ground transitions. The system block diagram is depicted in Fig. 12.1.

The drone is powered by four 2200 kV-rated Brushless DC (BLDC) motors, each controlled by a 30A Electronic Speed Controller (ESC). These high-RPM motors enable the platform to lift payloads up to 1 kg, excluding the weight of the rover. This payload supports a broad array of onboard sensors and communication hardware without compromising flight stability [11]. The drone's motion can be directed through pre-programmed mission paths using inertial measurements from onboard Inertial Measurement Unit (IMU) sensors, with the added flexibility of semi-autonomous operation. In urgent scenarios or communication delays, the system permits manual override by authorized ground operators for safety–critical interventions.

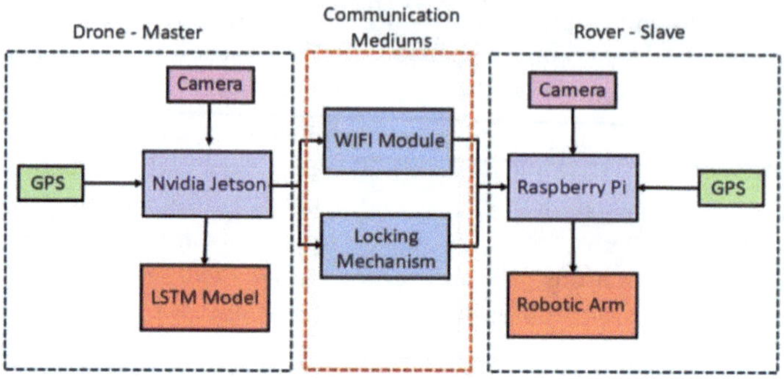

Fig. 12.1 Block diagram of Drone-Rover hybrid system

Calculations:

1. Estimate Propeller Speed

$$\text{RPM} = kV_{\text{BLDC}} \times V_{\text{Battery}} = 1000 \times 11.1 = 11,100\ \text{RPM} \tag{12.1}$$

$$N_{\text{BLDC}} = 11,100/60 \approx 185\ \text{RPS} \tag{12.2}$$

2. Thrust Calculation

   For a 10-inch 2-blade propeller, assume $C_{\text{Thrust}} = 0.1$ (Thrust Coefficient)

$$T = 0.1 \times 1.225 \times (185)^2 \times (0.254)^4 \approx 8.9\text{N} \tag{12.3}$$

   Thrust required ≈ 8.9N.

3. Thrust-to-Weight Ratio

   Drone-Rover Weight (W) = 1.8 kg

$$T_{\text{Total}}/W = 3.6/1.8 = 2.0 \tag{12.4}$$

where kV_{BLDC} = kV rating for BLDC, V_{Battery} = Voltage of Battery, RPM = revolution per minute,

RPS = Revolutions Per Second N_{BLDC} = Speed, C_{Thrust} = Thrust Coefficient, T = Required level of Thrust, W = Weight, T_{Total} = Total Thrust.

For ideal stable flying and good maneuverability Thrust-to-Weight Ratio is 2:1. This allows the drone to have longer flight time, better stability, more agility, and efficient hovering. To assess flight efficiency, a series of aerodynamic calculations were conducted. With a 10-inch, two-blade propeller rotating at an estimated 11,100 RPM (185 RPS), and using a thrust coefficient of 0.1, each motor can generate approximately 8.9 N of thrust. The total system thus yields 35.6 N of upward force, giving a thrust-to-weight ratio of 2:1 for the combined 1.8 kg drone-rover payload more than adequate for stable hovering, agile maneuvering, and quick directional changes, even in moderate wind conditions.

12.2.2 Flight Control and Navigation System

Central to the drone's autonomous flight capability is the Pixhawk 2.4.8 flight controller [21] which governs motion, stabilization, and navigation. The system integrates multiple sensor inputs, including GPS (connected via Inter-Integrated Circuit (I2C)), a 3-axis accelerometer, a gyroscope, and a barometer to ensure accurate spatial orientation and real-time adjustments during flight. Two primary flight modes are configured; namely a stabilized mode and an altitude hold mode. The stabilized mode allows for semi-manual or manual control while actively correcting for roll, pitch, and yaw deviations. The altitude hold mode helps maintain the same altitude. This helps maintain a fixed elevation. It is also ideal for sustained hover during data collection or search operations. A built-in fail-safe protocol is activated under critical conditions, such as loss of transmitter–receiver communication beyond a predefined threshold and battery voltage falling below 30%. If either condition is met, the system attempts an automatic Return to Launch (RTL) maneuver. However, in cases where battery charge is insufficient for a full aerial return, the drone attaches and signals the ground rover to autonomously navigate back to base while carrying the drone. This helps conserve power and preserve mission integrity. This is done by using bi-directional communication signals between the drone and the rover using the Jetson Nano [22] and the Raspberry Pi [9].

12.2.3 Power and Communication Architecture

A unified power architecture is employed to support both the aerial and ground modules, using a 12 V, 4200 mAh Lithium Polymer (LiPo) battery [23]. Under typical conditions, this provides up to 20 min of continuous drone flight. When operating in conjunction with the rover, average mission durations extend in between 30 and 40 min, while the rover can function independently for over 1 h. This enables the hybrid platform to conduct extended missions without the need for immediate recharging or human intervention. The drone maintains a dual-channel communication infrastructure; a long-range Radio Frequency (RF) transmitter/receiver pair with a range of 10 km to 20 km. It handles control and telemetry between drone and rover, and a Wi-Fi/Message Queuing Telemetry Transport (MQTT) module streams environmental and visual data to a cloud server for live monitoring, alert generation, and historical logging. This system ensures that even in the event of temporary connection loss, essential data is preserved locally and synchronized once the connectivity is restored.

12.3 Development of Rover

The rover component of the hybrid system serves as the ground-based operational unit, engineered for rugged, all-terrain mobility along coastal environments. Its primary function is to autonomously navigate beaches, rocky shorelines, and even narrow cave-like structures where aerial units are rendered ineffective due to space constraints or collision risks. Designed with a six-wheeled differential drive architecture, the rover demonstrates high stability, low energy consumption, and superior maneuverability in confined or obstacle-dense areas. One of the most advantageous features of this architecture is its ability to decouple from the drone and operate independently, thereby reducing overall power expenditure during extended missions. The rover's low-power profile allows for sustained performance even when battery levels drop below optimal. This makes it especially effective as a fallback system when the drone cannot return due to depleted energy reserves, or when missions require prolonged terrestrial engagement. Its integration with the drone ensures that both systems can re-couple autonomously, enabling recovery and redeployment cycles with minimal human intervention.

12.3.1 Motor Control and Localization

To ensure precise movement and terrain adaptability, the rover utilizes a L298N motor driver module, which interfaces with six high-torque Direct Current (DC) motors. Each L298N driver module supports up to 2A current and 36 V input voltage, thus allowing for both directional and speed control of the wheels. Two motor drivers are used in parallel configuration. The first one handles four motors (two on each side), while the second controls the remaining two. The motor driver, though commonly used in micro-controller projects, was deemed insufficient due to its lower current-handling capacity and was avoided to prevent thermal failures underload. The rover incorporates front-mounted and rear-view vision cameras to assist in terrain perception and obstacle avoidance. These real-time video feeds are processed locally via an onboard AI module, supporting intelligent path-finding and dynamic rerouting when unexpected terrain conditions are encountered.

12.3.2 Micro-controller and Sensor Integration

At the core of the rover's control architecture is the ESP32 micro-controller by Espressif Systems chosen for its robust dual-core processing capabilities, low-power consumption, and integrated Wi-Fi/Bluetooth connectivity. This unit acts as the central processing hub, coordinating motor control, sensor data acquisition, and communication with both the drone and external cloud platforms. The ESP32 receives command signals from the FS-i6 receiver and transmits motor control signals to the L298N modules. It also serves as the primary data logger and I/O controller for the suite of onboard sensors, which include:

i. Digital Humidity and Temperature Sensor (DHT11)—for real-time temperature and humidity readings.
ii. Air Quality Gas Sensor (MQ135)—for monitoring air quality and detecting harmful gases.
iii. Barometric Pressure Sensor (BMP280)—for atmospheric pressure and altitude estimation.
iv. Anemometer module—for measuring local wind speed, critical for drone launch/landing operations.

These sensors provide a real-time environmental context that not only supports mission-critical decision-making but also enables adaptive control of the aerial unit. For instance, high wind speeds detected on the ground can automatically delay drone reattachment and redeployment. Collected data is transmitted live via Internet of Things (IoT) platforms such as the Blynk or Arduino IoT Cloud. Additionally, all datasets are redundantly stored in local Comma Separate Value (CSV) files on a Secure Digital (SD) card, allowing for later retrieval, archival, and trend analysis using spreadsheet applications such as Microsoft Excel.

12.3.3 Transmitter and Receiver System

The rover's remote control and communication rely on the Fly Sky FS-i6 which is a 2.4 Ghz radio frequency transmitter and receiver system. This model supports 10 channels, with the first four mapped to joystick control and the remaining six configured as auxiliary channels for custom behaviors. The VRA and VRB knobs are present in the FSi6 transmitter to transmit pulses ranging from 1000 to 2000 us which are utilized for fine-tuning rover speed, while SWA to SWD switches that are present in the transmitter for toggling different flight modes/commands such as manual override, return-to-base, or environmental scan. The effective communication range of this setup is up to 1.5 km in unobstructed conditions, reducing to 800–1000 m in environments with signal interference. This coverage allows the rover to function independently in remote coastal zones while maintaining constant connectivity with human operators or automated base stations.

12.3.4 Power System and Onboard Computation

The rover uses the same 4200 mAh 12 V Lithium Polymer (LiPo) battery used by the aerial drone. This is chosen as the operating voltage of almost all the components are 12 V with capacity of 4200 mAh. Due to its efficient power usage, the rover can remain active for over an hour on a single charge. Power distribution is regulated to ensure that all components, sensors, motors, micro-controller, cameras, and onboard computing, receive stable voltage. Therefore, the risk of thermal instability or power drops during long missions is minimal. For advanced visual processing, the rover is equipped with an NVIDIA Jetson Nano module [22]. This GPU-enabled single-board computer is used to handle video feeds from dual onboard cameras. It performs real-time object detection, terrain classification, and map generation. Its deep learning capabilities, powered by models such as YOLOv8, support autonomous navigation and real-time hazard identification, essential for intelligent waste retrieval and search-and-rescue operations.

12.4 Proposed Methodology of Drone-Rover Hybrid System

The proposed drone-rover hybrid system represents an integrated aerial-terrestrial robotic platform engineered for autonomous exploration, real-time environmental monitoring, and remote data acquisition. The system employs a dual-mode operational framework integrating drone and rover subsystems, allowing smooth transitions between aerial and terrestrial mobility for effective operation in challenging and hazardous terrains. At its core, the system operates under both manual and autonomous control paradigms, facilitated via a FlySky FS-i6 transmitter and receiver system. Channels 1–4 are dedicated to drone navigation, while channels 5 and 6, mapped to the analog knobs VRA and VRB that govern the rover's movement. This division allows the operator to control each unit independently or in synchronized tandem, ensuring mission versatility.

12.4.1 Communication and Control Configuration

The FS-i6 receiver-transmitter pair allows for a range of up to 1 km under ideal line-of-sight conditions. A critical operational condition is that the Pulse Width Modulation (PWM) control must be disabled before the rover commands can be executed. This may ensure that both mobility modes remain isolated during task-specific transitions. Motor signals are mapped to PWM ranges between 1000 and 2000, where:

$$\begin{aligned} 1500 &= \text{Neutral(stationary rover)}, \\ 1500\text{–}2000 &= \text{Forward motion(increasing speed)}, \\ 1500\text{–}1000 &= \text{Reverse motion(increasing speed in reverse direction)}. \end{aligned}$$

This signal mapping enables smooth and gradual motion control which is ideal for maneuvering over uneven or delicate terrain where abrupt motion changes could lead to instability or mechanical stresses.

12.4.2 Conceptual Design

The system's operational workflow is illustrated in the structured flowchart depicted in Fig. 12.2. Upon powering on, all modules, including a Globa Positioning System (GPS), Inertia Measurement Unit (IMU) sensors, image processors, and environmental sensors, undergo automatic calibration. This step ensures that positional and sensory data is accurate, minimizing error propagation throughout the mission. Once calibrated, the system begins capturing GPS coordinates, dual-camera images, and weather sensor data. These inputs are essential for location tracking, environmental assessment, and terrain mapping. In case the GPS lock is not achieved, the system enters a fail-safe loop, re-initializing the calibration and sensor stack. When the GPS is available, localization data is forwarded for processing to the ESP32 microcontroller and Jetson Nano unit which collaboratively handles incoming data streams. The ESP32 manages sensor fusion and command routing, while the Jetson Nano applies machine learning models to camera input for object detection, path prediction, and navigation adjustment. The drone adjusts altitude, pitch, and yaw using data from GPS, gyroscope, and accelerometer. It executes aerial path planning and scanning maneuvers. The rover moves along the lines of the terrain classification. Accordingly, it executes forward/reverse motion, obstacle navigation, and location stabilization. All mission data including video, sensor telemetry, and location logs are streamed live to the operator via cloud platforms or stored locally for post-processing. The collected data (e.g., air quality, pressure, humidity, GPS tracks, and images) are analyzed to predict environmental changes and are uploaded to cloud servers for remote accessibility. Simultaneously, a CSV log file is maintained on an onboard SD card.

12.4.2.1 Sea Level and Climate Monitoring

The proposed drone-rover hybrid system is designed to autonomously perform sea level and climate monitoring by combining aerial sensing, AI-based prediction, and wireless data transmission to a central control station or cloud repository. As illustrated in Fig. 12.3, the Jetson Nano, located on the drone, acts as the computational hub for the drone, receiving GPS coordinates and environmental readings from onboard sensors, while simultaneously collecting visual data from the drone's camera. The captured images are processed to estimate waves heights and detect anomalies in sea conditions using machine learning techniques. The Jetson Nano runs a trained multidimensional Long Short-Term Memory (LSTM) model for wave height forecasting, enabling short-term sea condition predictions that are crucial for

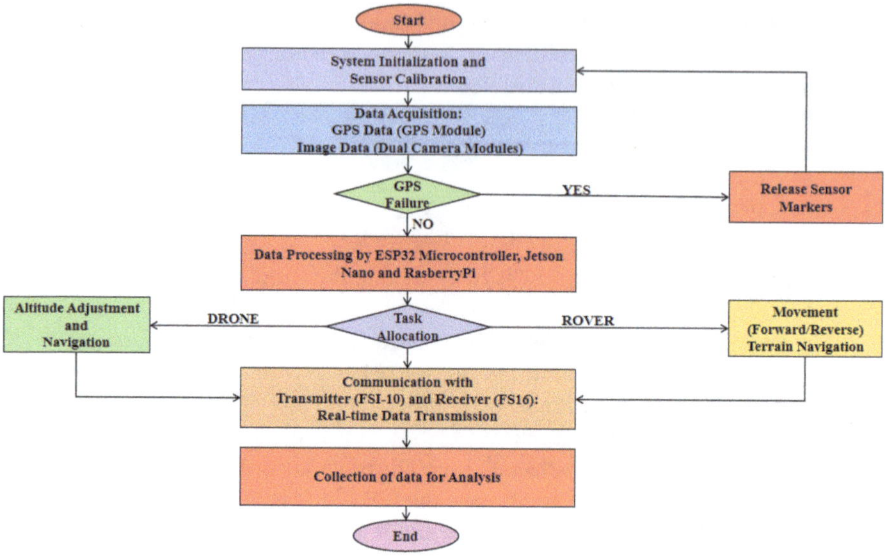

Fig. 12.2 Flowchart diagram of Drone-Rover hybrid system

maritime safety and planning. Once data is acquired and analyzed, it is transmitted through two separate channels.

High-frequency, short-range Radio Frequency (RF) transmission delivers real-time video and sensor data directly to the control station for monitoring, while a Long Range (LoRa) module handles long-range, low-power transmission of key environmental metrics to the cloud. This dual-path communication setup ensures both real-time oversight and asynchronous storage for long-term climate modeling.

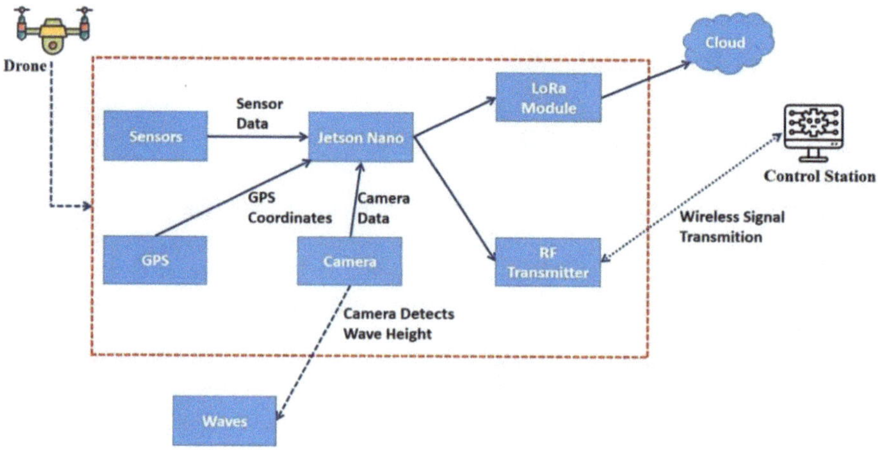

Fig. 12.3 Architecture for sea level and climate monitoring

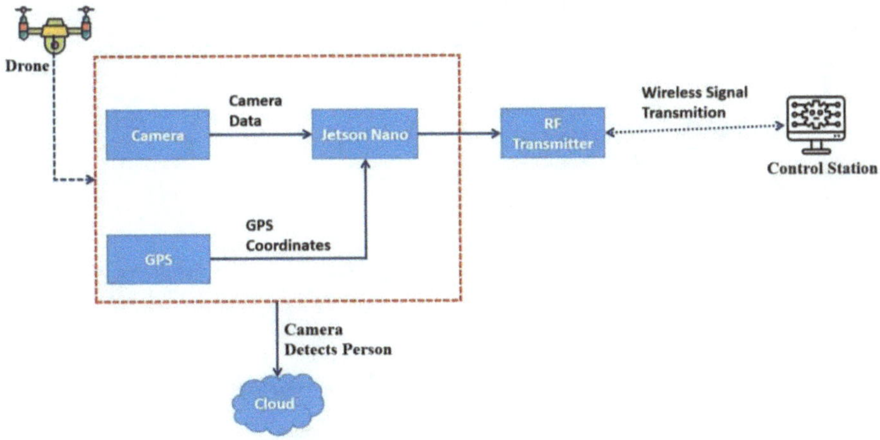

Fig. 12.4 Rescue detection and response architecture

By synchronizing the GPS data with visual and environmental inputs, the system creates geo-tagged datasets of sea surface behavior. These datasets help detect trends in wave dynamics, surface anomalies, and the influence of atmospheric pressure and temperature shifts on local sea conditions. Over time, this enables predictive analytics for storm surge events or rising sea levels.

12.4.2.2 Sea Monitoring Rescue Aid

The proposed drone-based system enhances rescue operations by autonomously identifying human presence at sea and transmitting real-time location data to a control station. As illustrated in Fig. 12.4, the drone leverages onboard intelligence to support immediate and accurate responses in rescue scenarios.

Aerial surveillance is carried out by a camera-equipped drone that streams visual data to a Jetson Nano embedded module. This compact AI processor runs a YOLOv8 object detection model capable of distinguishing people even in complex sea environments. Since GPS and visual feed are combined, geo-tagging of detected individuals is possible. Once a person is identified, the Jetson Nano associates with the detection of the GPS coordinates and passes this information to an RF transmitter module which relays the rescue alert wirelessly to a control station, which can then dispatch a response unit.

12.4.2.3 Trash and Jellyfish Collection

The YOLOv8 model runs on an onboard Jetson Nano, which processes camera feed in real time, as illustrated in Fig. 12.5. Once a detection is confirmed, the system marks

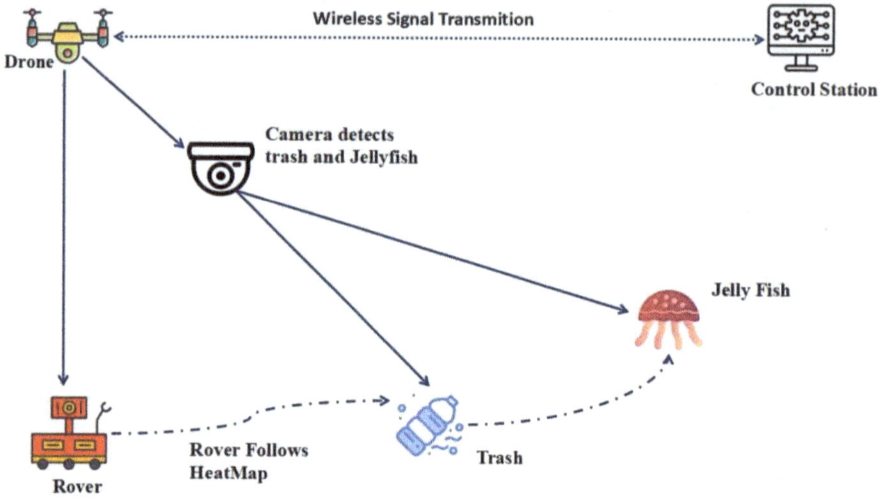

Fig. 12.5 Coordinated Drone-Rover cleanup operation

the object's location using GPS coordinates and visual metadata. These coordinates are logged into a local geospatial database and are used to generate dynamic heat maps, indicating concentration zones of trash or jellyfish. This mapping is critical for route optimization, prioritizing zones with higher pollution density for ground intervention. The heat map is updated incrementally as the drone covers new areas or reassesses known locations.

Upon localization, the drone transmits the precise GPS coordinates and object classification to a rover operating in the vicinity. This transmission occurs over a short-range communication protocol, RF or Wi-Fi Direct. It therefore enables the rover to receive immediate instructions without requiring internet connectivity. The control station also receives this data via wireless signal transmission for monitoring, logging, and potential human override. The rover is equipped with a GPS module and basic image recognition for final confirmation. Therefore, it can autonomously navigate along the marked coordinates. Depending on the identified object type, it either collects the trash using a grabber mechanism or flags jellyfish sightings for further assessment.

12.4.3 System Design and Algorithms

The proposed hybrid system is designed with a modular architecture that enables coordinated aerial and terrestrial operations. At the core of this architecture lies a multi-layered control system responsible for task allocation, navigation, perception, and decision-making. The aerial drone functions as the primary surveillance unit,

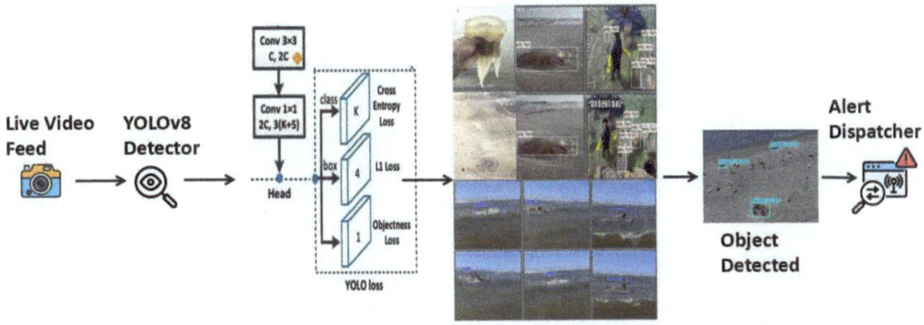

Fig. 12.6 Architecture diagram for YOLOv8 object detection

scanning coastal zones for anomalies such as floating debris, jellyfish blooms, or abnormal wave patterns. It utilizes YOLOv8 for object detection and classification in real time, leveraging the computational capabilities of the Jetson Nano. This model enables the drone-rover hybrid to perceive and interpret its surroundings by detecting and classifying objects critical to its mission.

The system integrates the detection pipeline into a unified architecture for feature extraction, bounding box regression, and classification. The detection head outputs bounding box coordinates, class scores, and object probabilities across multiple scales, as illustrated in Fig. 12.6 on some images from the dataset. The key goal is to support autonomous navigation, obstacle avoidance, and environmental monitoring by detecting three primary object classes namely Human, Trash, and Jellyfish. These objects are relevant to applications such as search and rescue, marine pollution analysis, and biological monitoring. Concurrently, wave height predictions are performed using a multidimensional LSTM neural network model trained on historical and real-time sensor data, Fig. 12.7. It therefore enables anticipatory response to hazardous sea conditions and predictive analytics for environmental conditions, particularly ocean wave height forecasting, which plays a vital role in flight and navigation planning.

The wave height prediction task is framed as a multivariate time series regression problem, where historical wave height data is used to forecast future values. The preprocessing pipeline involves data extraction and normalization, where only relevant features, such as significant wave height, are extracted and scaled using a MinMaxScaler, mapping all values to the [0, 1] range, and the sliding window technique in which time series data are split into overlapping sequences. The latter allows the LSTM model to learn temporal dependencies through a fixed-length input window of past values. The model architecture, see Fig. 12.8, is designed to capture both short-term fluctuations and long-term wave patterns.

For advanced autonomous navigation and target interaction, the hybrid system incorporates a sensor-driven position tracking architecture, as in Fig. 12.9. This subsystem computes real-time spatial parameters including the rover's current position (X, Y), orientation angle (Theta), and distance and bearing to target coordinates. The architecture begins with input from positional sensors, which are passed

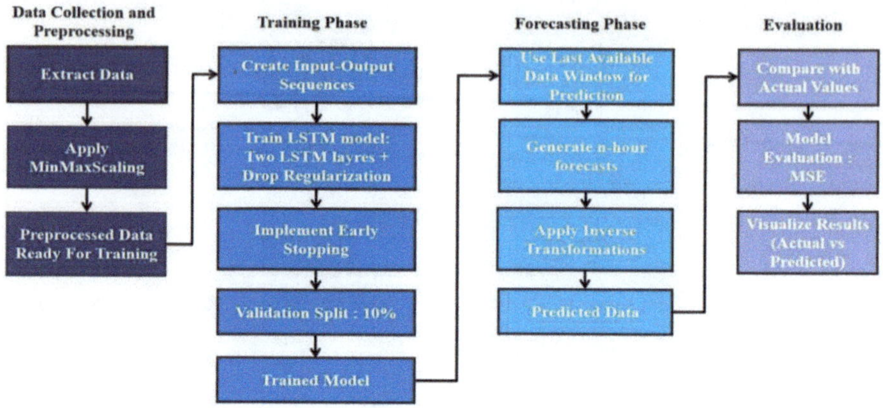

Fig. 12.7 Flowchart of LSTM network

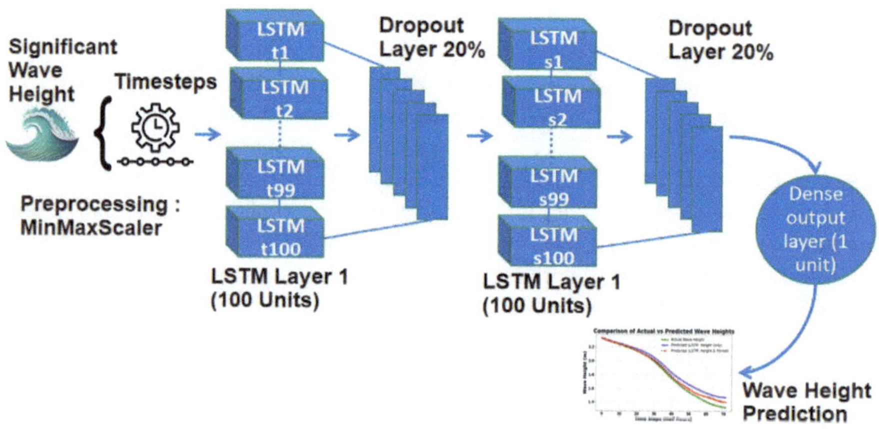

Fig. 12.8 Architecture diagram for wave height prediction

through Zero Order Hold (ZOH) blocks to stabilize discrete-time analysis. These values are normalized and reshaped for internal computation, allowing for real-time distance calculation, dynamic bearing angle computation, and an adaptive orientation correction.

To simulate real-world delays such as sensor latency and actuator lag, the architecture employs transport delay and Braitenberg filtering mechanisms. A Braitenberg filtering mechanism is a reactive control strategy in which sensor signals are directly filtered and mapped to actuators, producing emergent behaviors without explicit planning or learning. These ensure that positional updates and commands reflect actual system dynamics rather than idealized or instantaneous responses. Outputs of this subsystem include updated rover coordinates, angle and distance to next waypoint, and navigation commands for the rover's path controller. This enables smooth path

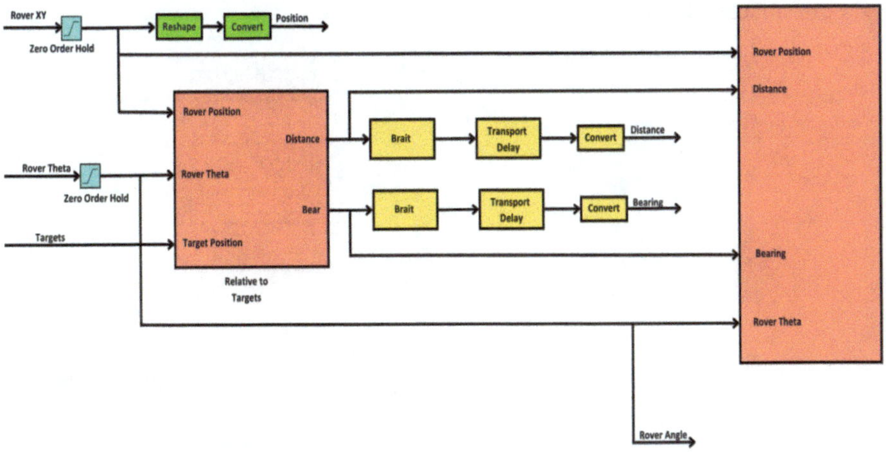

Fig. 12.9 Camera and image processing

tracking and high-precision navigation in scenarios such as GPS-denied environments, search and rescue missions, and hazardous terrain mapping. Environmental data from sensors, such as wave height, wind speed, and temperature, are continuously streamed to a cloud-based database, where they are visualized and analyzed for patterns indicative of ecological hazards. If predefined thresholds like sudden sea level rise or jellyfish proximity are crossed, the system triggers prerecorded audio alerts through shoreline speakers. This integration of AI algorithms, environmental sensing, and dual-domain mobility results in a highly adaptive system suitable for dynamic coastal conditions and mission-critical environmental monitoring tasks.

12.4.3.1 Autonomous Waste Retrieval Protocol

For the application of autonomous coastal waste management, a two-part hybrid system made up of an aerial drone and a ground-based rover is used. As shown in Fig. 12.10, the drone is equipped with a Jetson Nano, that performs real-time aerial surveillance to detect marine debris and jellyfish near the shoreline, while the rover, powered by a Raspberry Pi [9], is responsible for ground-level waste retrieval through GPS-guided navigation and actuation. The drone functions as the primary surveillance unit. It captures real-time imagery using an onboard camera and processes this data on Jetson Nano using YOLOv8. The system calculates the corresponding geographic coordinates using GPS data and camera geometry. Once marine debris or jellyfish is detected from aerial imagery, the drone computes the GPS location of the object and packages these data into a lightweight American Standard Code for Information Interchange (ASCII) encoded string. This message is transmitted directly via a LoRa module operating in P2P mode (Long Range Peer to Peer). On the ground, the rover's Raspberry Pi [9] is connected to a compatible LoRa

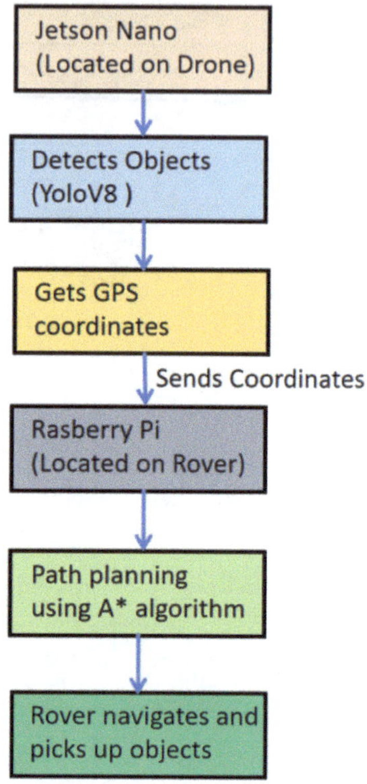

Fig. 12.10 Architecture of autonomous waste retrieval

receiver module configured to operate with identical parameters (e.g., frequency, spreading factor, and bandwidth). The extracted GPS coordinates are input into the rover's navigation and task-planning system, to calculate the most efficient path reaching the targeted debris or jellyfish to be collected.

12.4.3.2 Real-Time Climate Data Logging and Autonomous Coastal Alert System

The local control station equipped with a speaker system functions as a node for receiving, interpreting, and broadcasting warnings without requiring permanent internet access in the coastal areas. The aerial drone is equipped with a Jetson Nano board capable of executing onboard inference using deep learning models such as YOLOv8. During each surveillance flight, the drone captures high-resolution visual data and sensor readings (e.g., GPS, temperature, altitude, barometric pressure). Object detection tasks include:

i. Identification of marine debris floating near the coast
ii. Recognition of Jellyfish clusters, especially during spawning season

Fig. 12.11 Prototype of the designed drone

iii. Monitoring of wave patterns and tidal level changes, inferred from visual and environmental sensor data

Once the object and/or anomaly is detected, the drone immediately logs: The Type of hazard, Latitude and Longitude (from GPS), and Timestamp. This data is then serialized and transmitted via LoRa to the local control station. Once the Local station has received the data, based on their own unique parameters, they identify hazards and send the appropriate message through their broadcasting system. The development of the hybrid drone-rover platform involved an iterative process of design, fabrication, and integration to achieve robust environmental monitoring and autonomous deployment in coastal areas. The aerial module, such as the quadrotor drone shown in Fig. 12.11, is engineered with four 2200 kV BLDC motors, each regulated by a 30A ESC, enabling it to generate a combined thrust of 35.6 N sufficient for stable lift of a 1.8 kg payload.

This allows it to carry, deploy, and retrieve the rover using a modular coupling mechanism. Aerodynamic calculations confirm a thrust-to-weight ratio of 2.0, providing stability and agility during flight. The Pixhawk 2.4.8 flight controller governs navigation with dual-mode control semi-autonomous stabilization and altitude hold alongside a fail-safe return-to-launch feature. For extended missions, the drone utilizes a 4200 mAh 12 V LiPo battery delivering up to 20 min of aerial operation, while real-time telemetry and environmental data are relayed through a Wi-Fi/MQTT-enabled communication architecture.

The ground-based rover, Fig. 12.12, is built with a six-wheeled differential drive and driven by high-torque Direct Current (DC) motors interfaced via dual L298N motor drivers. It is equipped with dual front and rear cameras, and sensor modules (DHT11, MQ135, BMP280, anemometer) managed by an ESP32 micro-controller. Its embedded Jetson Nano handles AI-based object detection and navigation using YOLOv8. Operating independently or in tandem with the drone, the rover can

Fig. 12.12 Prototype of designed 6 wheeled rover

Fig. 12.13 Prototype of integrated Drone-Rover hybrid

autonomously navigate complex terrains, perform data logging, and execute return-to-base functions. Both modules share a common power and control infrastructure, with seamless transitions facilitated by the FlySky FS-i6 transmitter/receiver system. Overall, the drone-rover hybrid prototype shown in Fig. 12.13, demonstrates a scalable, modular, and resilient robotic system optimized for dynamic and remote environmental monitoring tasks.

12.5 Experimental Results

This section presents the results of the two core modules of the proposed hybrid drone-rover platform: wave height prediction using a Multidimensional LSTM model and real-time object detection using YOLOv8. Performance is evaluated using standard metrics, and integration outcomes are discussed in the context of autonomous coastal monitoring tasks.

12.5.1 Wave Height Prediction Using Multidimensional LSTM

The wave prediction model employs a stacked LSTM architecture with two sequential layers of 100 units each, interleaved with dropout layers with rate of 0.2 to prevent over-fitting. A final dense layer outputs the predicted wave height. The model is trained using the Adam optimizer with mean squared error (MSE) as the loss function. Training configuration comprises of data split: 70% training, 30% testing (10% of testing reserved for validation), 1000 epochs, 32 batch size, patience of 10 epochs for early stopping, historical wave height and wave period time series taken at 30 min intervals as input features, and preprocessing using MinMaxScaler normalization. A recursive prediction strategy is employed, where each predicted value is recursively used as input for multi-step forecasting.

The model's performance across different forecasting horizons (3 h to 72 h) is shown in Fig. 12.14, which illustrates the close correspondence between observed and predicted values. The results indicate the model's capability to capture both high-frequency variations and long-term trends in wave behavior. These predictions serve as inputs to the environmental awareness module, enabling preemptive actions such as altitude adjustment or rover deployment delays in response to rising wave conditions (Table 12.1).

The LSTM model trained on historical wave data demonstrated strong predictive capability. The model achieved a mean squared error (MSE) of 0.0035 and a mean absolute error (MAE) of 0.046 m on the test set at 48 h forecast. As shown in Fig. 12.13, the predicted values closely followed actual wave height patterns, confirming that the model effectively captured both short-term fluctuations and broader wave trends.

12.5.2 Real-Time Object Detection with YOLOv8

To complement temporal prediction, spatial perception is achieved through a YOLOv8 model trained for oriented object detection. The training dataset consists of approximately 1200 images gathered from multiple Roboflow datasets and manually annotated drone footage with three classes for human, trash and jellyfish. The training configuration consists of 300 epochs, 8 batch size, input size of 640 $\times$ 640 pixels, an SGD optimizer with cosine annealing schedule and for loss functions, both Cross-Entropy (for classification) and Binary Cross-Entropy (for objectiveness). During inference on the Jetson Nano, YOLOv8 achieved the following results: mAP

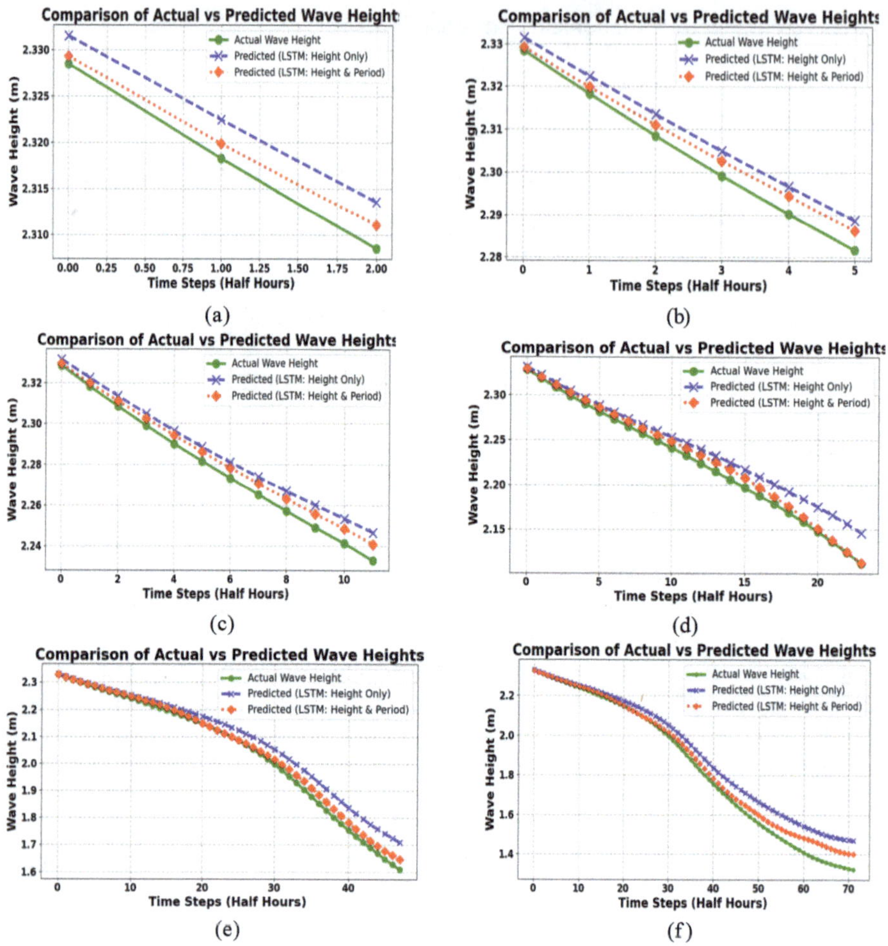

Fig. 12.14 Observed versus predicted wave heights for forecast of **a** 3 h, **b** 6 h, **c** 12 h, **d** 24 h, **e** 48 h, **f** 72 h

Table 12.1 LSTM and multi-LSTM statistical results for wave height prediction

Prediction time (Hours)	LSTM				Multi-LSTM			
	MAE	MSE	RMSE	R^2	MAE	MSE	RMSE	R^2
3	0.0170	0.0003	0.0170	0.8463	0.0080	0.0001	0.0082	0.9964
6	0.0113	0.0025	0.0113	0.8705	0.0089	0.0002	0.0091	0.9260
12	0.0138	0.0090	0.0153	0.9383	0.0062	0.0007	0.0070	0.9885
24	0.0259	0.0094	0.0343	0.9721	0.0112	0.0002	0.0126	0.9967
48	0.0464	0.0035	0.0594	0.9716	0.0462	0.0029	0.0537	0.9794
72	0.0583	0.0042	0.0645	0.9665	0.0464	0.0033	0.0571	0.9768

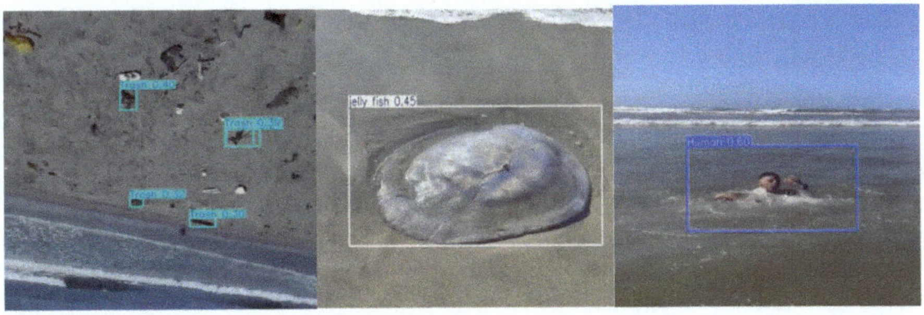

(a) Trash (b) Jelly Fish (c) Human

Fig. 12.15 Object detection using YOLOv8

at 0.5 of 89.7%, precision of 93.3%, recall of 88.5% and inference speed of 24 FPS. Figure 12.15 illustrates sample detection of trash, jellyfish, and humans under challenging visual conditions such as water glare and partial occlusion. Detected objects were assigned unique IDs for frame-to-frame tracking, and a custom rule-based logic triggered hazard alerts and cloud logging upon detection of specific threats like biohazards or plastic waste.

12.5.3 System-Level Integration and Field Evaluation

The hybrid system was tested in both simulated and real coastal scenarios. The integrated pipeline demonstrated reliable operation across all core tasks, including autonomous drone deployment, wave height forecasting, real-time object detection, and rover navigation. The flight duration per mission was ~ 19 min while rover runtime was ~ 60 min per full charge. The deployment and retrieval success rate was 97% and Jetson Nano average processing latency less than 100 ms per frame. Rover localization accuracy had ~ 15 cm average error. The system's performance validates its applicability for dynamic coastal monitoring, offering an effective combination of temporal foresight (via multi-LSTM) and spatial awareness (via YOLOv8), coordinated through an onboard AI hub.

12.6 Discussion

The experimental results confirm that the proposed hybrid drone-rover system can address key challenges in autonomous coastal monitoring, including environmental variability, dynamic object detection, and real-time decision-making.

12.6.1 Synergistic Performance of Temporal and Spatial Modules

The integration of a multidimensional LSTM model for wave prediction with a YOLOv8-based object detection pipeline enhances the system's situational awareness. The LSTM model achieved low MSE and MAE values, demonstrating its reliability in predicting short- and medium-term wave behavior. This foresight is critical for proactive mission planning, such as adjusting drone altitude during takeoff or delaying rover deployment in anticipation of unstable marine conditions. In parallel, YOLOv8's high detection precision and real-time inference capability provided accurate recognition of both anthropogenic and biological objects under complex visual conditions. The combined use of temporal and spatial modules, unified by the Jetson Nano-based AI hub, allowed for real-time adaptability. This design supports a more nuanced and robust understanding of the environment, compared to standalone systems. One notable strength is the system's adaptability to varying environmental conditions. The inclusion of data augmentation in the YOLOv8 training phase allowed for resilience to changes in lighting, scale, and object orientation, which are typical of coastal scenes.

12.6.2 Limitations and Future Enhancements

While the system demonstrated strong baseline performance, several limitations were observed. The object detection dataset of about 1200 images may constrain generalization to rare or novel objects. Expanding the dataset with more diverse environmental samples could further improve robustness. Object detection performance drops under poor lighting or water-reflective conditions, with false positives for semi-transparent objects like jellyfish. Integrating advanced edge AI modules such as the Jetson Orin Nano could significantly improve real-time processing capabilities, enabling more complex deep learning models for refined object detection and environmental classification. Although recursive LSTM forecasting performed well up to 48–72 h, prediction error accumulation becomes noticeable at longer horizons. Techniques such as scheduled sampling or encoder-decoder architecture could mitigate drift. The Jetson Nano performed well within current demands but may become a bottleneck with more complex multi-modal fusion tasks. Upgraded localization techniques, including sensor fusion of GPS, IMU, and visual SLAM, can improve precision in GPS-denied areas. The system can also benefit from a more intelligent task allocation algorithm, allowing dynamic decision-making based on terrain analysis, battery levels, and environmental hazards. Future versions could explore edge TPU or Xavier-based platforms for enhanced processing power. Despite these limitations, the system provides a strong foundation for autonomous, AI-driven environmental monitoring. Its modular design supports future extensions such as underwater sensing, chemical pollutant detection, or reinforcement learning-based adaptive planning. Incorporating solar charging or wireless docking stations would also ensure longer mission durations in remote or maritime environments.

12.7 Conclusion

This work introduces and validates an integrated drone-rover hybrid system capable of autonomous operation across both aerial and ground domains for coastal monitoring and environmental response. The combination of advanced sensing modules, a flexible control framework, and onboard intelligence driven by YOLOv8 detection and multidimensional LSTM forecasting enables the platform to perceive, interpret, and react to environmental conditions in real time. Through this integration, the system achieves a high degree of operational reliability and situational awareness across diverse coastal scenarios. Performance assessments demonstrate that the hybrid platform delivers precise object identification, low-latency computation, and consistent wave height prediction accuracy, confirming its suitability for complex coastal tasks. The synergy between the drone's wide-area coverage and the rover's terrain adapted mobility supports a broad range of applications, including debris tracking, jellyfish detection, environmental surveillance, and search and rescue assistance. These findings show that the multi-domain architecture offers clear advantages over conventional single-platform solutions, particularly in dynamic and unpredictable coastal environments. Although the system performs effectively, further advancements such as enlarging the training dataset, improving perception under challenging lighting conditions, and increasing onboard processing capabilities would enhance overall resilience and adaptability. Even so, the present study provides a strong technical foundation for future work in hybrid robotic platforms. Overall, the developed system marks a meaningful step toward autonomous, scalable, and environmentally responsive coastal monitoring solutions. Its demonstrated capabilities indicate significant potential for future deployment in disaster readiness, marine ecosystem protection, and long-term environmental observation.

References

1. Nicholls, R.J., Cazenave, A.: Sea-level rise and its impact on coastal zones. Science **328**(5985), 1517–1520 (2010)
2. Halpern, B.S., et al.: A global map of human impact on marine ecosystems. Science **319**(5865), 948–952 (2008)
3. Galgani, F., Hanke, G., Maes, T.: Global distribution, composition and abundance of marine litter. In Marine Anthropogenic Litter, pp. 29–56. Springer, Cham (2015)
4. Whitehead, K., Hugenholtz, C.H.: Remote sensing of the environment with small unmanned aircraft systems (UASs), part 1: a review of progress and challenges. J. Unmanned Veh. Syst. **2**(3), 69–85 (2014)
5. Panigrahi, S., Maski, P., Thondiyath, A.: Real-time biodiversity analysis using deep-learning algorithms on mobile robotic platforms. PeerJ Comput. Sci. **9**, e1502 (2023)

6. Hashim, A.S., Mohamad Tamizi, M.S.: Development of Drone for search and rescue operation in Malaysia flood disaster. Int. J. Eng. Technol. **7**, 9–12 (2018)
7. Kaggle.: Ocean Wave Prediction with LSTM (2022)
8. Siegwart, R., Nourbakhsh, I.R., Scaramuzza, D.: Introduction to autonomous mobile robots. MIT press (2011)
9. Kant, K., Mahankali, S., Bharadwaj, A., Sundaram, S.: An autonomous hybrid Drone-Rover vehicle for weed removal and spraying applications in agriculture. In: Proceeding 2023 IEEE International Conference on Agrosystem Engineering, Technology and Applications (AGRETA) (2023). https://doi.org/10.1109/AGRETA57740.2023.10262416
10. Kumar, S., et al.: An autonomous hybrid Drone-Rover vehicle for weed removal and spraying applications in agriculture. ResearchGate Preprint (2023)
11. Kalantari, A., et al.: Drivocopter: a concept Hybrid Aerial/Ground vehicle for long-endurance mobility. 2020 IEEE Aerospace Conference, Big Sky, MT, USA, pp. 1–10 (2020). https://doi.org/10.1109/AERO47225.2020.9172782
12. Kumarasan, T., Murugan, D., Benedictvinoth, M.: Drone-enabled wildlife monitoring system: revolutionizing conservation efforts. J.J. Coll. Eng. Technol. (2020). https://doi.org/10.48047/nq.2020.18.8.nq20245
13. Ergene, Y.: Analysis of Unmanned Systems in Military Logistics. Naval Postgraduate School (U.S.) (2016)
14. Emimi, M., et al.: The current opportunities and challenges in drone technology. Int. J. Electr. Eng. Sustain. **1**(3), 74–89 (2023). Accessed 6 Mar 2025
15. Arumugam, A., et al.: Deep learning-based wave height prediction using LSTM networks. Environ. Model. Softw. **150** (2022)
16. Akdag, M.: Ocean Wave Prediction with LSTM. Kaggle Notebook (2022). Available: https://www.kaggle.com/code/akdagmelih/ocean-wave-prediction-with-lstm
17. Keylabs.ai.: Real-Time Object Detection with YOLOv8 (2023). Available: https://keylabs.ai/blog/real-time-object-detection-with-yolov8
18. Singh, J.D., Valdenegro, M.: The Marine Debris Dataset for Forward-Looking Sonar Semantic Segmentation. arXiv preprint (2021). https://doi.org/10.48550/arXiv.2108.06800
19. Karnataka State Council for Science and Technology (KSCST).: Autonomous Drone-Rover for Environmental Monitoring (2024). Available: https://www.kscst.org.in/spp/47_series/47s_spp/Exhibition%20Projects/154_47S_BE_0455.pdf
20. Velvizhi, V.A., Anbarasan, M., Pradeep, G.S., Kaviarasu, K.S., Vishal, S.M.: Development and integration of a dual-mode Drone-Rover system for enhanced search and rescue operations. 2024 International Conference on Power, Energy, Control and Transmission Systems (ICPECTS), Chennai, India (2024)
21. Meier, L., Tanskanen, P., Fraundorfer, F., Pollefeys, M.: PIXHAWK: a system for autonomous flight using on board computer vision. 2011 IEEE International Conference on Robotics and Automation, pp. 2992–2997. Shanghai, China (2011). https://doi.org/10.1109/ICRA.2011.5980229
22. Utomo, P.A., Satyawan, A.S., Puspita, H., Wulandari, I.Y.: 360-degree image processing on NVIDIA Jetson Nano. Internet Things Artif. Intell. J. **4**(2), 15 (2024). https://doi.org/10.31763/iota.v4i2.722
23. Chang, K., Rammos, P., Wilkerson, S.A., Gadsden, S.A.: LiPo battery energy studies for improved flight performance of unmanned aerial systems. In: Proceeding SPIE Defense + Security (2016). https://doi.org/10.1117/12.2223352

Chapter 13
Securing the Internet of Marine Things: A Context-Aware Zero-Trust Framework with Post-quantum Cryptography and Swarm Intelligence

Ansh Raj Rath, Niharika Sharma, B. P. Sreeja, and V. Jothi Prakash

13.1 Introduction: The Connected Ocean and the Expanded Attack Surface

The systematic approach for developing the LC-ZTA model is depicted in Fig. 13.1. The framework initiates with an analysis of underwater acoustic physics and adversarial threats, such as the "Harvest Now, Decrypt Later" (HNDL) quantum threat [1]. It then progresses through the integration of multi-layered security—specifically Physical Layer Security (PLS) and Post-quantum Cryptography (PQC)—before validating the system using Digital Twin simulations [2].

The proposed research framework for the LC-ZTA model, illustrated in Fig. 13.1, outlines a systematic transition from identifying the physical and adversarial challenges of the IoMT to implementing a multi-layered, quantum-resilient, and context-aware zero-trust architecture.

A. R. Rath (✉) · N. Sharma · B. P. Sreeja
School of Artificial Intelligence, Amrita Vishwa Vidyapeetham, Coimbatore, India
e-mail: anshrajrath12@gmail.com

B. P. Sreeja
e-mail: bp_sreeja@cb.amrita.edu

V. J. Prakash
Department of Information Technology, Karpagam College of Engineering, Coimbatore, India
e-mail: jothiprakash.v@kce.ac.in

T. A. Tran et al. (eds.), *Big Data Analytics for Internet of Marine Things*,
Internet of Things, https://doi.org/10.1007/978-3-032-28171-5_13

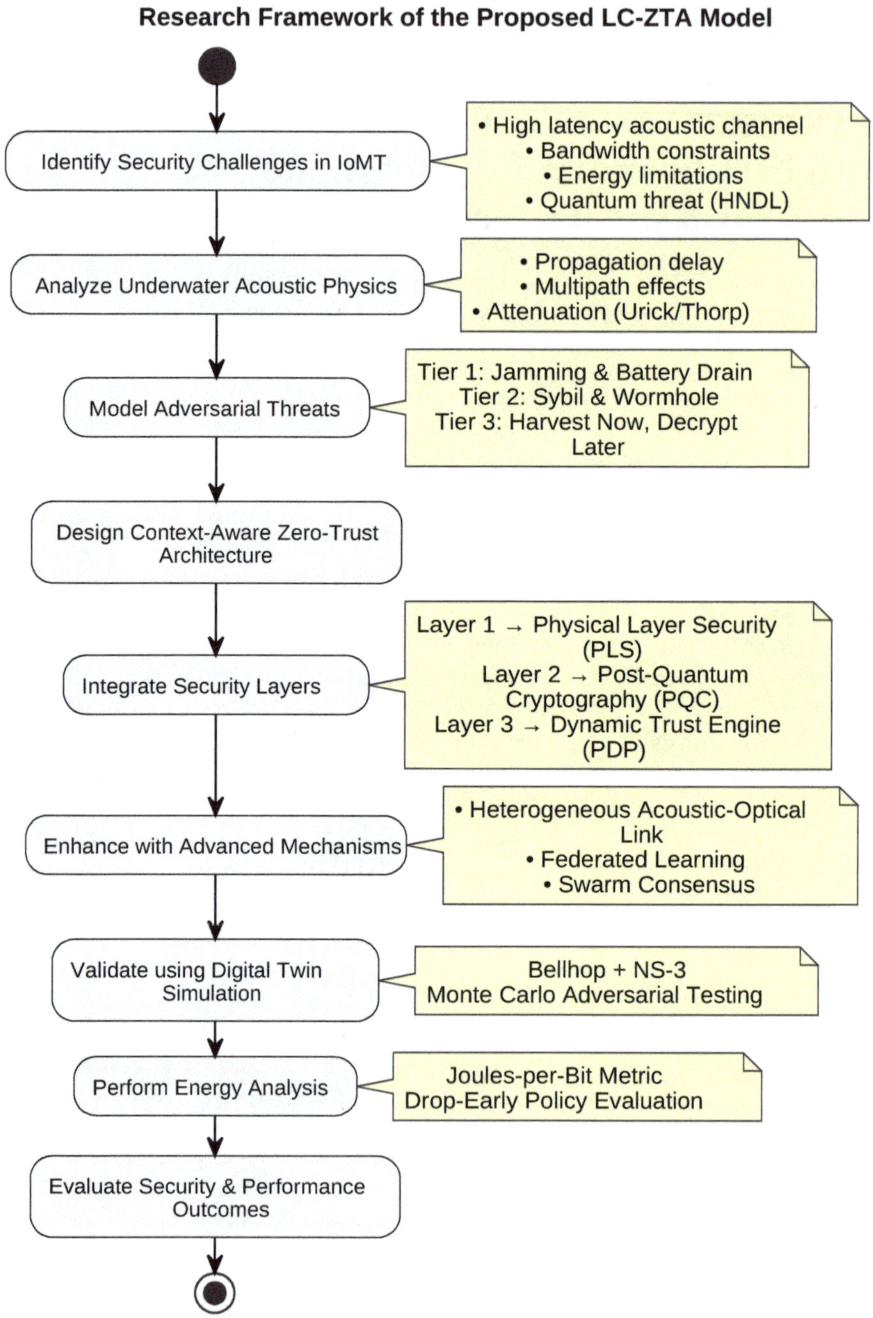

Fig. 13.1 Research framework of the proposed LC-ZTA model, illustrating the progression from challenge identification to security validation

13.1.1 The Paradigm Shift: Internet of Marine Things (IoMT)

The oceans still are the least connected and least explored domain of the current digital world. Over the past decade, a collective convergence of advanced sensing technologies, autonomous robotics, and hydro-acoustic communication has started reshaping this domain, giving rise to the Internet of Marine Things (IoMT) [3].

Unlike traditional IoT networks on land, the Internet of Underwater Things (IoMT) refers to a unique cyber-physical system designed to meet the challenges of the underwater environment. It integrates a heterogeneous network of connected nodes, including:

- **Autonomous Underwater Vehicles (AUVs)**: Smart mobile systems that perform a variety of jobs ranging from pipeline inspection, tactical reconnaissance to full bathymetric mapping.

  Recent Advancements: The major theme of the recent studies is incorporating Deep Reinforcement Learning (DRL) into AUVs to enable autonomous, real-time trajectory planning in uncharted terrains. Moreover, AUVs are frequently deployed as mobile "data mules" in 6G-supported underwater acoustic sensor networks (UASNs) to solve the problem of energy usage by static nodes [4].

  Restrictions: AUVs are severely limited in their operations due to battery life and the high latency of acoustic channels, which also makes it difficult for them to be controlled remotely in real time. From a security standpoint, because they operate independently and are isolated, their intelligent decision-making systems are very vulnerable to adversarial data injection and physical node capture. This isolated operational scenario necessitates continuous, zero-trust verification, which is also highly important [4].
- **Static Sensor Arrays**: Several arrays of sensors installed at a fixed location for keeping a continuous record of seismic activity, changes in chemical composition, and variations in ocean temperature.

  Recent Advancements: Based on recent studies, in order to break the energy limitation of the Internet of Underwater Things (IoUT), edge computing and energy-aware fuzzy routing protocols integration are emphasized as the main features in the latest articles. Thanks to these technologies, static arrays can analyze environmental information on-site and therefore, do not have to send the raw data which consume a lot of energy to power the acoustic communication lines [5].

  Limitations: Their immobility makes them very susceptible to environmental biofouling, physical tampering, and localized signal jamming. Since these nodes are data collection points that work continuously and require very few hardware updates through their long deployment lifecycles, they are very attractive to "harvest now, decrypt later" quantum attacks, which necessitates the use of lightweight yet quantum-resistant cryptographic solutions [5].

- **Surface Gateways**: Floating entities like buoys or unmanned surface vehicles that help in bridging the underwater acoustic-based networks with terrestrial or satellite RF networks.

  Recent Advancements: Architectures nowadays are orienting hybrid multi-modal communication (mashing acoustic and optical links) and Integrated Sensing and Communication (ISAC) networks. They empower gateways to tune spectrum usage closely and be compatible with Low Earth Orbit (LEO) satellite constellations for global data exfiltration [6].

  Limitations: Running power-consuming dual-communication links in rough surface weather conditions leads to rapid depletion of energy. More importantly, being the central chokepoint for all network traffic, surface gateways constitute a single point of failure. They are highly vulnerable to Distributed Denial of Service (DDoS) attacks and traffic analysis, thus decentralized anomaly detection mechanisms rather than gateway-heavy processing are essential for network survivability [6].

While modern AUVs exhibit extended autonomy and swarm coordination capabilities, this newly introduced inter-connectivity also leads to a greater adversarial attack surface. New vulnerabilities that can compromise vital offshore infrastructure or neutralize strategic defense assets are brought about by the shift from isolated assets to networked and interconnected fleets.

13.1.2 The Inefficiency of Terrestrial Security Framework

Conventionally, underwater networks operated under the concept of "security by isolation" or "security by obscurity." The inherent hostility of the subsea environment and the proprietary nature of acoustic modulation schemes were deemed sufficient barriers to entry. This doctrine is now obsolete. The widespread availability of software-defined acoustic modems (SDAMs) and commercial off-the-shelf (COTS) hydrophones has lowered the barrier to accessing the underwater acoustic spectrum, allowing even non-state actors to carry out interception and signal injection attacks.

Efforts to transpose terrestrial security standards to the underwater domain have faltered due to the physics of the medium. Protocols like TLS/SSL involve a lot of handshake and certificate transfer. In a fiber-optic network, a 4 KB handshake is insignificant; but in unmanned underwater acoustic network (UWAN) with a data rate of 1 kbps and a 2-s propagation delay, this can lead to considerable latency (about 30 s) and energy drain [7].

Moreover, the trust model in the traditional Public Key Infrastructure (PKI) is not compatible with the nature of the AUV swarms, which are intermittently connected and decentralized, and in which communication with the central Certificate Authority (CA) is usually interrupted.

13.1.3 Architectural Pivot: Context-Aware Zero-Trust

This chapter outlines a clear deviation from perimeter-based security, moving forward with a Context-Aware Zero-Trust Architecture (ZTA).

- **Zero Trust**: "Never Trust, Always Verify." Any implicit trust that was previously based on network location or possession of cryptographic keys is now invalid and carries no value.
- **Contextual Binding**: Authentication is now advanced beyond cryptographic validity to include physical integrity. The system validates and confirms a transmission by correlating location of origin (via acoustic physics) with behavioral patterns using analytics [8].

By strictly combining security protocols with the stochastic properties of the acoustic channel, we construct a new and improved defense mechanism that now uses the hostile environment as a cryptographic asset.

13.1.4 Main Contributions of the Study

Acknowledging the critical security vulnerabilities and resource limitations that are present in underwater environments, this chapter proposes a novel Lightweight Context-Aware Zero-Trust Architecture. The main contributions of this study are the following:

- **Leading Physiocryptographic Verification**: We introduce a method that uses the random and harsh properties present in the underwater acoustic channel impulse response as a dynamic and valuable cryptographic asset. This transitions authentication from static key possession based dependency to a continuous physical layer verification.
- **Combination of Post-quantum Cryptography (PQC)**: We direct and integrate lightweight PQC protocols that are specifically tailor made for resource constrained use cases. In our case, the marine nodes, protects the long-lifecycle underwater assets from emerging "harvest now, decrypt later" quantum computing threats.
- **Decentralized Threat Detection via Federated Learning**: We then customize the framework with localized Federated Learning models, that enables the nodes to collectively collaborate and identify zero-day anomalies and malicious behavioral patterns all without transmitting any raw, bandwidth heavy data to surface gateways.
- **Bio-inspired Swarm Consensus**: We also implement a robust consensus mechanism that grants the autonomous underwater swarms to maintain and preserve data integrity and also the ability to continue their operation even when isolated or under attacks like localized denial-of-service attacks.

13.1.5 *Significance to the Internet of Marine Things (IoMT)*

The central significance of this study is in its direct resolution of the problem that is security versus efficiency tradeoff which is present within the Internet of Marine Things (IoMT). Traditional perimeter-based security model that are used as a norm in terrestrial systems are completely incompatible with IoMT due to their highly mobile, fragmented, and easily compromised nature of these marine nodes. Decentralizing the threat intelligence involved and establishing continuous context-aware verification now ensures that IoMT networks can reliably support critical operations that range from continuous bathymetric mapping to tactical maritime surveillance. This framework redefines IoMT security that used to be a reactive and computationally heavy burden into a highly proactive, lightweight, and quantum-resilient system which ensures long-term survivability in such hostile aquatic environments.

13.2 Physics of the Hostile Channel: A Security Perspective

To understand the failure of the standard protocol and the efficiency of the proposed architecture, we must analyze the underwater acoustic channel, which is widely regarded as the most challenging communication medium in existence.

13.2.1 *The Temporal Attack Surface: Propagation Delay*

The speed of sound in water ($c \approx 1,500$ m/s) is five orders of magnitude lower than the speed of light ($c \approx 3 \times 10^8$ m/s).

$$T_{\text{prop}} = \frac{d}{c} \tag{13.1}$$

A transmission distance (d) of 3 km induces a propagation delay $\left(T_{\text{prop}}\right)$ of 2 s [9]. A traditional three-way handshake (SYN, SYN-ACK, ACK) has a minimum idle delay of 4 s, before even accounting for processing or serialization overhead.

Security Implication: This prohibitive latency creates a "temporal attack surface." An adversary can intercept a command (e.g., actuation of a valve), allow the legitimate execution, and replay the signal post-execution. Traditional networks make use of accurate timestamping (T_{send}) to discard late packets (i.e., if$|T_{\text{current}} - T_{\text{send}}| > \delta$). Meanwhile, in underwater networks, clock synchronization is inherently compromised by variable propagation delays which causes high false-positive rates due to strictly enforced timestamp checking [10].

13.2.2 Spectral Efficiency and Attenuation Constraints

The phenomenon of acoustic wave attenuation is dependent on frequency, which makes it essential to utilize low-frequency bands (10 kHz–50 kHz) in long-distance data transmission. Transmission loss ($T\,L$) is calculated via the Urick equation [11]:

$$TL(d, f) = k \cdot \log(d) + d \cdot \alpha(f) \times 10^{-3} \tag{13.2}$$

Let k be the geometric spreading coefficient and (f) the absorption coefficient as follows from Thorp's formula [12]:

$$\alpha(f) = \frac{0.11f^2}{1+f^2} + \frac{44f^2}{4100+f^2} + 2.75 \times 10^{-4} f^2 + 0.003 \tag{13.3}$$

Security Implication: The above-mentioned constraints result in a large bandwidth restriction. A typical example of a post-quantum cryptographic (PQC) signature, like SPHINCS+, is tens of kilobytes in size. The transfer of such a large payload would monopolize the acoustic channel for several minutes, thus resulting in a denial-of-service attack. Therefore, it is essential to focus on Lightweight Cryptography (LWC) and Key Encapsulation Mechanisms (KEM).

13.2.3 Stochastic Multipath Propagation

The underwater acoustic channel can be considered a waveguide where the signal is reflected from the sea surface as well as from the ocean floor. This results in a Channel Impulse Response (CIR) that lasts for tens of milliseconds, thus causing Inter-Symbol Interference (ISI).

$$(t) = \sum_{p=0}^{p-1} A_p(t)x\big(t - \tau_p(t)\big) + n(t) \tag{13.4}$$

Here, A_p and τ_p are the amplitude and delay of the p-th multipath component, respectively.

Security Implication (Physiocryptography): Although ISI is compromising the data integrity, it also introduces a unique location-dependent characteristic that can be used to help from the security perspective. The multipath delay profile varies with source depth, receiver depth, and distance thus making it unique, which improves authentication. This property of the ISI uses this phenomenon for Physical Layer Authentication's advantage [8].

13.3 Adversarial Modeling in the Acoustic Domain

An efficient and ironclad architecture must be designed to resist possible and targeted threat vectors. Threats faced by the architecture are divided into three types:

13.3.1 Tier 1: Spectrum Denial and Resource Exhaustion

Attacks targeting availability:

- **Reactive Jamming**: The attacker actively listens to an intercepted channel waiting for a known pattern that is sent at the start of data packets (data preamble) by using energy detection. The attacker then sends noise bursts only when valid transmissions are occurring to maximize channel disruption with minimal energy expenditure to conserve resources at the attacker's side [13].
- **Sleep Deprivation (Battery Drain)**: The attacker persistently sends requests to initiate a handshake with a node, aiming to keep the processor active or in a high power mode, which leads to considerable battery depletion. This results in mission failure due to drained batteries [14].

13.3.2 Tier 2: Identity Deception and Routing Attacks

Attacks aimed at compromising integrity and trust.

- **Sybil Attacks**: A malicious node creates multiple fake identities to manipulate the consensus mechanisms like swarm voting and also affect the network's decision-making [15].
- **Wormhole Attacks**: The attackers relay packets through a low-latency external link (not part of the target network, for example, a radio link), making distant nodes appear to be neighbors. This allows the attackers to later hijack routing paths [15].

13.3.3 Tier 3:"Harvest Now, Decrypt Later" (Quantum Threat)

A novel strategic risk associated with data longevity is utilized. Attackers have now started to actively archive encrypted network traffic. Upon the maturity of cryptographically relevant quantum computers (CRQC), these archives will be decrypted using Shor's Algorithm. The need to adopt post-quantum cryptography (PQC) for long-lifecycle data immediately arises in order to maintain its confidentiality [1].

13.4 Proposed Architecture: Context-Aware Zero-Trust

The shift from perimeter-focused defense to a Zero-Trust Architecture (ZTA) is driven by the decentralization of naval forces. Our proposed Lightweight Context-Aware Zero-Trust Architecture (LC-ZTA) builds upon existing identity authentication by considering the physics of the hostile underwater channel as a security strength rather than a weakness.

13.4.1 Architectural Adherence to NIST SP 800-207

In order to ensure alignment with industry standards, the LC-ZTA is mapped to the logical components described in NIST SP 800-207 as follows [16]: An overview of the proposed architecture is given in Fig. 13.2

1. **Policy Enforcement Point (PEP)**: The PEP is integrated into the acoustic modem firmware or FPGA and acts as a low-level gatekeeper. It performs real-time signal analysis and implements "hard drops" of packets that have anomalous physical properties (e.g., Doppler shifts or RSS spikes) in order to shield the upper network stack from exhaustion.
2. **Policy Decision Point (PDP)**: The PDP is found inside a Secure Enclave (TrustZone) on the main controller. The PDP is the central decision-making authority for the system. The PDP calculates a dynamic Trust Score (Tk) through the integration of physical proof of verification, cryptographic validity, and behavioral heuristics.
3. **Dynamic Trust Algorithm**: The trust value is calculated through a weighted multi-factor formula:

$$T_k = w_1 \cdot P_{\text{physics}} + w_2 \cdot P_{\text{crypto}} + w_3 \cdot P_{\text{behavior}} \tag{13.5}$$

   Any packet that is below the adaptive threshold ($T_{\text{threshold}}$) is purged immediately. This is to maintain the integrity of the nodes.

13.4.2 Layer 1: Physical Layer Security (PLS)

The first line of defense relies on the characteristics of the underwater environment.

13.4.2.1 RSS-Distance Consistency

The system correlates the Received Signal Strength (P_{rx}) with the GPS location (d_{gps}) claimed in the payload.

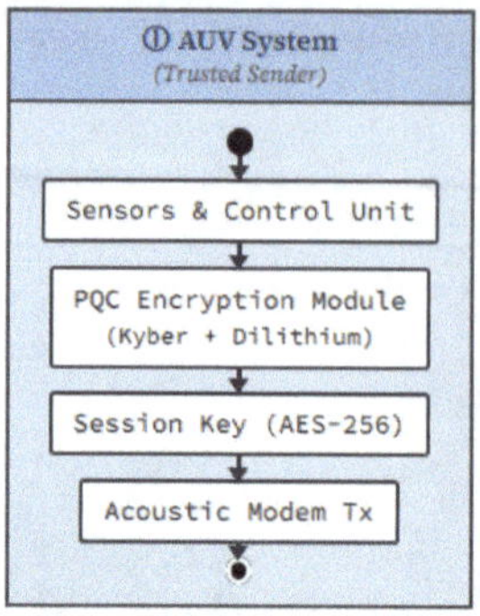

(a) AUV System (Trusted Sender)

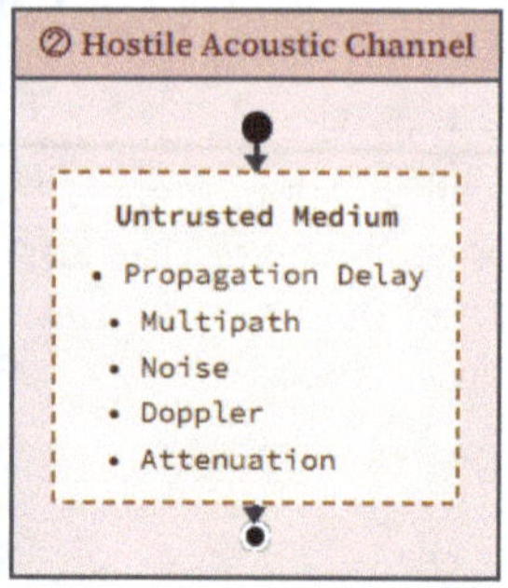

(b) Hostile Acoustic Channel

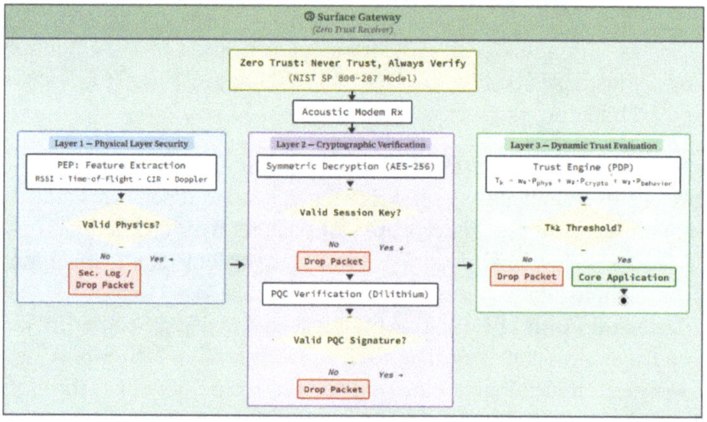

(c) Surface Gateway with Zero-Trust Verification

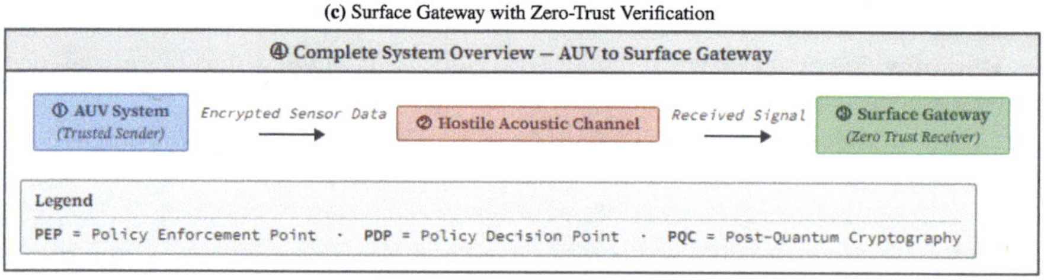

(d) Complete System Overview — AUV to Surface Gateway

Fig. 13.2 Architecture of the proposed secure underwater communication framework. **a** The **AUV System (Trusted Sender)** collects sensor data and performs post-quantum cryptographic protection using Kyber and Dilithium, establishes an AES-256 session key, and transmits encrypted data through an acoustic modem. **b** The **Hostile Acoustic Channel** represents the untrusted underwater medium affected by propagation delay, multipath propagation, noise, Doppler effects, and attenuation. **c** The **Surface Gateway** follows a zero-trust model (NIST SP 800-207), performing physical layer feature extraction (RSSI, time-of-flight, CIR, Doppler), AES-256 decryption, Dilithium signature verification, and dynamic trust evaluation before forwarding validated packets to the core application. **d** The **Complete System Overview** illustrates the end-to-end communication pipeline from the AUV to the surface gateway under the zero-trust architecture. *Legend* PEP—Policy Enforcement Point; PDP—Policy Decision Point; PQC—Post-quantum Cryptography

$$P_{rx}^{\text{est}} = P_{tx} - TL\big(d_{gps}, f\big) \pm \Delta_{\text{stochastic}} \tag{13.6}$$

Inconsistencies between the calculated and actual signal strength help detect Sybil attacks (where signal strength is unrealistically high) or Wormhole tunneling (where signal strength is unrealistically low).

13.4.2.2 Bathymetric Fingerprinting

The receiver matches the signal's Channel Impulse Response (CIR) to a bathymetric database in the receiver. If a signal does not contain the multipath reverberations characteristic of the sender's claimed location, it is labeled as a spoofing attack.

13.4.3 Layer 2: Post-quantum Cryptography (PQC)

Because of the prolonged operational life of sea-based assets, it is mandatory to protect against potential quantum threats.

13.4.3.1 Cryptographic Suite: CRYSTALS-Kyber and Dilithium

The suite uses standardized NIST algorithms, namely ML-KEM (Kyber-512) for establishing ephemeral keys and ML-DSA (Dilithium) for digital signatures to provide non-repudiation and firmware integrity.

13.4.3.2 PQC Segmentation Protocol

Since PQC signatures are relatively large, a PQC Segmentation Protocol is implemented using Forward Error Correction (FEC), such as Reed-Solomon codes. This facilitates effective key reconstruction even in environments with high interference, eliminating the need for the energy-consuming process of re-transmission requests.

13.4.4 Layer 3: Energy-Centric "Drop Early" Policy

In order to ensure the optimal longevity of the mission as well as resource usage, we use a tiered validation system that is carefully designed based on the computational complexity of each step. This will enable us to rapidly eliminate threats with low probability while ensuring that computationally complex steps are only performed for security validation. The following steps are checks are performed:

1. **Header/Format Check**: At this point, a preliminary check with a low computational cost is carried out to preserve the basic integrity of the data. Any packets that fail this check, i.e., have formatting errors or structural inconsistencies, are immediately dropped.
2. **Physics Check (PEP)**: After the header check, a low-cost Physics Check (PEP) is carried out utilizing Digital Signal Processing (DSP) algorithms. This stage aims to identify spoofed or tampered packets by analyzing physical characteristics of the signal, such as inconsistencies in Received Signal Strength or anomalies in bathrymetric fingerprinting. Sophisticated spoofing attacks are dropped here before decryption, thus saving energy.
3. **Symmetric Decryption**: After the preliminary checks, the payload is decrypted incurring a medium computational cost using a standard and fast algorithm like AES-256. This phase is essential in determining whether the party actually has possession of the shared secret key. Without the correct key, decryption will produce a random stream of unreadable characters, and the packet is dropped immediately.
4. **PQC Verification**: The most computationally intensive phase, Post-Quantum Cryptography (PQC) verification, is only carried out during session establishment. Absolute proof of identity is required to execute critical commands, and hence, PQC verification is employed in this case as it offers long-term resistance to quantum attacks.

13.5 Advanced Advancements: Future-Proofing IoMT

As the Internet of Marine Things moves from remote and isolated sensors to huge, highly interconnected autonomous swarms traditional single channel networks will inevitably become deprecated due to the increased data load. To future proof the present day IoMT against both bandwidth exhaustion and upcoming quantum threats, this framework uses three advanced mechanisms.

13.5.1 Heterogeneous Acoustic-Optical Link Layer (HAOC)

The future of IoMT mandates the need to abandon the forced choice that happens between communication distance and data speed. Through the implementation of a dynamic and dual-band HAOC approach, the framework completely overcomes the acoustic bandwidth bottleneck faced by traditional single channel networks:

- **Optical Band (Blue-Green Laser)**: Deployed exclusively for short range and high-speed communications (<100 m) needed during critical swarm docking, localized data offloading, and close quarters coordination. Crucial for future-proofing as its high bandwidth (>10 Mbps) acts as the primary reason for the rapid and bulk exchange of huge Post-Quantum Cryptography (PQC) keys and localized

Federated Learning model updates which are operations that are impossible to perform over slow acoustic links. Furthermore, the rapid attenuation of light in water fundamentally secures these high value exchanges from long-range acoustic eavesdropping [17].

- **Acoustic Band**: Dedicated only for long-distance (>10 km), low bitrate Command and Control (C2) and emergency telemetry. In the future proofed framework, the acoustic band is no longer overwhelmed with heavy data payloads. Instead it uses the PQC keys securely that have been preestablished during previous optical sessions to maintain an unbreakable cryptographic connection across vast oceanic expanses. This makes sure that continuous swarm survivability and node verification can happen even in highly turbid waters where optical links temporarily fail [17].

13.5.2 Federated Learning for Distributed Anomaly Detection

To counter the bandwidth overhead that comes with centralized logging, nodes use Federated Learning (FL). Nodes train an autoencoder on local traffic patterns and send only the gradients of the model (compressed vectors) to the leader of a swarm of nodes. The leader combines the gradients to update a global model, which allows the whole swarm to detect new jamming signatures without exfiltrating data. In the context of the Internet of Marine Things (IoMT), the integration of FL provides the following significant impacts [18]:

- **Bypassing Acoustic Bandwidth Limits**: Traditional and centralized intrusion detection needs streaming huge raw network logs to a surface server. In the IoMT, this is practically impossible to carry out using low bitrate underwater acoustic channels [3, 19]. FL compresses this data exchange into lightweight model updates, thus making advanced threat detection feasible in the ocean [18].
- **Extending Node Life cycles**: Acoustic data transmission is the most energy intensive operation that can be performed by an underwater node [6]. By processing raw data locally and transmitting only minimized gradient updates, FL greatly reduces the acoustic communication time, conserving battery life of the nodes and also extending the deployment lifecycle of these autonomous swarms [4, 18].
- **Decentralized Swarm Survivability**: If a surface gateway is jammed or destroyed in a traditional IoMT architecture, the underwater network security falls apart as it becomes blind to cyber threats [7, 14]. Due to FL that decentralizes threat intelligence, an isolated swarm can work together to identify and quarantine a compromised peer completely by itself without relying on surface-level command and control [5, 18].

13.5.3 Bio-inspired Swarm Consensus

Based on Reynolds' Flocking (Boids) theory, an "Immune System" for the swarm is designed. Nodes observe the physical kinematics of neighboring nodes; if a node demonstrates impossible motion (indicative of sensor hijacking), the swarm creates a digital "repulsion field" [20] to exclude the node from the consensus process. In the context of the Internet of Marine Things (IoMT), adding this bio-inspired consensus mechanism has the following vital reasoning and operational impacts:

- **Bridging Cyber and Physical Security**: In the IoMT, an attacker might successfully steal a node's cryptographic keys, but they can't easily simulate complex underwater kinematics [8]. By tying network consensus to verifiable physical behavior, this mechanism makes sure that trust is based on the real physical environment of the ocean, immediately overcoming advanced spoofing or GPS-faking attacks [9, 20].
- **Autonomous Self-healing**: Because acoustic communication with surface command centers is ridden by high latency and frequent disconnections [10], IoMT swarms need to be able to survive independently. This biological "immune system" allows the swarm to instantly quarantine compromised nodes and reorganize its formation on the go, ensuring no disruptions in mission execution without waiting for external instructions [4, 5].
- **Zero-Overhead Threat Mitigation**: Traditional continuous authentication protocols heavily drain the limited battery life of underwater nodes [7, 15]. Though observing the kinematics of neighboring nodes uses the standard navigational sensors that are already active on AUVs [2]. This provides a highly effective, continuous security layer that imposes virtually no additional computational or energy overhead.

13.6 Digital Twin Modeling for Predictive Security

We are performing validation on the proposed architecture through High-Fidelity Cyber-Physical Emulation. In the context of the Internet of Marine Things (IoMT), the essential meaning of this digital twin model is to transition underwater security from a reactive mode of defense to a proactive, predictive defense. By continuously mirroring the unpredictable ocean environment, the system is able to anticipate physical vulnerabilities much ahead of any possible exploitation by attackers.

13.6.1 The Cyber-Physical Twin

A virtual replica that uses both the **Bellhop** acoustic ray-tracing model and the **NS-3** network simulator [21]. This Twin recreates the environmental variables (thermoclines, salinity) that affects the vehicles in real life scenarios to accurately calculate the Signal-to-Noise Ratio (SNR) and multipath expectations for any given coordinate [5].

- **Contribution to the IoMT**: In IoMT networks, distinguishing between a benign acoustic channel fade and an active cyber jamming attack is very difficult. By establishing a real time, high-fidelity expected physical acoustic behavior as the baseline allows the IoMT nodes to instantly recognize artificial signal deviations as security threats rather than falsely perceive it as natural environmental noise [2].

13.6.2 Predictive Adversarial Simulation

The Digital Twin executes Monte Carlo simulations to predict vulnerabilities. For instance, if the digital twin predicts that the vehicle will be entering a "communication shadow zone," the system can automatically reroute the asset to a safe depth [2].

- **Contribution to the IoMT**: Attackers continuously wait to utilize physical communication blind spots to isolate nodes, execute node capture attacks, or launch localized spoofing without triggering network wide alarms. By modeling these adversarial opportunities together with physical oceanography, the IoMT can predict and adapt its swarm topology well in advance to handle these challenges. This makes sure that no asset is left physically isolated or vulnerable, thus bridging the gap between environmental hazards and cyber threats.

13.7 Energy Analysis: The Joules-Per-Bit Metric

In the energy-limited underwater domain, security has to be energy-efficient.

$$E_{\text{total}} + E_{tx} + E_{rx} + E_{\text{compute}} \tag{13.7}$$

where the elements of the equation are defined as follows:

- E_{total}: The total energy needed by the underwater node to undergo a single secure communication event.
- E_{tx}: The energy drawn by the acoustic or optical modem which is used to broadcast data through the water.
- E_{rx}: The energy required to continue receiving by keeping the receiving hydrophones and processing circuits active.

- $E_{compute}$: The energy consumed by the local processor present on the node to execute and perform tasks like Post-Quantum Cryptography (PQC) verification, Federated Learning updates, and consensus algorithms.

The analysis shows that although PQC verification results in computational overhead, the energy cost in millijoules is negligible compared to the $\approx$ 5 Joules per kilobyte acoustic transmission energy cost (E_{tx}). The "Drop Early" policy provides a net positive energy budget by removing malicious traffic at the PEP layer, thus avoiding battery waste on illegitimate data floods [17].

13.8 Conclusion

The Internet of Marine Things (IoMT) is a frontier area where the ambitions of the digital world clash with the hostility of the physical world. This chapter proposed a Lightweight Context-Aware Zero-Trust Architecture that delivers several critical results that need to be adapted for next generation maritime security:

1. **Decentralized Threat Mitigation**: By removing obsolete perimeter-based security the architecture removes all the centralized points of failure. Through the integration of Federated Learning nodes can now detect anomalies locally, resulting in a massive reduction in the bandwidth required to perform continuous threat monitoring.
2. **Zero-Overhead Physical Authentication**: The study shows that exploiting the random and noisy nature of the underwater acoustic channel gives a highly secure and unspoofable authentication mechanism. This result demonstrates that the hostile environment can be weaponized for security, thus lowering the computational overhead usually required for performing node verification.
3. **Future Proofed Data Integrity**: By enforcing Post-Quantum Cryptography (PQC), the architecture now has mathematically robust defense to combat emerging "harvest now, decrypt later" attacks, thus ensuring the long-term confidentiality of long-lifecycle marine assets.

Significance and Impact on the IoMT: The ultimate significance of these results is in proving that robust security in the IoMT does not have to come at the compromise of operational efficiency. By combining bio-inspired adaptability, digital twinning, Physiocryptography, and PQC, this proposed framework impacts the IoMT ecosystem in three critical ways:

- **Eliminating the Security-Efficiency Tradeoff**: It proves that marine networks too can achieve quantum resistant, zero-trust security all while not draining the finite battery life or overwhelming the low bandwidth acoustic channels of underwater nodes.
- **Enabling Sustainable Scalability**: By transitioning underwater security from a theoretical, computationally heavy idea into a lightweight reality, this framework allows autonomous marine swarms to scale effectively without network collapse.

- **Ensuring Long-Term Mission Resilience**: It gives us the foundational robustness that is required to deploy critical IoMT assets for long term, unattended missions. This guarantees that the marine networks can survive and continuously operate in unpredictable, harsh aquatic environments against both physical node failures and upcoming cyber threats.

References

1. Adeola, Y., Charles, E.: Addressing the Quantum Cybersecurity Challenge: The Harvest Now, Decrypt Later Threat (2025)
2. Martynova, L., Pashkevich, I., Bykova, V.: Development of a digital twin of an autonomous underwater vehicle to assess the effectiveness of searching for bottom objects. In: 2024 International Russian Smart Industry Conference (SmartIndustryCon), pp. 243–248. Sochi, Russian Federation (2024)
3. Jahanbakht, M., Xiang, W., Hanzo, L., Rahimi Azghadi, M.: Internet of underwater things and big marine data analytics—a comprehensive survey. IEEE Commun. Surv. Tutorials **23**(2), 904–956 (2021)
4. Ding, Q., Ye, L., Chen, H., Liu, H., Liang, A., Cui, W.: AUV Intelligent Decision-Making System Empowered by Deep Learning: Evolution, Challenges and Future Prospects. Preprints (2025)
5. Jiang, B., Feng, J., Cui, X., Wang, J., Liu, Y., Song, H.: Security and reliability of internet of underwater things: architecture, challenges, and opportunities. ACM Comput. Surv. **57**(3), (2024)
6. Hazarika, A., Rahmati, M.: Underwater acoustic integrated sensing and communication: a spatio-temporal freshness for intelligent resource prioritization. J. Mar. Sci. Eng. **13**(9), (2025)
7. Jiang, S.: On securing underwater acoustic networks: a survey. IEEE Commun. Surv. Tutorials **21**(1), 729–752 (2019)
8. Kahraman, Y.: Physical Layer Security for Underwater Acoustic Channels (2017)
9. Campagnaro, F., et al.: Replay-attack countermeasures for underwater acoustic networks. In: Global Oceans 2020: Singapore—U.S. Gulf Coast, pp. 1–9. Biloxi, MS, USA (2020)
10. Syed, A.A., Heidemann, J.: Time synchronization for high latency acoustic networks. In: Proceedings IEEE INFOCOM 2006. 25TH IEEE International Conference on Computer Communications, pp. 1–12. Barcelona, Spain (2006)
11. Rossing, T.D., Fletcher, N.H.: Underwater sound. In: Principles of Vibration and Sound. Springer, New York, NY (2004)
12. Thorp, W.H.: Analytic description of the low-frequency attenuation coefficient. J. Acoust. Soc. Am. **42**, 270–270 (1967)
13. Mounica, G., Deepthi, P.: Reactive jammers in wireless sensor networks: the trigger service. Int. J. Comput. Sci. Trends Technol. (IJCST) **5** (2014)
14. Raymond, D.R., Midkiff, S.F.: Denial-of-service in wireless sensor networks: Attacks and defenses. IEEE Pervasive Comput. **7**(1), 74–81 (2008)
15. Yang, G., Dai, L., Wei, Z.: Challenges, threats, security issues and new trends of underwater wireless sensor networks. Sensors **18**(11), 3907 (2018)
16. Rose, S., Borchert, O., Mitchell, S., Connelly, S.: Zero trust architecture. Technical Report NIST Special Publication 800-207, National Institute of Standards and Technology (2020)
17. Mala, D.J., Rastogi, A.K., Raj, A., Manori, V.: Hybrid kem-qkd integration with parameter optimization for secure underwater networks via netsquid simulations. IEEE Access **13**, 157081–157103 (2025)

18. Meftah, A., Do, T.N., Kaddoum, G., Talhi, C.: Federated learning-enabled jamming detection for stochastic terrestrial and non-terrestrial networks. IEEE Trans. Green Commun. Network. **9**(1), 271–290 (2024)
19. Kao, C.-C., Lin, Y.-S., Geng-De, W., Huang, C.-J.: A comprehensive study on the internet of underwater things: applications, challenges, and channel models. Sensors **17**(7), 1477 (2017)
20. Gu, Z., Zhao,H., Chen, H., et al.: UAV Swarm GNSS Spoofing Defense Method Based on Consensus Artificial Potential Field. Preprint (2024)
21. Guerra, F., Casari, P., Zorzi, M.: World ocean simulation system (WOSS): a simulation tool for underwater networks with realistic propagation modeling. In: WUWNet'09: Proceedings of the 4th International Workshop on Underwater Networks (2009)

Index

T. A. Tran et al. (eds.), *Big Data Analytics for Internet of Marine Things*, Internet of Things, https://doi.org/10.1007/978-3-032-28171-5